Student Solutions Manual

AF816874

Intermediate Algebra
Everyday Explorations

FIFTH EDITION

Alice Kaseberg
Lane Community College (Emeritus)

Greg Cripe
Spokane Falls Community College

Peter Wildman
Spokane Falls Community College

Prepared by

Shawna Cosner
Spokane Falls Community College

Becky Cosner
Gonzaga Community Colleges of Spokane,
Gonzaga University

BROOKS/COLE
CENGAGE Learning

Australia • Brazil • Japan • Korea • Mexico • Singapore • Spain • United Kingdom • United States

BROOKS/COLE
CENGAGE Learning

For product information and technology assistance, contact us at
Cengage Learning Customer & Sales Support,
1-800-354-9706

For permission to use material from this text or product, submit all requests online at **www.cengage.com/permissions**
Further permissions questions can be emailed to
permissionrequest@cengage.com

ISBN-13: 978-1-133-36397-2
ISBN-10: 1-133-36397-0

Brooks/Cole
20 Davis Drive
Belmont, CA 94002-3098
USA

Cengage Learning is a leading provider of customized learning solutions with office locations around the globe, including Singapore, the United Kingdom, Australia, Mexico, Brazil, and Japan. Locate your local office at: **www.cengage.com/global**

Cengage Learning products are represented in Canada by Nelson Education, Ltd.

To learn more about Brooks/Cole, visit **www.cengage.com/brookscole**

Purchase any of our products at your local college store or at our preferred online store **www.cengagebrain.com**

Printed in the United States of America
1 2 3 4 5 6 7 16 15 14 13 12

Contents

CHAPTER 0

Review

Exercises 0.1: Number Sense

1. integer **3.** irrational **5.** natural numbers **7.** natural numbers

9. **a.** $4+(-12)=-8$

$4-(-12)=4+12=16$

$(4)(-12)=-48$

$4\div(-12)=\dfrac{4}{-12}=\dfrac{4\cdot 1}{4\cdot(-3)}=-\dfrac{1}{3}$

b. $\dfrac{1}{2}+\dfrac{1}{3}=\dfrac{1\cdot 3}{2\cdot 3}+\dfrac{1\cdot 2}{3\cdot 2}=\dfrac{3}{6}+\dfrac{2}{6}=\dfrac{5}{6}$

$\dfrac{1}{2}-\dfrac{1}{3}=\dfrac{3}{6}-\dfrac{2}{6}=\dfrac{1}{6}$

$\left(\dfrac{1}{2}\right)\left(\dfrac{1}{3}\right)=\dfrac{1\cdot 1}{2\cdot 3}=\dfrac{1}{6}$

$\left(\dfrac{1}{2}\right)\div\left(\dfrac{1}{3}\right)=\dfrac{1}{2}\cdot\dfrac{3}{1}=\dfrac{3}{2}$

c. $\dfrac{-2}{3}+\dfrac{1}{6}=\dfrac{-2\cdot 2}{3\cdot 2}+\dfrac{1}{6}=\dfrac{-4}{6}+\dfrac{1}{6}=\dfrac{-3}{6}=-\dfrac{1}{2}$

$\dfrac{-2}{3}-\dfrac{1}{6}=\dfrac{-4}{6}-\dfrac{1}{6}=-\dfrac{5}{6}$

$\left(\dfrac{-2}{3}\right)\left(\dfrac{1}{6}\right)=\dfrac{-2\cdot 1}{3\cdot 6}=\dfrac{-2}{18}=\dfrac{-2}{2\cdot 9}=-\dfrac{1}{9}$

$\left(\dfrac{-2}{3}\right)\div\left(\dfrac{1}{6}\right)=\dfrac{-2}{3}\cdot\dfrac{6}{1}=\dfrac{-12}{3}=-4$

d. $0.75+(-0.2)=0.55$

$0.75-(-0.2)=0.75+0.2=0.95$

$(0.75)(-0.2)=-0.15$

$(0.75)\div(-0.2)=-3.75$

e. $-0.45+(-1.5)=-1.95$

$-0.45-(-1.5)=-0.45+1.5=1.05$

$(-0.45)(-1.5)=0.675$

$(-0.45)\div(-1.5)=0.3$

f. $\dfrac{-5}{8}+\dfrac{-10}{9}=\dfrac{-5\cdot 9}{8\cdot 9}+\dfrac{-10\cdot 8}{8\cdot 9}=\dfrac{-45}{72}+\dfrac{-80}{72}=\dfrac{-125}{72}=-1\dfrac{53}{72}$

$\dfrac{-5}{8}-\dfrac{-10}{9}=\dfrac{-45}{72}-\dfrac{-80}{72}=\dfrac{-45}{72}+\dfrac{80}{72}=\dfrac{35}{72}$

$\left(\dfrac{-5}{8}\right)\left(\dfrac{-10}{9}\right)=\dfrac{-5\cdot -10}{8\cdot 9}=\dfrac{50}{72}=\dfrac{2\cdot 25}{2\cdot 36}=\dfrac{25}{36}$

$\left(\dfrac{-5}{8}\right)\div\left(\dfrac{-10}{9}\right)=\dfrac{-5}{8}\cdot\dfrac{9}{-10}=\dfrac{-5\cdot 9}{8\cdot 2\cdot -5}=\dfrac{9}{16}$

11. a. $8-(-6)+(-2)=8+6-2=14-2=12$

b. $15+(-9)-(-8)=15-9+8=6+8=14$

c. $-3.6-(-4.8)+(4.3)=-3.6+4.8+4.3=1.2+4.3=5.5$

d. $-8.3+(-3.8)-(-2.1)=-8.3-3.8+2.1=-12.1+2.1=-10$

e. $4\frac{1}{4}-5\frac{1}{2}+2\frac{1}{4}=4\frac{1}{4}+2\frac{1}{4}-5\frac{1}{2}=6\frac{1}{2}-5\frac{1}{2}=1$

f. $6\frac{1}{4}+4\frac{1}{2}-3\frac{1}{2}=6\frac{1}{4}+1=7\frac{1}{4}$

13. a. $\sqrt{-9}$ is undefined on the set of real numbers **b.** $\sqrt{9}=3$ **c.** $-\sqrt{9}=-3$

15. a. $-(-3)$; the opposite of negative three **b.** -2; negative two

c. $5-(-3)$; negative three subtracted from five

17. a. $19\cdot 92+19\cdot 8=19(92+8)=19(100)=1900$

b. $20\left(\frac{1}{5}+0.5\right)=20\left(\frac{1}{5}\right)+20(0.5)=4+10=14$

c. $299\cdot 5=5(300-1)=1500-5=1495$

d. $107\cdot 29-7\cdot 29=29(107-7)=29(100)=2900$

e. $8\left(4\frac{1}{4}\right)=8\left(4+\frac{1}{4}\right)=32+2=34$

19. a. $A=\pi(122\div 2)^2$

$A=\pi(61)^2$

$A=\pi(3721)$

$A\approx 11689.9\,cm^2$

b. $S=2\pi\cdot 2^2+2\pi(2)(4.5)$

$S=2\pi\cdot 4+2\pi(9)$

$S=8\pi+18\pi$

$S=26\pi\approx 81.7\,in^2$

c. $d=(-5)^2-4(6)(-7)$

$d=25-4(6)(-7)$

$d=25+168$

$d=193$

d. $x=9-4(9-2)+8\div 4-2\cdot 5$

$x=9-4(7)+2-10$

$x=9-28+2-10$

$x=-27$

21. a. $m = \dfrac{4-8}{6-(-2)}$

$m = \frac{-4}{8}$

$m = -\frac{1}{2}$

b. $c = \sqrt{3^2 + 4^2}$

$c = \sqrt{9+16}$

$c = \sqrt{25}$

$c = 5$

c. $F = \dfrac{9}{5}(100) + 32$

$F = \frac{900}{5} + 32$

$F = 180 + 32$

$F = 212$

d. $C = \dfrac{5}{9}(212-32)$

$C = \dfrac{5}{9}(180)$

$C = \dfrac{900}{9}$

$C = 100$

e. $h = \dfrac{4}{125}(-20)^2$

$h = \dfrac{4}{125}(400)$

$h = \dfrac{1600}{125}$

$h = 12.8$

23. a. $d = \sqrt{[6-(-2)]^2 + [3-(-3)]^2}$

$d = \sqrt{(8)^2 + (6)^2}$

$d = \sqrt{64+36}$

$d = \sqrt{100}$

$d = 10$

b. $m = \dfrac{3-(-3)}{6-(-2)}$

$m = \dfrac{6}{8}$

$m = \dfrac{3}{4}$

c. $x = \dfrac{-(-2) + \sqrt{(-2)^2 - 4(3)(-5)}}{2(3)}$

$x = \dfrac{2 + \sqrt{4+60}}{6}$

$x = \dfrac{2 + \sqrt{64}}{6}$

$x = \dfrac{2+8}{6}$

$x = \dfrac{10}{6}$

$x = \dfrac{5}{3}$

d. $d = \{|4.05 - 7.12| + |7.51 - 7.12| + |9.80 - 7.12|\} \div 3$

$d = \{|-3.07| + |0.39| + |2.68|\} \div 3$

$d = (3.07 + 0.39 + 2.68) \div 3$

$d = 6.14 \div 3$

$d = 2.04\overline{6}$

25. a. $S = \frac{100}{2}\{2(1) + [(100) - 1](2)\}$

$S = 50[2 + (99)2]$

$S = 50(2 + 198)$

$S = 50(200)$

$S = 10{,}000$

b. $S = \frac{100}{2}\{2(2) + [(100) - 1](2)\}$

$S = 50[4 + (99)2]$

$S = 50(4 + 198)$

$S = 50(202)$

$S = 10{,}100$

c. $S = \frac{100}{2}\{2(5) + [(100) - 1](5)\}$

$S = 50[10 + (99)5]$

$S = 50(10 + 495)$

$S = 50(505)$

$S = 25{,}250$

27. a. $D = \sqrt{[6-(-2)]^2 + [3-(-3)]^2}$

$D = \sqrt{(8)^2 + (6)^2}$

$D = \sqrt{64+36}$

$D = \sqrt{100}$

$D = 10$

b. $D = \sqrt{(14-2)^2 + [2-(-3)]^2}$

$D = \sqrt{(12)^2 + (5)^2}$

$D = \sqrt{144+25}$

$D = \sqrt{169}$

$D = 13$

c. $D = \sqrt{[(-4)-11)]^2 + [3-(-5)]^2}$

$D = \sqrt{(-15)^2 + (8)^2}$

$D = \sqrt{225+64}$

$D = \sqrt{289}$

$D = 17$

29. a. $W = 35.74 + 0.6215T - 35.75\left(V^{0.16}\right) + 0.4275T\left(V^{0.16}\right)$

$W = 35.74 + 0.6215(25) - 35.75\left(5^{0.16}\right) + 0.4275(25)\left(5^{0.16}\right)$

$W \approx 18.9^\circ$

The result is the same to one decimal place.

b. $W = 35.74 + 0.6215(25) - 35.75\left(10^{0.16}\right) + 0.4275(25)\left(10^{0.16}\right)$

$W \approx 15.1^\circ$

c. $W = 35.74 + 0.6215(25) - 35.75\left(15^{0.16}\right) + 0.4275(25)\left(15^{0.16}\right)$

$W \approx 12.6^\circ$

d. $W = 35.74 + 0.6215(25) - 35.75\left(20^{0.16}\right) + 0.4275(25)\left(20^{0.16}\right)$

$W \approx 10.8^\circ$

$W = 35.74 + 0.6215(25) - 35.75\left(25^{0.16}\right) + 0.4275(25)\left(25^{0.16}\right)$

$W \approx 9.3^\circ$

$W = 35.74 + 0.6215(25) - 35.75\left(30^{0.16}\right) + 0.4275(25)\left(30^{0.16}\right)$

$W \approx 8.1^\circ$

The results are the same to one decimal place.

e. $W = 35.74 + 0.6215(-5) - 35.75\left(5^{0.16}\right) + 0.4275(-5)\left(5^{0.16}\right)$

$W \approx -16.4^\circ$

$W = 35.74 + 0.6215(-5) - 35.75\left(10^{0.16}\right) + 0.4275(-5)\left(10^{0.16}\right)$

$W \approx -22.1^\circ$

$W = 35.74 + 0.6215(-5) - 35.75\left(15^{0.16}\right) + 0.4275(-5)\left(15^{0.16}\right)$

$W \approx -25.8^\circ$

31. a. True. Property, because whether you are driving in the right lane or in the left lane, the driver would be in center of the road.

b. True. The number 5 can be written as 1 + 1 + 1 + 1 + 1, so $5\frac{1}{2}$ can be written as $5 + \frac{1}{2}$

Exercises 33 through 37 are writing exercises and possible answers are given.

33. The division can be changed to multiplication by the reciprocal of the number.

35. For subtraction, the symbol would be placed between two numbers. To represent a negative number, the symbol would be placed in front of the number. To represent the opposite of a number, the symbol would be placed in front of parentheses with a number inside.

37. Subtraction and Division are not commutative. The agreement is that the numbers will be subtracted or divided in the order in which the numbers are listed.

Exercises 0.2: Solving Equations using Symbolic Representations

1. **a.** In $2x + 3$, the variable is x, the numerical coefficient is 2, and the constant term is 3.

b. The variable is r and the numerical coefficient is π. There is no constant term.

c. The variable is x, the numerical coefficient is -1 (implied by the $-$ in front of x), and the constant term is 4.

d. The variable is x, the numerical coefficient on x^2 is 1, the numerical coefficient on x is 1, and the constant term is -1.

3. **a.** $-2(3 - 4x) = (-2)(3) + (-2)(-4x) = -6 + 8x$

b. $-a(b - c) = (-a)(b) + (-a)(-c) = -ab + ac$

c. $x(x^2 - 2x - 3) = (x)(x^2) + (x)(-2x) + (x)(-3) = x^3 - 2x^2 - 3x$

5. **a.** There are 3 (x^2) tiles, 5 (x) tiles, and 6 (1) tiles; total is $3x^2 + 5x + 6$.

b. There are 4 (a^2) tiles, 3 (ab) tiles, and 2 (b^2) tiles; total is $4a^2 + 3ab + 2b^2$.

7. **a.** $6ac - 3ab + 2ca + 7bc = 6ac + 2ac - 3ab + 7bc = ac(6 + 2) - 3ab + 7bc = 8ac - 3ab + 7bc$

b. $x(4 + 2x) - x(x - 4) = 4x + 2x^2 - x^2 + 4x = 2x^2 - x^2 + 4x + 4x = x^2 + 8x$

9. **a.** $-4 - 6 = -4 + (-6) = -10$

b. $-5 - (-3) = -5 + 3 = -2$

c. $-\frac{1}{2} - (-\frac{3}{2}) = -\frac{1}{2} + \frac{3}{2} = 1$

d. $100 \cdot \frac{1}{4} = 100 \div 4 = 25$

e. $\frac{5}{8} \div \frac{5}{8} = \frac{5}{8} \cdot \frac{8}{5} = 1$

f. $-6 \div \frac{1}{3} = -6 \cdot 3 = -18$

11. **a.** $-3 + (3) = 0$

b. $5 \cdot (\frac{1}{5}) = 1$

c. $-2 + (3) = 1$

d. $8 \cdot (0) = 0$

e. $\frac{7}{8} \cdot \left(\frac{8}{7}\right) = 1$

f. $\frac{3}{8} + \left(-\frac{3}{8}\right) = 0$

13. **a.** $4 = 4$ is an Identity.

b. $8 = 3$ is a Contradiction.

15. **a.** $x - 2 = 2x - x$ is a Contradiction.

b. $\frac{x}{4} = x(\frac{1}{2} - 0.25)$ is an Identity ($\frac{1}{2} - \frac{1}{4} = \frac{1}{4}$).

17. **a.**
$$x - 8 = -4$$
$$x - 8 + 8 = -4 + 8$$
$$x = 4$$

b.
$$x + 3 = -6$$
$$x + 3 - 3 = -6 - 3$$
$$x = -9$$

c. $x - 6 = -8$
$x - 6 + 6 = -8 + 6$
$x = -2$

d. $\frac{x}{2} = -6$
$2 \cdot \frac{x}{2} = -6 \cdot 2$
$x = -12$

e. $\frac{3x}{4} = 21$
$(\frac{4}{3})(\frac{3x}{4}) = (\frac{4}{3})(21)$
$x = 28$

f. $\frac{3x}{8} = 12$
$(\frac{8}{3})(\frac{3x}{8}) = (\frac{8}{3})(12)$
$x = 32$

g. $-\frac{2}{3}x = 24$
$(-\frac{3}{2})(-\frac{2}{3}x) = (-\frac{3}{2})24$
$x = -36$

h. $-\frac{1}{4}x = 8$
$(-4)(-\frac{1}{4}x) = (-4)(8)$
$x = -32$

i. $\frac{7}{8}x = 35$
$(\frac{8}{7})(\frac{7}{8}x) = (\frac{8}{7})35$
$x = 40$

19. a. $2x - 4 = 8$
$2x - 4 + 4 = 8 + 4$
$2x = 12$
$2x \div 2 = 12 \div 2$
$x = 6$

b. $2x - 4 = 6$
$2x - 4 + 4 = 6 + 4$
$2x = 10$
$2x \div 2 = 10 \div 2$
$x = 5$

c. $2x - 4 = -6$
$2x - 4 + 4 = -6 + 4$
$2x = -2$
$2x \div 2 = -2 \div 2$
$x = -1$

d. $2x - 4 = 0$
$2x - 4 + 4 = 0 + 4$
$2x = 4$
$2x \div 2 = 4 \div 2$
$x = 2$

21. a. $\frac{5}{9}(F - 32) = -25$
$\frac{9}{5} \cdot \frac{5}{9}(F - 32) = -25 \cdot \frac{9}{5}$
$F - 32 = -45$
$F - 32 + 32 = -45 + 32$
$F = -13$

b. $\frac{5}{9}(F - 32) = 10$
$\frac{9}{5} \cdot \frac{5}{9}(F - 32) = 10 \cdot \frac{9}{5}$
$F - 32 = 18$
$F - 32 + 32 = 18 + 32$
$F = 50$

c. $\frac{5}{9}(F - 32) = 5$
$\frac{9}{5} \cdot \frac{5}{9}(F - 32) = 5 \cdot \frac{9}{5}$
$F - 32 = 9$
$F - 32 + 32 = 9 + 32$
$F = 41$

d. $\frac{5}{9}(F - 32) = -30$
$\frac{9}{5} \cdot \frac{5}{9}(F - 32) = -30 \cdot \frac{9}{5}$
$F - 32 = -54$
$F - 32 + 32 = -54 + 32$
$F = -22$

23. $3x + 3 = 0$
$3x + 3 - 3 = 0 - 3$
$3x = -3$
$3x \div 3 = -3 \div 3$
$x = -1$

25. $\frac{x}{3} - 1 = 4$
$\frac{x}{3} - 1 + 1 = 4 + 1$
$\frac{x}{3} = 5$
$3 \cdot \frac{x}{3} = 5 \cdot 3$
$x = 15$

27.
$\frac{3x}{4} + 5 = 23$
$\frac{3x}{4} + 5 - 5 = 23 - 5$
$\frac{3x}{4} = 18$
$\frac{4}{3} \cdot \frac{3x}{4} = 18 \cdot \frac{4}{3}$
$x = 24$

29.
$\frac{3}{8}x - 4 = 8$
$\frac{3}{8}x - 4 + 4 = 8 + 4$
$\frac{3}{8}x = 12$
$\frac{8}{3} \cdot \frac{3}{8}x = 12 \cdot \frac{8}{3}$
$x = 32$

31.
$x - 2(4 - x) = -17$
$x - 8 + 2x = -17$
$3x - 8 = -17$
$3x - 8 + 8 = -17 + 8$
$3x = -9$
$3x \div 3 = -9 \div 3$
$x = -3$

33.
$x - 3(4 + x) = 6$
$x - 12 - 3x = 6$
$-2x - 12 = 6$
$-2x - 12 + 12 = 6 + 12$
$-2x = 18$
$-2x \div -2 = 18 \div -2$
$x = -9$

35.
$x + 3(5 - x) = -9$
$x + 15 - 3x = -9$
$-2x + 15 = -9$
$-2x + 15 - 15 = -9 - 15$
$-2x = -24$
$-2x \div -2 = -24 \div -2$
$x = 12$

37.
$8 - 3(9 - x) = -14.5$
$8 - 27 + 3x = -14.5$
$3x - 19 = -14.5$
$3x - 19 + 19 = -14.5 + 19$
$3x = 4.5$
$3x \div 3 = 4.5 \div 3$
$x = 1.5$

39.
$14 - 9(x + 2) = 45.5$
$14 - 9x - 18 = 45.5$
$-9x - 4 = 45.5$
$-9x - 4 + 4 = 45.5 + 4$
$-9x = 49.5$
$-9x \div -9 = 49.5 \div -9$
$x = -5.5$

41.
$37.45 = x + 0.07x$
$37.45 = 1.07x$
$37.45 \div 1.07 = 1.07x \div 1.07$
$35 = x$
$x = 35$

43.
$46.17 = x + 0.15x + 0.065x$
$46.17 = 1.215x$
$46.17 \div 1.215 = 1.215x \div 1.215$
$38 = x$
$x = 38$

45.
$2(x + 3) = 1 + 4x$
$2x + 6 = 1 + 4x$
$2x + 6 - 2x - 1 = 1 + 4x - 2x - 1$
$5 = 2x$
$5 \div 2 = 2x \div 2$
$2.5 = x$
$x = 2.5$

47. $4-(x-2)=2(x+6)$

$4-x+2=2x+12$

$6-x=2x+12$

$6-x+x-12=2x+12+x-12$

$-6=3x$

$-6\div 3=3x\div 3$

$-2=x$

$x=-2$

49. $2-3(x+1)=x-3$

$2-3x-3=x-3$

$-1-3x=x-3$

$-1-3x+3x+3=x-3+3x+3$

$2=4x$

$2\div 4=4x\div 4$

$\frac{1}{2}=x$

$x=\frac{1}{2}$

51. $D=rt$

$$\frac{D}{r}=\frac{rt}{r}$$

$$t=\frac{D}{r}$$

53. $y=mx+b$

$y-mx=mx+b-mx$

$b=y-mx$

55. $$V=\frac{4\pi r^3}{3}$$

$$\frac{3V}{4\pi}=\frac{3}{4\pi}\cdot\frac{4\pi r^3}{3}$$

$$r^3=\frac{3V}{4\pi}$$

57. $pN=1$

$$\frac{pN}{N}=\frac{1}{N}$$

$$p=\frac{1}{N}$$

59. $A=\frac{1}{2}h(a+b)$

$$\frac{2A}{h}=\frac{2}{h}\left(\tfrac{1}{2}h(a+b)\right)$$

$$\frac{2A}{h}-a=a+b-a$$

$$b=\frac{2A}{h}-a$$

61. $a_n=a_1+(n-1)d$

$a_n-(n-1)d=a_1+(n-1)d-(n-1)d$

$a_n-(n-1)d=a_1$

$a_1=a_n-(n-1)d$

63. $$A=\frac{a+b+c}{3}$$

$$3A=\frac{3(a+b+c)}{3}$$

$3A-b-c=a+b+c-b-c$

$3A-b-c=a$

65. $S=\frac{1}{2}n(a_1+a_n)$

$$\frac{2S}{n}=\frac{2}{n}\left(\tfrac{1}{2}n(a_1+a_n)\right)$$

$$\frac{2S}{n}-a_n=a_1+a_n-a_n$$

$$a_1=\frac{2S}{n}-a_n$$

67. $E = \dfrac{T_h - T_c}{T_h}$

$ET_h = \dfrac{T_h(T_h - T_c)}{T_h}$

$ET_h + T_c = T_h - T_c + T_c$

$T_c + ET_h - ET_h = T_h - ET_h$

$T_c = T_h - ET_h$

69. $V = 344 + 0.6(T - 20)$

$V - 344 = 344 + 0.6(T - 20) - 344$

$\dfrac{V - 344}{0.6} = \dfrac{0.6(T - 20)}{0.6}$

$\dfrac{V - 344}{0.6} + 20 = T - 20 + 20$

$T = \dfrac{V - 344}{0.6} + 20$

71. $I = \dfrac{E}{R}$

$I \cdot \dfrac{R}{I} = \dfrac{E}{R} \cdot \dfrac{R}{I}$

$R = \dfrac{E}{I}$

73. $T = \dfrac{0.16V}{S_e}$

$T\left(\dfrac{S_e}{T}\right) = \left(\dfrac{0.16V}{S_e}\right)\left(\dfrac{S_e}{T}\right)$

$S_e = \dfrac{0.16V}{T}$

75. $S = \dfrac{a_1}{r-1}$

$S\left(\dfrac{r-1}{S}\right) = \left(\dfrac{a_1}{r-1}\right)\left(\dfrac{r-1}{S}\right)$

$r - 1 = \dfrac{a_1}{S}$

$r - 1 + 1 = \dfrac{a_1}{S} + 1$

$r = \dfrac{a_1}{S} + 1$ or $r = \dfrac{a_1}{S} + \dfrac{S}{S} = \dfrac{a_1 + S}{S}$

77. $D = \left(\dfrac{D}{t}\right)t$, D = D

79. $A = \frac{1}{2}\left(\dfrac{2A}{a+b}\right)(a+b)$

$A = \frac{1}{2}(2A)$; A = A

81. $E = \dfrac{\dfrac{T_c}{1-E} - T_c}{\dfrac{T_c}{1-E}}$

$E = \dfrac{\dfrac{T_c}{1-E} - \dfrac{T_c(1-E)}{1-E}}{\dfrac{T_c}{1-E}}$

$\Rightarrow E = \dfrac{\dfrac{T_c - T_c(1-E)}{1-E}}{\dfrac{T_c}{1-E}}$

$E = \dfrac{T_c(1-(1-E))}{1-E} \cdot \dfrac{1-E}{T_c}$, E = E

Exercises 0.3: Solving Equations with Visual and Numeric Representations

1.

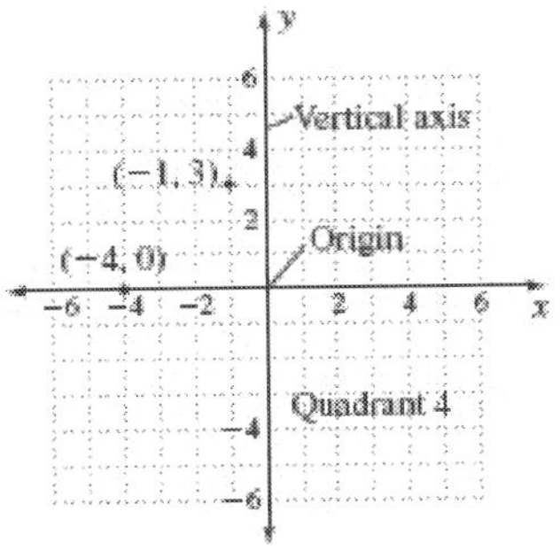

3. **a.** (75, 167), J **b.** (10, 50), E
c. (37, 98.6), G **d.** (100, 212), K

5. **a.** B = (−18, 0) **b.** A, (−40, −40)

7. Independent variable, number of trips; dependent variable, cost of riding local transit.

9. Independent variable, adjusted gross income; dependent variable, income tax rate.

11. Visualize x columns of 2 dots with one dot missing from the last column.

13. a.

Input, x	1	2	3	4
Output, y	3	6	9	12

b. The rule is $y = 3x$.

15. a.

Input, x	1	2	3	4	5
Output, y	2	5	8	11	14

b. We can estimate the rule as $y = 3x + ?$. When $x = 1$, $y = 3(1) - 1$ or $y = 2$.
The rule is $y = 3x - 1$.

17.

x	y = 3x − 2
−3	3(−3) − 2 = −11
−2	3(−2) − 2 = −8
−1	3(−1) − 2 = −5
0	3(0) − 2 = −2
1	3(1) − 2 = 1
2	3(2) − 2 = 4
3	3(3) − 2 = 7

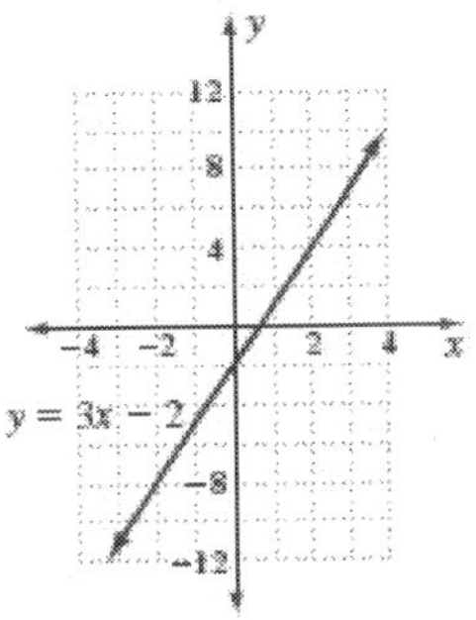

19.

x	y = 4 – 2x
–3	$4-2(-3)=10$
–2	$4-2(-2)=8$
–1	$4-2(-1)=6$
0	$4-2(0)=4$
1	$4-2(1)=2$
2	$4-2(2)=0$
3	$4-2(3)=-2$

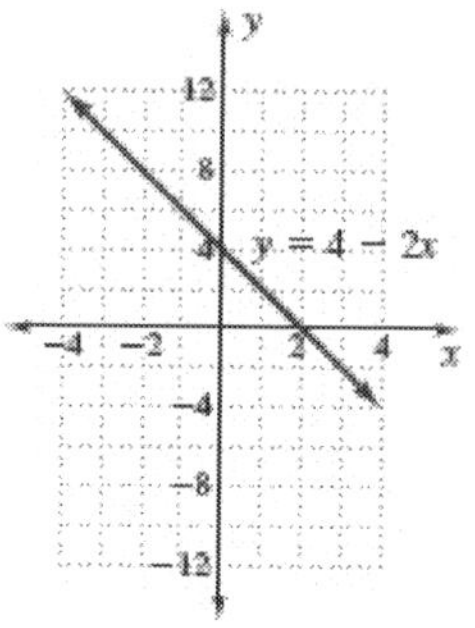

21.

x	$y=2x-x^2$
–3	$2(-3)-(-3)^2=-15$
–2	$2(-2)-(-2)^2=-8$
–1	$2(-1)-(-1)^2=-3$
0	$2(0)-(0)^2=0$
1	$2(1)-(1)^2=1$
2	$2(2)-(2)^2=0$
3	$2(3)-(3)^2=-3$

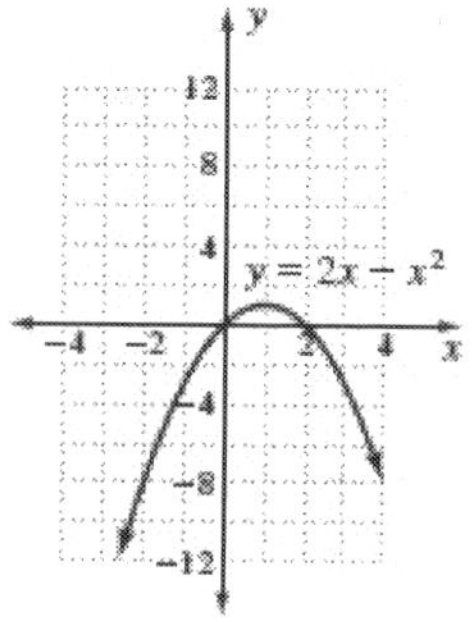

23. We chose the axes limits by examining the table.

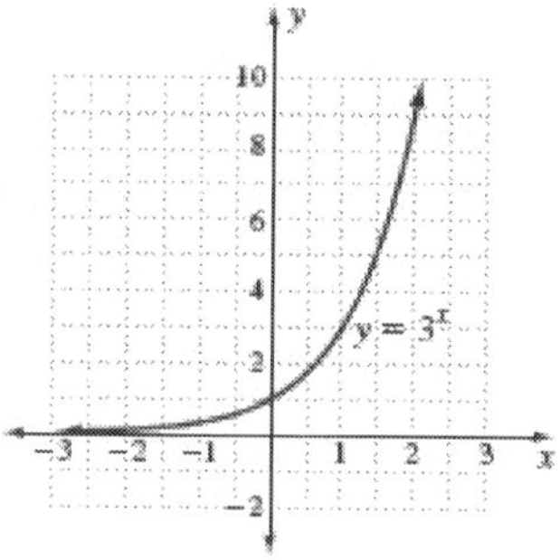

25. We chose the axes limits by examining the table.

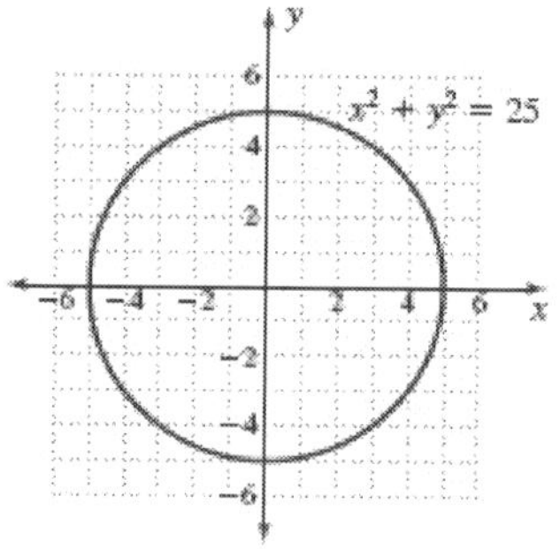

27. We chose the axes limits by examining the table.

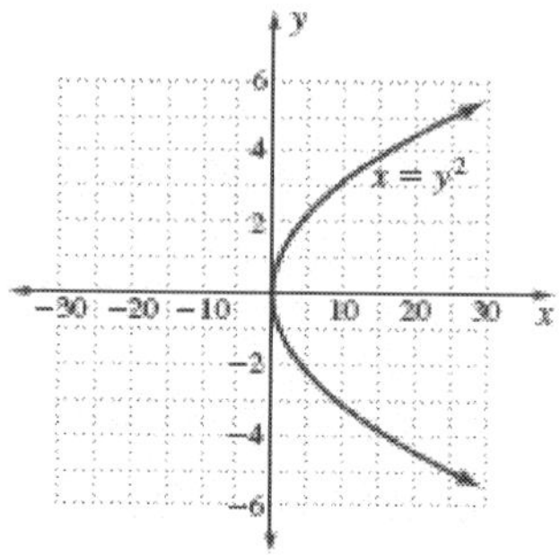

29. On the input axes, use \$0 to \$1000 with a scale of \$100; on the output axes, use \$0 to \$100 with a scale of \$10.

31. On the input axes, use 0 to 24 hours with a scale of 2 hours; on the output axes, use low tide to high tide with a scale of 10% of the difference.

33.

x	**y = 1 – 0.25x**
a. –4	1 – 0.25(–4) = 2
0	1 – 0.25(0) = 1
b. 1	1 – 0.25(1) = 0.75
2	1 – 0.25(2) = 0.5

35. a. $\{0, \frac{1}{2}\}$ **b.** ≈{1.3, –0.8}

c. {–1.5, 2} **d.** no real number solutions, { }

37. a. x = 0 or x = –1 **b.** x = –3 or x = 2

c. no real number solution **d.** x = –0.5

39. a. x = –2 or x = 2 **b.** no real number solution

c. x = –3 or x = 3 **d.** x = 0

41. a. x = –1 or x = 0 **b.** x = –2 or x = 1

c. no real number solution **d.** x = –3 or x = 2

For Exercises 43 through 47, use calculator graphs and tables to solve for x. Answers are given for reference only.

43. x = 4 **45. a.** x = 1.5 **b.** x = 3.5 **47. a.** x = –2 **b.** x = –3

49.

Weight (lb)	**Index** $I = W(704.5) \div H^2$
100	$(100)(704.5) \div (62)^2 = 18.3$
110	$(110)(704.5) \div (62)^2 = 20.2$
120	$(120)(704.5) \div (62)^2 = 22.0$
130	$(130)(704.5) \div (62)^2 = 23.8$
140	$(140)(704.5) \div (62)^2 = 25.7$

51.

Weight (lb)	**Index** $I = W(704.5) \div H^2$
130	$(130)(704.5) \div (70)^2 = 18.7$
140	$(140)(704.5) \div (70)^2 = 20.1$
150	$(150)(704.5) \div (70)^2 = 21.6$
160	$(160)(704.5) \div (70)^2 = 23.0$
170	$(170)(704.5) \div (70)^2 = 24.4$

53. a. W = 130 lb **b.** W = 110 lb **55. a.** W = 160 lb **b.** W = 170 lb

57. Solving the equation 3x + 4 = –2 is *finding the independent variable.*

59. Locating y, given the graph and x, is *finding the dependent variable.*

61. Locating x, given the graph and y, is *finding the independent variable.*

63. Find the output for input 2.

65. Find the x-coordinate for the point of intersection of y = 3x + 4 and y = 7.

Exercises 0.4: Inequalities, Line Graphs, and Intervals

1. **a.** $\frac{1}{2}+\frac{1}{2}=1;\ \frac{1}{2}\cdot\frac{1}{2}=\frac{1}{4};\ 1>\frac{1}{4}$

b. $\frac{1}{2}\div\frac{1}{2}=\frac{1}{2}\cdot\frac{2}{1}=1;\ \frac{1}{2}+\frac{1}{2}=1;\ 1=1$

c. $\frac{1}{4}+\frac{1}{4}=\frac{1}{2};\ \frac{1}{2}\cdot\frac{1}{2}=\frac{1}{4};\ \frac{1}{2}>\frac{1}{4}$

d. $\frac{1}{2}-\frac{1}{4}=\frac{1}{4};\ \frac{7}{8}-\frac{3}{8}=\frac{1}{2};\ \frac{1}{4}<\frac{1}{2}$

e. $0.3\cdot 0.5=0.15;\ 0.4\cdot 0.4=0.16;\ 0.15<0.16$

f. $1.5-3.5=-2.0;\ -2.5-(0.5)=-3.0;\ -2.0>-3.0$

3. $-3<x<4$ is $x>-3$ and $x<4$

5. Neither $x>4$ nor $x<1$ is appropriate for a compound inequality.

7. $x>-1$ and $x\le 5$ is $-1<x\le 5$

9. $-1\le x<1$ is $x\ge -1$ and $x<1$

11. $3<x$ and $4>x$ is $3<x<4$

	Inequality	Interval	Words	Line Graph
13.	$-3<x<5$	$(-3, 5)$	**Set of numbers greater than −3 and less than 5**	
15.	$-4<x\le 2$	$(-4, 2]$	**Set of numbers greater than −4 and less than or equal to 2**	
17.	$x>5$	$(5, \infty)$	**Set of numbers greater than 5**	see text
19.	$x<-2$	$(-\infty, -2)$	**Set of numbers less than −2**	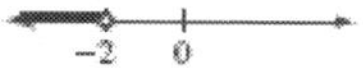
21.	$x\le -3$	$(-\infty, -3]$	**Set of numbers less than or equal to −3**	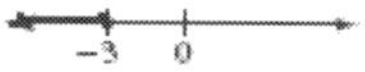
23.	$x\ge 4$	$[4, \infty)$	**Set of numbers greater than or equal to 4**	

25. $y=\$4$ for $0<x\le 2$; $y=\$4+\$0.50(x-2)$ for $x>2$

27. $y=\$20$ for $0<x\le 3$; $y=\$20+\$5(x-3)$ for $x>3$, x rounded up to the next integer.

29. $y=\$65$ for $0<x\le 100$; $y=\$65+\$0.15(x-100)$ for $x>100$

31. $y=\$85$ for $0<x\le 10$; $y=\$85+\$4.75(x-10)$ for $x>10$

33. Answers may vary. Here is one example: 3.14159261 and 3.14159269.

35. **a.** interval **b.** ordered pair **c.** ordered pair **d.** interval
e. interval **f.** ordered pair **g.** interval **h.** ordered pair

37. The inequality is correct because $3<4$ and $4=4$.

Exercises 0.5: Solving Systems of Equations

1.

$y = 3x - 4$

$3y - 2x = 9$

$3(3x - 4) - 2x = 9$

$9x - 12 - 2x = 9$

$7x - 12 = 9$

$7x = 21 \quad \Rightarrow x = 3$

$y = 3(3) - 4 \quad \Rightarrow y = 5$

$3(5) - 2(3) \stackrel{?}{=} 9$

$15 - 6 \stackrel{?}{=} 9$. It checks.

3.

$0.6x + y = 3$

$1.4x + y + 3 = 0$

$y = 3 - 0.6x$

$y = -1.4x - 3$

$3 - 0.6x = -1.4x - 3$

$-0.6x + 1.4x = -3 - 3$

$0.8x = -6 \quad \Rightarrow x = -7.5$

$y = 3 - 0.6(-7.5) \quad \Rightarrow y = 7.5$

Solution check will be left to the student for the remainder of this section.

5.

$x + 2y = 9.4$

$3x - 5y = 4$

$x = 9.4 - 2y$

$3(9.4 - 2y) - 5y = 4$

$28.2 - 6y - 5y = 4$

$28.2 - 11y = 4$

$28.2 - 4 = 11y$

$24.2 = 11y \quad \Rightarrow y = 2.2$

$x = 9.4 - 2(2.2) \quad \Rightarrow x = 5$

7.

$y = 1.8x - 6.6$

$y + 1.4x = 3$

$y = 3 - 1.4x$

$1.8x - 6.6 = 3 - 1.4x$

$1.8x + 1.4x = 3 + 6.6$

$3.2x = 9.6 \quad \Rightarrow x = 3$

$y = 1.8(3) - 6.6 \quad \Rightarrow y = -1.2$

9.

$3y + 2x = -2$ (1)

$3y - 5x = 26$ (2)

$-1(3y - 5x) = -1(26)$ (2)

$-3y + 5x = -26$ (2)

$3y + 2x = -2$ (1) *add*

$0y + 7x = -28$

$7x = -28 \quad \Rightarrow x = -4$

$3y + 2(-4) = -2$

$3y - 8 = -2$

$3y = 6 \quad \Rightarrow y = 2$

11.

$5x = 14.7 + 6y$ (1)

$2x + 6y = -4.2$ (2)

$5x - 6y = 14.7$ (1) *add*

$7x + 0y = 10.5$

$7x = 10.5 \quad \Rightarrow x = 1.5$

$5(1.5) = 14.7 + 6y$

$7.5 = 14.7 + 6y$

$7.5 - 14.7 = 6y$

$-7.2 = 6y \quad \Rightarrow y = -1.2$

13. $2x + 4y = 0 \ (1)$
$3x - 2y = 24 \ (2)$
$2(3x - 2y) = 2(24) \ (2)$
$6x - 4y = 48 \ (2)$
$2x + 4y = 0 \ (1)$ *add*
$8x + 0y = 48$
$8x = 48 \ \Rightarrow x = 6$
$2(6) + 4y = 0$
$12 + 4y = 0$
$4y = -12 \ \Rightarrow y = -3$

15. $C + D = 2$
$2C + D = -11$ *subtract*
$-C = 13$
$\mathbf{C = -13}$
$-13 + D = 2$
$\mathbf{D = 15}$

17. $4E = F$
$3E + 4F = 34.2$ *substitute*
$3E + 4(4E) = 34.2$
$3E + 16E = 34.2$
$19E = 34.2 \ \Rightarrow E = 1.8$
$4(1.8) = F \ \Rightarrow F = 7.2$

19. $3G + 2H = 0$
$H + G = 0.8$
$H = 0.8 - G$ *substitute*
$3G + 2(0.8 - G) = 0$
$3G + 1.6 - 2G = 0 \ \Rightarrow G = -1.6$
$H + (-1.6) = 0.8 \ \Rightarrow H = 2.4$

21.

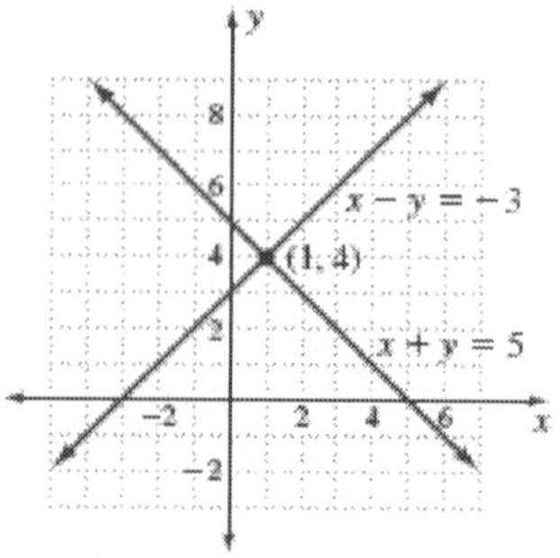

(1, 4)

23.

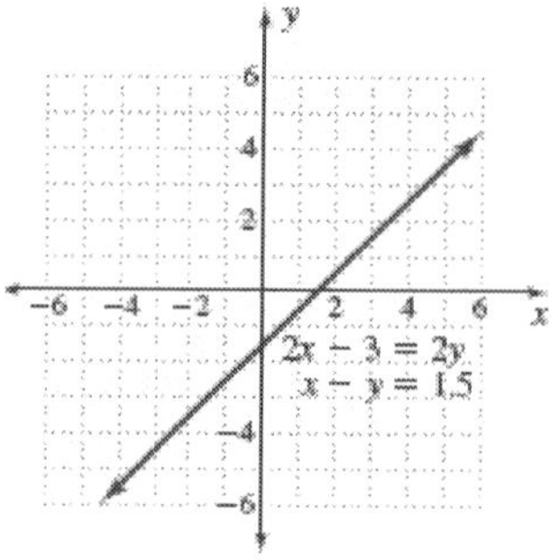

Coincident lines
$x = 1.5 + y; \ 2(1.5 + y) - 3 = 2y$
$3 + 2y - 3 = 2y \quad 2y = 2y$; identity

25.

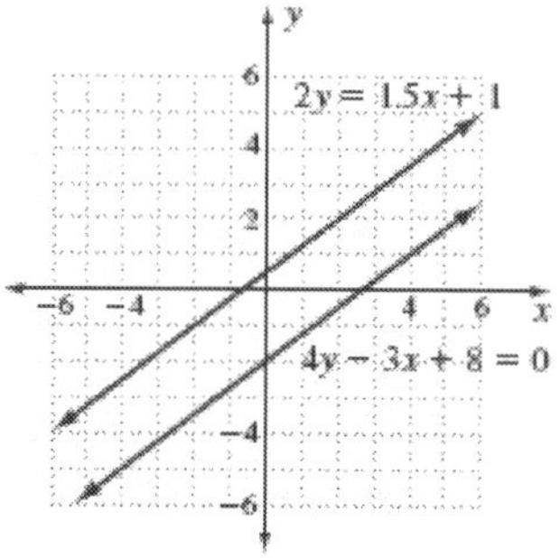

Parallel lines
$2(2y) = 2(1.5x + 1)$
$4y = 3x + 2 \quad 4y = 3x - 8$
$3x + 2 = 3x - 8 \quad 2 = -8$; contradiction

27.

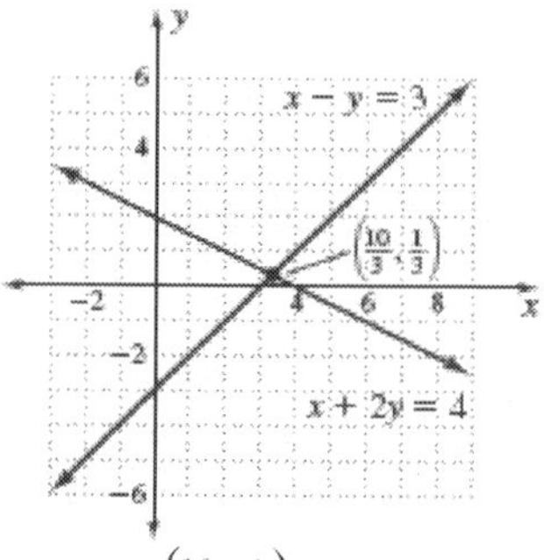

$\left(\frac{10}{3}, \frac{1}{3}\right)$

29.

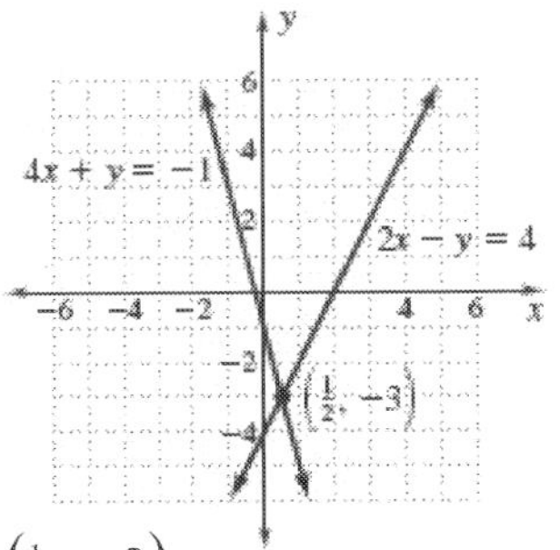

$\left(\frac{1}{2}, -3\right)$

31.

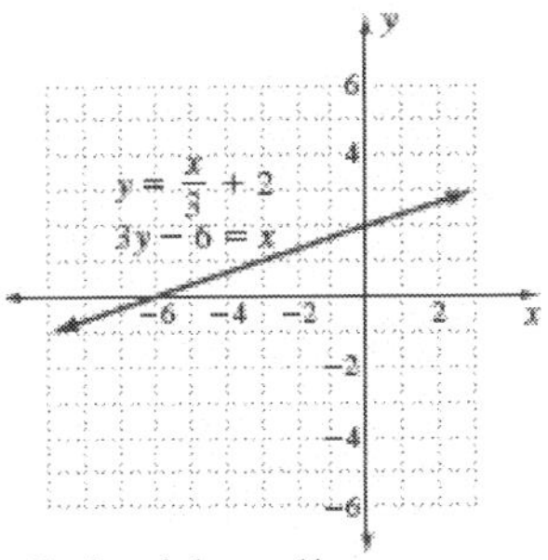

Coincident lines; $3\left(\frac{x}{3}+2\right) - 6 = x$

$\Rightarrow\ x + 6 - 6 = x \quad \Rightarrow\ x = x$; identity

33. Package 1 = 250 + 0.10x
Package 2 = 350 + 0.08x
250 + 0.10x = 350 + 0.08x
0.02x = 100
x = \$5000
250 + 0.10(5000) = \$750

35. Package 3 = 550 + 0.04x
Package 4 = 750 + 0.04x
550 + 0.04x = 750 + 0.04x
0 = 200; contradiction
Parallel graphs; pkg 4 is always better (note y-intercept value).

37. Increasing the percent of sales is the same as increasing the slope: the graph becomes steeper.

39. Packages 3 and 5 have the same fixed amount of \$550. Therefore, they have the same vertical axis intercept.

41. length + width = 30

length = width + 18
(width + 18) + width = 30
2width + 18 = 30
2width = 12 $\Rightarrow$ **width = 6"**
length = 6 + 18 $\Rightarrow$ **length = 24"**

Check: $24 + 6 \stackrel{?}{=} 30$

$30 \stackrel{?}{=} 30$ It checks.

43. $2l + 2w = 228$

$w = \frac{l}{2} - 3$

$2l + 2\left(\frac{l}{2} - 3\right) = 228$

$2l + l - 6 = 228$

$3l = 234 \quad \Rightarrow$ ***l* = 78 ft**

$w = \dfrac{78}{2} - 3 \quad \Rightarrow$ ***w* = 36 ft**

Check:

$2(78) + 2(36) \stackrel{?}{=} 228$

$156 + 72 \stackrel{?}{=} 228$

$228 \stackrel{?}{=} 228$ It checks.

45. box + choc = 45
box = choc – 40

(choc – 40) + choc = 45

2choc – 40 = 45

2choc = 85

choc = \$42.50

choc = \$42.50

box = 42.50 – 40

box = \$2.50

Exercises 0.6: Verbal Representations

1. The difference between x and seven.

3. The quotient of x and seven.

5. The difference between seven and x.

7. The quotient of seven and x.

9. The quotient of seven and the sum of x and two.

11. The quotient of the sum of x and two and the difference between x and two.

13. $10 - 2x = y$

15. $y = \frac{1}{2}x + 10$

17. $\frac{x}{15} + 2 = y$

19. $y = 7x - 14$

21. $y = 1.29x - 0.45$

23. $y = 1.25 - 0.05x$

25. $y = 125x + 85$

27. a. $\frac{1}{6} = 0.1\overline{6} \Rightarrow 0.1\overline{6} \cdot 100 = 16.\overline{6}\% = 16\frac{2}{3}\%$

b. $\frac{3}{5} = 0.60 \Rightarrow 0.60 \cdot 100 = 60\%$

c. $20 \cdot 100 = 2000\%$

d. $0.05 \cdot 100 = 5\%$

29. a. $20\% = \frac{20}{100} = 0.2$

b. $12.5\% = \frac{12.5}{100} = 0.125$

c. $0.005\% = \frac{0.005}{100} = 0.00005$

d. $500\% = \frac{500}{100} = 5$

31. x = season ticket cost, y = daily cost, 5% = 0.05; $y = 0.05x$

33. x = wages, y = amount of tax, 6.2% = 0.062; $y = 0.062x$

35. x = amount of loan, y = one borrowing cost, 1% = 0.01; $y = 0.01x$

37. x = amount of wages, y = net pay, 7.65% = 0.0765, 20% = 0.20, $n\% = \frac{n}{100}$;
$= x - (0.0765x + 0.20x + \frac{n}{100}x)$

39. Quantity: \$1600, \$12,000; Rate: 5% and 8% annual interest

41. Quantity: 20 hr, 15 hr; Rate: \$6.50 and \$7.25 per hour

43. Quantity: 100 mL, 1000 mL; Rate: 8% and 0% solution

For GPA calculations: A = 4 pts, B = 3 pts, C = 2 pts, D = 1 pt and F = 0

45.

Credit hours	Grade point	Total points
4	4	(4)(4) = 16
3	3	(3)(3) = 9
3	2	(3)(2) = 6
1	3	(1)(3) = 3
Total = 11	**GPA = 34 ÷ 11 = 3.09**	Total = 34

$4A + 3B + 3C + 1B = \text{total points}$

$4(4) + 3(3) + 3(2) + 1(3) = 34$

$4 + 3 + 3 + 1 = 11$

$\text{GPA} = 34 \div 11 = 3.09$

47.

Credit hours	Grade points	Total points
75	3.08	75(3.08) = 231
X	4.0	4x
75 + x	3.25	3.25(75 + x) = 231 + 4x

Solving for x: $3.25(75 + x) = 231 + 4x$

$243.75 + 3.25x = 231 + 4x$

$12.75 = 0.75x$

$x = 17$ credit hours of A

49.

$	Rate	Total
x	0.06	0.06x
75,000 – x	0.09	0.09(75,000 – x)
75,000	0.08	0.08(75,000)

$0.06x + 0.09(75{,}000 - x) = 6{,}000$

$0.06x + 6{,}750 - 0.09x = 6{,}000$

$-0.03x = -750$

$x = 25{,}000$

$25,000 should be invested at 6% and $50,000 at 9%.

51.

$	Rate	Total
C	0.21	0.21c
B	0.11	0.11b
10,000	0.13	0.13(10,000)

$b + c = 10{,}000$ and $b = 10{,}000 - c$

$0.11b + 0.21c = 1{,}300$

$0.11(10{,}000 - c) + 0.21c = 1{,}300$

$1{,}100 - 0.11c + 0.21c = 1{,}300$

$0.10c = 200 \Rightarrow$ **c = $2,000**

$b = 10{,}000 - 2{,}000 \Rightarrow$ **b = $8,000**

53.

$	Rate	Total
x	0.045	0.045x
23,600 – x	0.085	0.085(23,600–x)
23,600		1,782

$0.045x + 0.085(23{,}600 - x) = 1{,}782$

$0.045x + 2006 - 0.085x = 1782$

$224 = 0.04x \Rightarrow$ **$x = 5{,}600$**

$5,600 is invested at 4.5% and $18,000 at 8.5%.

55. $A + B = 1{,}000$

$0.15A + 0.10B = 0.12(1{,}000)$

$0.15A + 0.10(1{,}000 - A) = 120$

$0.15A + 100 - 0.10A = 120$

$0.05A = 20 \Rightarrow$ **$A = 400$ lb**

$400 + B = 1{,}000 \Rightarrow$ **$B = 600$ lb**

57.

Liters	%	Total
x	0.03	0.03x
y	0.20	0.20y
1,000	0.064	0.064(1,000)

$x + y = 1{,}000$ and $y = 1{,}000 - x$

$0.03x + 0.20y = 64$

$0.03x + 0.20(1{,}000 - x) = 64$

$0.03x + 200 - 0.20x = 64$

$136 = 0.17x \Rightarrow$ **$x = 800$ L**

$y = 1{,}000 - 800 \Rightarrow$ **$y = 200$ L**

59. $x + y = 200$

$0x + 0.24y = 0.10(200) \Rightarrow$ **$y \approx 83.3$ gal carbamide peroxide**

$x + 83.3 = 200 \Rightarrow$ **$x \approx 116.7$ gal anhydrous glycerol**

61.

Liters	%	Total
x	0.50	0.50x
y	0.00	0y
0.5	0.05	0.05(0.5)

$x + y = 0.5$

$0.50x = 0.025 \Rightarrow$ **$x = 0.05$ L**

$0.05 + y = 0.5 \Rightarrow$ **$y = 0.45$ L**

63.

mL	%	Total
x	0.50	0.50x
y	0.00	0y
100	0.20	0.20(100)

$x + y = 100$

$0.50x = 20$

$x = 40$ mL of dextrose solution

$40 + y = 100$

$y = 60$ mL of distilled water

Exercises 0.7: Mathematical Thinking: Problem Solving

Exercises 1 and 3 are writing exercises, and answers will vary.

5. We might ask: What is the total value of the money? The reader will need to supply the following facts: dimes are \$0.10 and quarters are \$0.25.

7. We might ask: What is measure of each angle? The reader will need to supply the following facts: the sum of the three interior angles in a triangle is 180°.

9. We might ask: What is the perimeter? What are the measures of the other two sides? The reader will need to supply the following fact: opposite sides of a rectangle are equal.

11.

Change in x		1	1	1
x	1	2	3	4
y	2	5	8	11
Change in y		3	3	3

The changes in x and y are constant. Working backwards in the table, when $x = 0$, $y = 2 - 3 = -1$. Using the change in y as the coefficient on x, the equation is $y = 3x - 1$.

13.

Change in x		1	1	1
x	1	2	3	4
y	3	5	7	9
Change in y		2	2	2

The changes in x and y are constant. Working backwards in the table, when $x = 0$, $y = 3 - 2 = 1$. Using the change in y as the coefficient on x, the equation is $y = 2x + 1$.

15.

Change in x		1	1	1
x	1	2	3	4
y	2	4	6	8
Change in y		2	2	2

The changes is x and y are constant. Working backwards in the table, when $x = 0$, $y = 2 - 2 = 0$. Using the change in y as the coefficient on x, the equation is $y = 2x$.

17.

x	y	xy = 4
1	4	(1)(4) = 4
2	2	(2)(2) = 4
3	$\frac{4}{3}$	(3)($\frac{4}{3}$) = 4

19.

x	y	4x = y
1	4	(4)(1) = 4
2	8	(4)(2) = 8
3	12	(4)(3) = 12

21.

x	y	x + y = 4
1	3	(1) + (3) = 4
2	2	(2) + (2) = 4
3	1	(3) + (1) = 4

23. Guesses will vary. Equation is y = 1.07x + \$24.95. When y = \$350, x = \$303.79.

25. Guesses will vary. Equation is y = (\$74(3) + \$0.50x)0.95. When y = \$431.30, x = 464 miles.

27. Guesses will vary. Equations are x = first, x – \$50,000 = second, and x – \$30,000 = third. x + (x – \$50,000) + (x – \$30,000) = \$1,000,000; shares are \$360,000, \$310,000, and \$330,000.

29.

Input	1	4	5	8	n
Output	9	**12**	13	16	$n + 8$

31.

Input	0	2	4	5	n
Output	0	1	**2**	2.5	$\frac{n}{2}$

33.

Input	–1	0	1	$\frac{1}{2}$	n
Output	1	0	**1**	$\frac{1}{4}$	n^2

35.

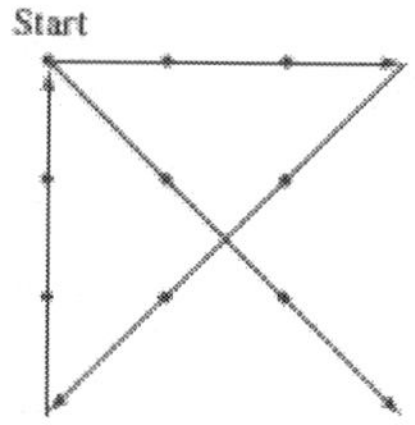

37. The four digits as a single number is the difference between 10^4 and the product of 30 and 293.

Chapter 0: Review Exercises

1. Associative properties for addition and multiplication, commutative properties for addition and multiplication and distributive property for multiplication over addition.

2. Integers, irrational numbers, natural numbers, rational numbers and whole numbers.

3. Absolute value, braces, brackets, fraction bar, parentheses and square root.

4. Axes, coordinate plane, ordered pair, origin, quadrants, scale, x- and y-axes, perpendicular, x-coordinate and y-coordinate.

5. Assumption and condition.

6. Constant, expression, numerical coefficient, term, variable, like terms, constant terms, greatest common factors and principal square root.

7. Sum, difference, product and quotient.

8. Equivalent equations, inverse order of operations, solution set, addition property of equations, multiplication property of equations, substitution method, elimination method, solution to the system and system of equations.

9. Input-output rule, independent variable and dependent variable.

10. Additive inverses, opposites, multiplicative inverses and reciprocals.

11. Evaluate, simplify, factor, and solve an equation.

12. Solution set, identity and contradiction.

13. Inequality, compound inequality, interval and line graph.

14. Coincident lines and parallel lines

15. a. $-3+(-5)-(-6)=-3+(-5)+6=-8+6=-2$

b. $-\frac{1}{2}+\frac{1}{4}-\frac{3}{4}-(-\frac{1}{4})=-\frac{1}{2}+\frac{1}{4}+\frac{1}{4}-\frac{3}{4}=-\frac{1}{2}+\frac{1}{2}-\frac{3}{4}=-\frac{3}{4}$

c. $2.25+8.50-3.75=10.75-3.75=7.0$

d. $6.2-8.6+1.8=6.2+1.8-8.6=8.0-8.6=-0.6$

16. a. $2\cdot 17\cdot 50=2\cdot 50\cdot 17=100\cdot 17=1700$

b. $\frac{1}{2}(-7)(-6)=\frac{1}{2}(42)=21$

c. $-4(-8)(-5)=-4(40)=-160$

d. $\frac{3}{5}(\frac{11}{6})(\frac{15}{121})=\frac{3(11)(3)(5)}{5(3)(2)(11)(11)}=\frac{3}{2(11)}=\frac{3}{22}$

17. a. $-4^2 = -(4)(4) = -16$

b. $(-4)^2 = (-4)(-4) = 16$

c. $0 - 4^2 = -(4)(4) = -16$

d. $8 - 6 \div 3 + 3 \cdot 4 + \frac{1}{2}(8 - 3) = 8 - 2 + 12 + \frac{1}{2}(5) = 6 + 12 + 2\frac{1}{2} = 20\frac{1}{2}$

e. $\dfrac{3-6}{2-(-7)} = \dfrac{-3}{2+7} = -\dfrac{3}{9} = -\dfrac{1}{3}$

f. $\dfrac{-(-2)-\sqrt{(-2)^2-4(5)(-3)}}{2(5)} = \dfrac{2-\sqrt{4+60}}{10} = \dfrac{2-\sqrt{64}}{10} = \dfrac{2-8}{10} = \dfrac{-6}{10} = -\dfrac{3}{5}$

g. $\sqrt{(2-(-7))^2+(3-6)^2} = \sqrt{9^2+(-3)^2} = \sqrt{81+9} = \sqrt{90} = 3\sqrt{10} \approx 9.5$

h. $\dfrac{|1.25-2.83|+|2.56-2.83|+|4.68-2.83|}{3} = \dfrac{1.58+0.27+1.85}{3} = \dfrac{3.7}{3} = \dfrac{37}{30} \approx 1.2$

18. a. $\frac{1}{2} + 96.43 + 3 \div 6 = \frac{1}{2} + 96.43 + \frac{1}{2} = 96.43 + 1 = 97.43$

b. $9 \cdot 59 + 59 = 59(9 + 1) = 59(10) = 590$

c. $8(8\frac{3}{8}) = 8(8) + 8(\frac{3}{8}) = 64 + 3 = 67$

d. $3.57(39) + 6.43(39) = 39(3.57 + 6.43) = 39(10) = 390$

e. $0.0333333\overline{3} + \frac{3}{10} = 0.033333\overline{3} + 0.300 = 0.33333\overline{3} = \frac{1}{3}$

f. $(225 + 127) + (576 - 127) = 225 + 576 - 127 + 127 = 225 + 576$

19. a. $-2^2 + 5 - 3(4 - 5) = -4 + 5 - 3(-1) = -4 + 5 + 3 = 4$

b. $3\{6 - 2[8 - 3(9 - 12) + 3] - 1\} = 3\{6 - 2[8 - 3(-3) + 3] - 1\} = 3\{6 - 2[8 + 9 + 3] - 1\}$
$= 3\{6 - 2[20] - 1\} = 3\{6 - 40 - 1\} = 3\{-35\} = -105$

c. $2\pi(1.5 \text{ in.})^2 + 2\pi(1.5 \text{ in.})(4 \text{ in.}) = 2\pi(1.5 \text{ in.})(1.5 \text{ in.} + 4. \text{ in.}) = 2\pi(1.5 \text{ in.})(5.5 \text{ in.})$
$= 16.5\pi \text{ in.}^2 \approx 51.84 \text{ in.}^2$

d. $\frac{1}{60{,}000}(1500)^2 - \frac{3}{100}(1500) + 1000 = \frac{1}{60{,}000}(2250000) - \frac{3}{100}(1500) + 1000$

$37.5 - 45 + 1000 = 992.5$

c.
$$\frac{\frac{4}{3}\pi\left(\frac{30}{2\pi}\right)^3}{\frac{4}{3}\pi\left(\frac{9\frac{1}{4}}{2\pi}\right)^3} = \frac{\frac{30^3}{2^3\pi^3}}{\frac{\left(\frac{37}{4}\right)^3}{2^3\pi^3}} = \frac{30^3}{2^3\pi^3} \div \frac{\left(\frac{37}{4}\right)^3}{2^3\pi^3} = \frac{30^3}{2^3\pi^3} \cdot \frac{2^3\pi^3}{\left(\frac{37}{4}\right)^3} = 27000 \cdot \frac{16}{50653} \approx 34$$

The ratio is approximately 34 to 1.

20. a. $h = \dfrac{s\sqrt{3}}{2} = \dfrac{4\sqrt{3}}{2} = 2\sqrt{3} \text{ yd}$

b. $A = \pi\left(\dfrac{d}{2}\right)^2 = \pi\left(\dfrac{5}{2}\right)^2 = \dfrac{25}{4}\pi \text{ cm}^2 = 6.25\pi \text{ cm}^2$

21. a. The difference between three and the product of two and a number.

b. The product of three and the difference between a number and 5.

22. a. The sum of five and the product of four and a number.

b. The product of five and the sum of a number and three.

23. $\frac{x}{15}+8$

24. $6(5-x)$

25. $9 - 3(2x - 5y) - 4(5x - 3y) = 9 - 6x + 15y - 20x + 12y = -6x - 20x + 15y + 12y + 9$
$= -26x + 27y + 9$

26. $5(x-3)-3(2-3x)-5(3x-2)=5x-15-6+9x-15x+10=-x-11$

27. $(-2)^2 - (-5)^2 = 4 - 25 = -21$

28. $(-3)^2+2(-3)(-4)+(-4)^2=9+24+16=49$

29. $A = \pi(3.5\text{ in})^2$
$A = 12.25\pi\text{ in}^2$

30. $A=\frac{1}{2}(4.5\,\text{cm})(6.5\,\text{cm})=14.625\,\text{cm}^2$

31.

x	y = −x
−2	−(−2) = 2
−1	−(−1) = 1
0	−(0) = 0
1	−(1) = −1
2	−(2) = −2

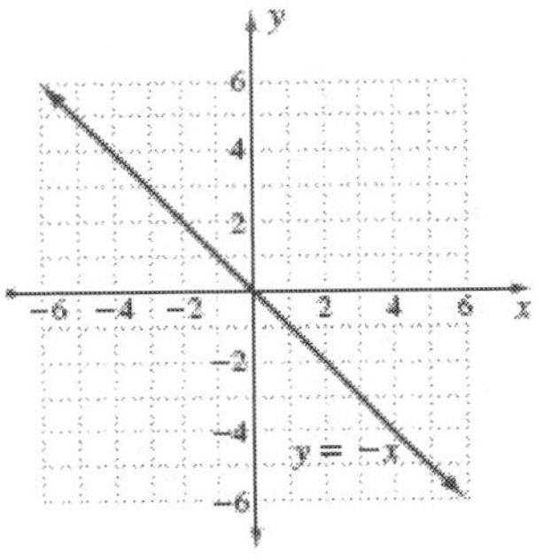

32.

x	y = 4 − 2x
−2	4 − 2(−2) = 8
−1	4 − 2(−1) = 6
0	4 − 2(0) = 4
1	4 − 2(1) = 2
2	4 − 2(2) = 0

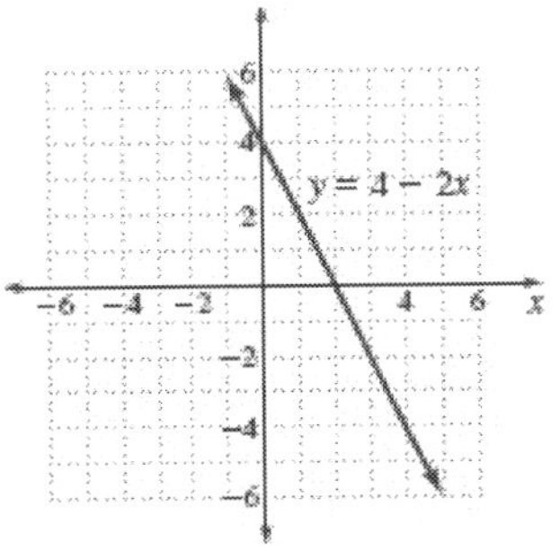

33.

x	y = 30 – 3.5x
–2	30 – 3.5(–2) = 37
–1	30 – 3.5(–1) = 33.5
0	30 – 3.5(0) = 30
1	30 – 3.5(1) = 26.5
2	30 – 3.5(2) = 23

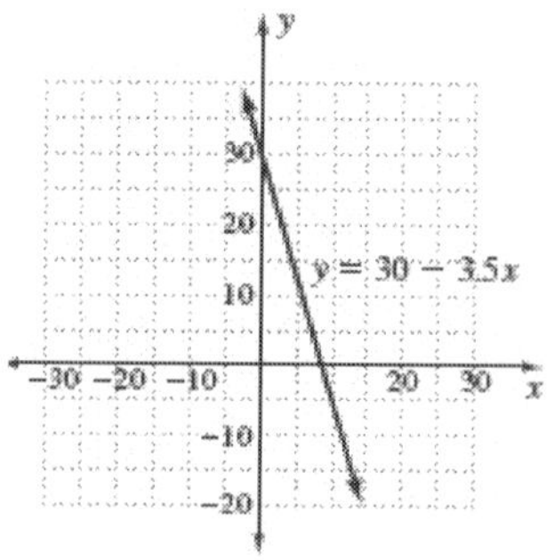

34.

x	y = 150x + 300
–2	150(–2) + 300 = 0
–1	150(–1) + 300 = 150
0	150 (0) + 300 = 300
1	150(1) + 300 = 450
2	150(2) + 300 = 600

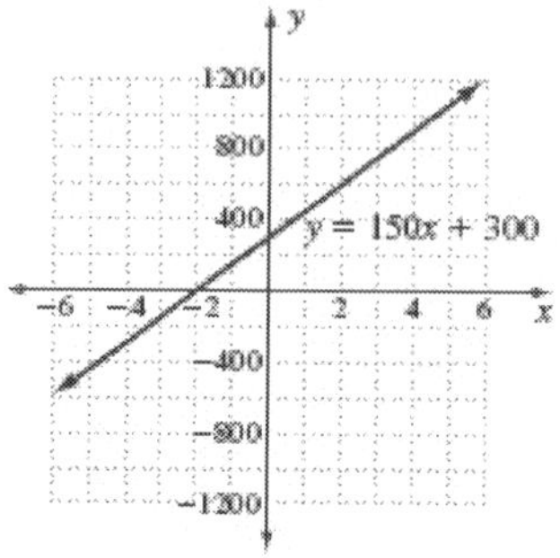

35.

$2x + 2(1.5) = 11$

$2x + 3 - 3 = 11 - 3$

$2x \div 2 = 8 \div 2$

$x = 4$

36.

$25 = \frac{1}{2}(2.5)(x)$

$25 \cdot 2 = \frac{1}{2}(2.5)(x) \cdot 2$

$50 \div 2.5 = (2.5)(x) \div 2.5$

$20 = x$ or $x = 20$

37.

$40 = \frac{5}{9}(F - 32)$

$\frac{9}{5}(40) = \frac{9}{5}[\frac{5}{9}(F - 32)]$

$72 + 32 = F - 32 + 32$

$104 = F$ or $F = 104$

38.

$-40 = \frac{5}{9}(F - 32)$

$-40 \cdot \frac{9}{5} = \frac{5}{9}(F - 32) \cdot \frac{9}{5}$

$-72 + 32 = F - 32 + 32$

$-40 = F$ or $F = -40$

39.

$20 + (n - 1)(-1.75) = 0.75$

$20 + (n - 1)(-1.75) - 20 = 0.75 - 20$

$\frac{(n-1)(-1.75)}{-1.75} = \frac{-19.25}{-1.75}$

$n - 1 + 1 = 11 + 1$

$n = 12$

40.

$60 + 3.75(n - 5) = 150$

$60 + 3.75n - 18.75 = 150$

$3.75n + 41.25 - 41.25 = 150 - 41.25$

$3.75n \div 3.75 = 108.75 \div 3.75$

$n = 29$

41.
$$I = Prt$$
$$\frac{I}{Pt} = \frac{Ptr}{Pt}$$
$$\frac{I}{Pt} = r$$
$$r = \frac{I}{Pt}$$

42.
$$A = \tfrac{1}{2}bh$$
$$A \cdot 2 = \tfrac{1}{2}bh \cdot 2$$
$$2A \div h = bh \div h$$
$$\frac{2A}{h} = b \text{ or } b = \frac{2A}{h}$$

43.
$$C = a + bY$$
$$C - a = a + bY - a$$
$$\frac{C-a}{Y} = \frac{by}{Y}$$
$$\frac{C-a}{Y} = b$$

44.
$$y = mx + b$$
$$y - mx = mx + b - mx$$
$$y - mx = b$$
$$b = y - mx$$

45. a.
$$3x + 1 = -5$$
$$3x + 1 - 1 = -5 - 1$$
$$3x = -6$$
$$x = -2$$

b.
$$3x + 1 = -2x + 1$$
$$3x + 1 - 1 + 2x = -2x + 1 - 1 + 2x$$
$$5x = 0$$
$$x = 0$$

46. a.
$$-2x + 1 = 3$$
$$-2x + 1 - 1 = 3 - 1$$
$$-2x = 2$$
$$x = -1$$

b.
$$\tfrac{1}{2}x - \tfrac{3}{2} = -2x + 1$$
$$\tfrac{1}{2}x - \tfrac{3}{2} + \tfrac{3}{2} + 2x = -2x + 1 + \tfrac{3}{2} + 2x$$
$$\tfrac{5}{2}x = \tfrac{5}{2}$$
$$x = 1$$

47. a.
$$\tfrac{1}{2}x - \tfrac{3}{2} = -4$$
$$\tfrac{1}{2}x - \tfrac{3}{2} + \tfrac{3}{2} = -4 + \tfrac{3}{2}$$
$$\tfrac{1}{2}x = -2.5$$
$$x = -5$$

b.
$$3x + 1 = \tfrac{1}{2}x - \tfrac{3}{2}$$
$$3x + 1 - \tfrac{1}{2}x - 1 = \tfrac{1}{2}x - \tfrac{3}{2} - 1$$
$$2.5x = -2.5$$
$$x = -1$$

48. $x^2 = -2x + 3$ is the intersection(s) of the two graphs; $x = -3$ and $x = 1$.

49. a. Number of quarters; parking time
b. Speed; distance traveled
c. Size of luggage; cost

50. E = escrow balance, x = total payments from account, $\frac{2}{12} = 0.167$;
E = 0.167x;
For E = \$367.40, \$367.40 = 0.167x
\$367.40 ÷ 0.167 = 0.167x ÷ 0.167
x = \$2200

51. a. $-8 < x \le -4$; $(-8, -4]$

b. $-\infty < x < \infty$; $(-\infty, \infty)$

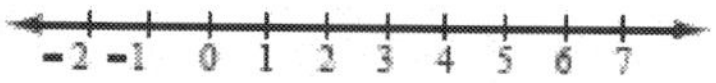

c. $-2 < x < 7$; $(-2, 7)$

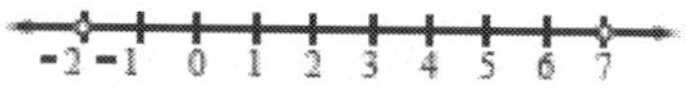

d. $x > -3$; $(-3, \infty)$

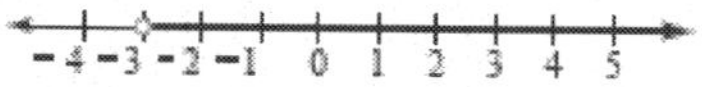

e. $x > 0$; $(0, \infty)$

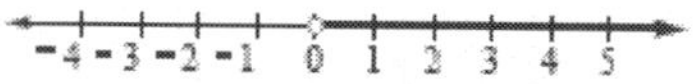

f. $x \ge 0$; $[0, \infty)$

52. a. Set of numbers less than or equal to -2: $x \le -2$, $(-\infty, -2]$

b. Set of numbers between -3 and 2, including 2: $-3 < x \le 2$, $(-3, 2]$

c. Set of numbers greater than -1: $x > -1$, $(-1, \infty)$

d. Set of numbers less than or equal to zero or numbers greater than 2: $x \le 0$ or $x > 2$, $(-\infty, 0]$ or $(2, \infty)$

53.

$x + 2y = -4$

$3x - 4y = 4$

$x = -4 - 2y$

$3(-4 - 2y) - 4y = 4$

$-12 - 6y - 4y = 4$

$-10y = 16$

$\mathbf{y = -1.6}$

$x + 2(-1.6) = -4$

$\mathbf{x = -0.8}$

54.

$2x - 3y = 13$

$4x - y = 11$

$y = 4x - 11$

$2x - 3(4x - 11) = 13$

$2x - 12x + 33 = 13$

$-10x = -20$

$\mathbf{x = 2}$

$4(2) - y = 11$

$\mathbf{y = -3}$

55.

$y = 2x - 1$

$\frac{1}{2}(y + 1) = x$

$\frac{1}{2}(y + 1)(2) = 2x$

$y + 1 = 2x$

$y = 2x - 1$; same as first equation.

Coincident lines

56.

$2x + 3y = 6$

$3y = 2(3 - x)$

$3y = 6 - 2x$

$3y + 2x = 6$; same as first equation.

Coincident lines

57.

$y = 3x - 1$

$y - 2 = 3x$

$y = 3x + 2$; same slope as first eqn.

Parallel lines

58.

$3x + 5y = 5$

$2x - 4y = 18$

$-1.5(2x - 4y) = -1.5(18)$

$-3x + 6y = -27$ *Add to first eqn.*

$11y = -22$; $\mathbf{y = -2}$

$3x + 5(-2) = 5$

$3x = 15$; $\mathbf{x = 5}$

59.

$5x - 102 = 9y$

$3.5x + 1.5y = 9$

$5x - 9y = 102$ (a)

$6(3.5x + 1.5y) = 6(9)$

$21x + 9y = 54$ (b)

Add (a) and (b).

$26x = 156$

$\mathbf{x = 6}$

$5(6) - 102 = 9y$

$-72 = 9y$

$\mathbf{y = -8}$

60.

$1.5x = 3.5y + 19$

$5.5x + 1.5y = 41$

$1.5x - 3.5y = 19$

$3.5(5.5x + 1.5y) = 3.5(41)$

$1.5(1.5x - 3.5y) = 1.5(19)$

$19.25x + 5.25y = 143.5$ (a)

$2.25x - 5.25y = 28.5$ (b)

Add (a) and (b).

$21.5x = 172$

$\mathbf{x = 8}$

$5.5(8) + 1.5y = 41$

$1.5y = -3$

$\mathbf{y = -2}$

61.

$y = -\frac{1}{3}x - \frac{1}{3}$

$2.5x - 4y = 20.5$

$2.5x - 4(-\frac{1}{3}x - \frac{1}{3}) = 20.5$

$2.5x + \frac{4}{3}x + \frac{4}{3} = 20.5$

$3\frac{5}{6}x = 19\frac{1}{6}$

$\mathbf{x = 5}$

$y = -\frac{1}{3}(5) - \frac{1}{3}$

$\mathbf{y = -2}$

62.

$-3x + 4y = 5$

$y = \frac{3}{4}x + 2$

$-3x + 4(\frac{3}{4}x + 2) = 5$

$-3x + 3x + 8 = 5$

$8 = 5$; contradiction

Parallel lines

63.

$0.3x - 0.4y = 1.4$

$0.2x - 0.5y = 0.7$

$-1.5(0.2x - 0.5y) = -1.5(0.7)$

$-0.3x + 0.75y = -1.05$

Add last eqn. to first eqn.

$0.35y = 0.35$

$\mathbf{y = 1}$

$0.3x - 0.4(1) = 1.4$

$0.3x = 1.8$

$\mathbf{x = 6}$

64.

$1.3x - 0.7y = 5$

$2x - 3y = 2.5$

$2x = 2.5 + 3y$

$x = 1.25 + 1.5y$

$1.3(1.25 + 1.5y) - 0.7y = 5$

$1.625 + 1.95y - 0.7y = 5$

$1.25y = 3.375$

$\mathbf{y = 2.7}$

$2x - 3(2.7) = 2.5$

$2x = 10.6$

$\mathbf{x = 5.3}$

65. $\frac{1}{2}x+\frac{3}{4}y=2$

$\frac{1}{3}x-\frac{1}{2}y=-1$

$2(\frac{1}{2}x+\frac{3}{4}y)=2(2)$

$-3(\frac{1}{3}x-\frac{1}{2}y)=-3(-1)$

x + 1.5y = 4

−x + 1.5y = 3 *add*

3y = 7

y = 2$\frac{1}{3}$

$\frac{1}{2}x+\frac{3}{4}(\frac{7}{3})=2$

0.5x + 1.75 = 2

0.5x = 0.25

x = 0.5

66. $\frac{1}{4}x-\frac{1}{2}y=-1$

$\frac{1}{2}x+\frac{3}{5}y=2$

$4(\frac{1}{4}x-\frac{1}{2}y)=4(-1)$

$-2(\frac{1}{2}x+\frac{3}{5}y)=-2(2)$

x − 2y = −4

−x − 1.2y = −4 *add*

−3.2y = −8

y = 2.5

0.25x − 0.5(2.5) = −1

0.25x − 1.25 = −1

0.25x = 0.25

x = 1

67. Solution is the intersection (−1, −5); y = 2x − 3 is the blue line; y = −x − 6 is the red line.

68. Solution is the intersection (−2, 2); y = 3x + 8 is the blue line; y = −x is the red line.

69.

Cr. Hrs.	Grade Pt.	Total
20	2.80	20(2.80) = 56.0
x	3.25	3.25x
20 + x	3.00	3.00(20 + x)

3.00(20 + x) = 56 + 3.25x

60 + 3x = 56 + 3.25x

4 = 0.25x; x = 16 credit hours

70.

Cr. Hrs.	Grade Pt.	Total
20	2.80	20(2.80) = 56.0
x	3.75	3.75x
20 + x	3.00	3.00(20 + x)

3(20 + x) = 56 + 3.75x

60 + 3x = 56 + 3.75x

4 = 0.75x; 5$\frac{1}{3}$ credit hours, or 6 if fractional hours are not available.

71.

Pounds	**Protein**	**Total**
C	0.13	0.13C
D	0.08	0.08D
1000	0.09	90

$C + D = 1000;\ C = 1000 - D$

$0.13C + 0.08D = 90$

$0.13(1000 - D) + 0.08D = 90$

$130 - 0.13D + 0.08D = 90 \quad 40 = 0.05D$ **D = 800 lb** $C = 1000 - 800;$ **C = 200 lb**

72.

mL	**%**	**Total**
W	0	0
L	0.25	0.25L
2000	0.03	60

$W + L = 2000$

$0.25L = 60$

L = 240 mL

$W + 240 = 2000$

W = 1760 mL

73. Let x = # of Jeanne's perfect games and y = # of Tish's perfect games.

$x = y + 2$

$x + y = 40$

$(y + 2) + y = 40$

$2y = 38;$ **y = 19**

$x = 19 + 2;$ **x = 21**

74. a. The algebra yielded a contradiction – **true.**

b. The graphs intersect in a unique point – **false.**

c. The algebra yielded an identity – **false.**

75. Tamika will carry the cat to the car and return for Kim. She will leave Kim in the car and carry the cat back to the house. Tamika will leave then the cat in the house and carry Ming to the car. She will leave Ming in the car and return to the house for the cat.

76. The 5 digits as a single number is the sum of 1953 and 19,247.

Chapter 0: Test

1. Another name for additive inverse is *opposite*.

2. An *ordered pair* is used to locate a point on the coordinate graph.

3. An *independent* variable is the input, x, or set of numbers on the horizontal axis.

4.

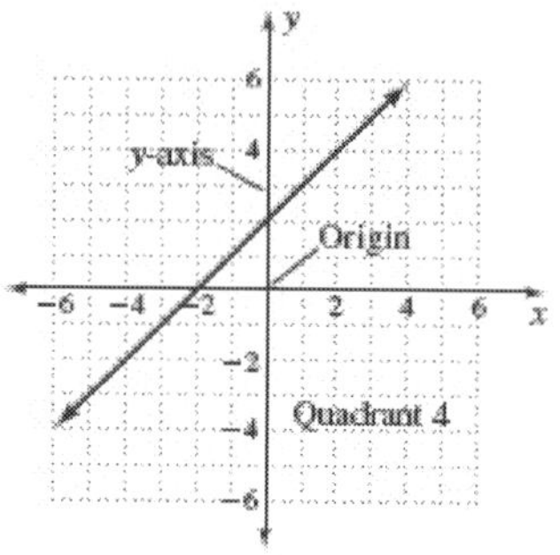

5. a. $-27 + (-3) = -30$

$(-27)(-3) = 81$

$-27 - (-3) = -27 + 3 = -24$

$(-27) \div (-3) = \frac{-27}{-3} = 9$

b. $-1.5 + 0.5 = -1$

$(-1.5)(0.5) = -\frac{3}{2} \cdot \frac{1}{2} = -\frac{3}{4} = -0.75$

$-1.5 - 0.5 = -1.5 + (-0.5) = -2$

$(-1.5) \div (0.5) = \frac{-3}{2} \div \frac{1}{2} = \frac{-3}{2} \cdot \frac{2}{1} = -3$

c. $1\frac{1}{2} + (-\frac{1}{4}) = \frac{3}{2} - \frac{1}{4} = \frac{6}{4} - \frac{1}{4} = \frac{5}{4} = 1\frac{1}{4}$

$(1\frac{1}{2})(-\frac{1}{4}) = (\frac{3}{2})(-\frac{1}{4}) = -\frac{3}{8}$

$1\frac{1}{2} - (-\frac{1}{4}) = \frac{3}{2} + \frac{1}{4} = \frac{6}{4} + \frac{1}{4} = \frac{7}{4} = 1\frac{3}{4}$

$(1\frac{1}{2}) \div (-\frac{1}{4}) = (\frac{3}{2}) \div (-\frac{1}{4}) = \frac{3}{2} \cdot (-\frac{4}{1}) = \frac{-12}{2} = -6$

6. a. $(348 + 295) + 105 = 348 + (295 + 105) = 348 + 400 = 748$; associative property for addition

b. $230 + 689 + 70 = 230 + 70 + 689 = 300 + 689 = 989$; commutative property for addition

7. a. $4\{2 - 3[4 + 5(2 - 5)] + 5\} = 4\{2 - 3[4 + 5(-3)] + 5\} = 4\{2 - 3[4 + (-15)] + 5\}$

$= 4\{2 - 3[-11] + 5\} = 4\{2 + 33 + 5\} = 4\{40\} = 160$

b. $-3^2 = -(3)(3) = -9$

c. $(-3)^2 = (-3)(-3) = 9$

d. $\frac{1}{2} \cdot 25(17 + 23) = \frac{1}{2} \cdot 25(40) = \frac{1}{2} \cdot 1000 = 500$

e. $3x - (8 - 3x) = 3x - 8 + 3x = 6x - 8$

f. $\frac{5 - \sqrt{4}}{2 - 3} = \frac{5 - 2}{2 - 3} = \frac{3}{-1} = -3$

8. The product of four and the difference between a number and three.

9. $A = \dfrac{a+b+c}{3}$

$3A = a + b + c$

$3A - a - c = b$

10. $A = \dfrac{h}{2}(a+b)$

$\dfrac{2A}{h} = a+b$

$\dfrac{2A}{h} - a = b$

11. a. graph crosses y = 0 (x-axis) at x = –5

b. graph intersects y = –2 at x = 2

c. graphs intersect at x = –2

12. a. $\frac{1}{3}x + \frac{5}{3} = 0$

$3(\frac{1}{3}x + \frac{5}{3}) = 3(0)$

$x + 5 = 0$

$x + 5 - 5 = 0 - 5$

$x = -5$

b. $-\frac{3}{4}x - \frac{1}{2} = -2$

$-4(-\frac{3}{4}x - \frac{1}{2}) = -4(-2)$

$3x + 2 = 8$

$3x + 2 - 2 = 8 - 2$

$3x = 6$

$\dfrac{3x}{3} = \dfrac{6}{3} \Rightarrow x = 2$

c. $\frac{1}{3}x + \frac{5}{3} = -\frac{3}{4}x - \frac{1}{2}$

$12(\frac{1}{3}x + \frac{5}{3}) = 12(-\frac{3}{4}x - \frac{1}{2})$

$4x + 20 = -9x - 6$

$4x + 20 - 20 + 9x = -9x - 6 - 20 + 9x$

$13x = -26$

$\dfrac{13x}{13} = -\dfrac{26}{13}$

$x = -2$

13. $A = \dfrac{1.1}{2}(1.4 + 2.6)\ \text{m}^2$

$A = 0.55(4.0)\ \text{m}^2$

$A = 2.2\ \text{m}^2$

14.

x	y
3	1
6	2
7	**$2\frac{1}{3}$**
8	**$2\frac{2}{3}$**
9	3

Extending the table gives x = 7, when y = $2\frac{1}{3}$.
Extending the table backwards, when x = 0, y = 0;
equation is y = $\frac{1}{3}$x.

15. a. C = cost, x = months; C = \$2875x + \$800x; C = \$3675x

b. \$250,000 = \$3675x

$x = \$250{,}000 \div \$3675$; x = 68 months or 5 years 8 months

c. Costs for apartment and care; amount of savings

d. No change in cost; no other costs or income

16. a. $x \le 5$; $(-\infty, 5]$

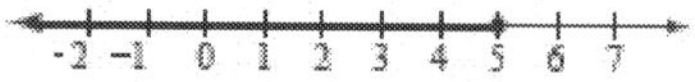

b. $-2 < x < 5$; $(-2, 5)$

c. $\Re$; $(-\infty, \infty)$

17.
$3x - 4y = 12$
$x - \frac{3}{2}y = -9$
$x = 1.5y - 9$ *substitute into 1st eqn.*
$3(1.5y - 9) - 4y = 12$
$4.5y - 27 - 4y = 12$
$0.5y = 39$; $\mathbf{y = 78}$
$x = 1.5(78) - 9$; $\mathbf{x = 108}$

18.
$5x - 2y = 5$
$5x = 2(2 + y)$
$5x - 2y = 4$ *subtract from 1st eqn.*
$0 = 1$; contradiction
No solution; parallel lines

19.
$2x - 3y = 6$
$4x + 5y = -32$
$-2(2x - 3y) = -2(6)$
$-4x + 6y = -12$ *add to 2nd eqn.*
$11y = -44$; $\mathbf{y = -4}$
$2x - 3(-4) = 6$
$2x = -6$; $\mathbf{x = -3}$

20.
$x + 2y = -8$; $x = -2y - 8$
$0.5x + y = -4$
$0.5(-2y - 8) + y = -4$
$-y - 4 + y = -4$
$-4 = -4$; identity
Infinite number of solutions
Coincident lines

21.
$C + P = 3922$
$C + 10 = P$
$C + (C + 10) = 3922$
$2C + 10 = 3922$
$2C = 3912$
$\mathbf{C = 1956}$
$1956 + 10 = P$; $\mathbf{P = 1966}$

22.
$0.16x = 0.014(400)$
$x + w = 400$
$0.16x = 5.6$
x = 35 gal
$35 + w = 400$
w = 365 gal

23. Formula is:

$45.59(3) + 0.149x + 0.0725(45.59(3) + 0.149x) + 0.97(3) \approx 149.60 + 0.16x$

x	50	60	**59**
149.60 + 0.16x	149.60 + 0.16(50) = 157.60	149.60 + 0.16(60) = 159.20	149.60 + 0.16(59) = **159.04**

≈59 clips

CHAPTER 1

Functions and Linear Equations

Exercises 1.1: Functions

1. $f(x) = 15x - 4$

3. Not a function

5. $f(x) = 25 - x^2$

7. Fails vertical-line test, all $x < 2$ have two outputs, not a function.

9. Each input has one output, function

11. Fails vertical-line test, each input has two outputs, not a function.

13. Exercise 8; $[-5, 5]$; $[0, 5]$

15. Exercise 9; $(-\infty, \infty)$; $[0, \infty)$

17. Function; one output for each input.

19. Function; one output for each input.

21. Not a function; 3 and 4 both have two outputs.

23. Function

25. Not a function

27. Not a function

29. x is any real number; $f(x) \geq 0$.

31. x is any real number; $h(x) \geq -2$.

33. x is any real number; $g(x) \leq 6$.

35.

x	$g(x) = 5 + 2(x - 3)$
−2	$5 + 2[(-2) - 3] = -5$
−1	$5 + 2[(-1) - 3] = -3$
0	$5 + 2[(0) - 3] = -1$
1	$5 + 2[(1) - 3] = 1$
2	$5 + 2[(2) - 3] = 3$
3	$5 + 2[(3) - 3] = 5$
4	$5 + 2[(4) - 3] = 7$

37.

x	$f(x) = x^2 - 2x - 3$
−2	$(-2)^2 - 2(-2) - 3 = 5$
−1	$(-1)^2 - 2(-1) - 3 = 0$
0	$(0)^2 - 2(0) - 3 = -3$
1	$(1)^2 - 2(1) - 3 = -4$
2	$(2)^2 - 2(2) - 3 = -3$
3	$(3)^2 - 2(3) - 3 = 0$
4	$(4)^2 - 2(4) - 3 = 5$

39.

x	$g(x) = 8 - x^2$
−2	$8 - (-2)^2 = 4$
−1	$8 - (-1)^2 = 7$
0	$8 - (0)^2 = 8$
1	$8 - (1)^2 = 7$
2	$8 - (2)^2 = 4$
3	$8 - (9)^2 = -1$
4	$8 - (4)^2 = -8$

41. a. $g(3) = (3)^2 + (3) - 2$
$g(3) = 9 + (3) - 2$
$g(3) = 10$

b. $g(1) = (1)^2 + (1) - 2$
$g(1) = 1 + (1) - 2$
$g(1) = 0$

c. $g(\square) = (\square)^2 + (\square) - 2$

d. $g(n) = (n)^2 + (n) - 2$
$g(n) = n^2 + n - 2$

e. $g(n - m) = (n - m)^2 + (n - m) - 2$
$g(n - m) = n^2 - 2nm + m^2 + n - m - 2$

43. 10 digits

45. 9 digits

47. a. $5 + 2(x - 3) = -3$
$5 + 2x - 6 = -3$
$2x - 1 + 1 = -3 + 1$
$2x \div 2 = -2 \div 2$
$x = -1$

b. $5 + 2(x - 3) = 5$
$5 + 2x - 6 = 5$
$2x - 1 + 1 = 5 + 1$
$2x \div 2 = 6 \div 2$
$x = 3$

49. a. From the table, $f(-1)$ and $f(3)$ both equal 0. The solution set is $\{-1, 3\}$.

b. $f(x) = 21$ does not appear on the table. We extend it to find $f(6) = (6)^2 - 2(6) - 3 = 21$. Noting the symmetry, we check $f(-4)$; $(-4)^2 - 2(-4) - 3 = 21$. The solution set is $\{-4, 6\}$.

51. a. From the table, $g(-2)$ and $g(2)$ both equal 4. The solution set is $\{-2, 2\}$.

b. From the table, $g(-1)$ and $g(1)$ both equal 7. The solution set is $\{-1, 1\}$.

53. $f(2)$ is the function $f(x)$ evaluated at $x = 2$.

55. Exterior house doors open into the house, movie theater doors open toward the outside, bedroom doors open into the room, bathroom doors open into the room, closet doors open outward, cupboard doors open outward, and doors on stairways open away from the staircase.

In the case of movie theater and stairway doors, the direction the door opens is mandatory for safety reasons. Closet and cupboard doors are a matter of space. Bedroom and bathroom doors are generally off a hallway, so opening into the room is safer for those in the hallway. Exterior house doors open inward for safety.

Exercises 1.2: Linear Functions

1. $f(x) = \frac{5}{9}(x - 32) = \frac{5}{9}x - \frac{160}{9}$; linear function

3. $C(x) = 2\pi x$; linear function

5. $f(x) = x^2 + 2x$; non-linear

7. $v(t) = gt + v_o$; linear function

9. $h(x) = ax^2 + bx + c$; non-linear

11. a. x-intercept: (–1, 0), y-intercept: (0, 2) **b.** x-intercept: (3, 0), y-intercept: (0, –2)

13. The function shown in **12b** decreases less rapidly.

15. The function shown in **11a** increases more rapidly.

17.
$f(x) = x + 4$
$0 = x + 4$
$-4 = x \Rightarrow$ x-intercept: (–4, 0)

$f(x) = x + 4$
$f(0) = 0 + 4$
$f(0) = 4 \Rightarrow$ y-intercept: (0, 4)

19.
$g(x) = 2x + 5$
$0 = 2x + 5$
$-5 = 2x$
$-\frac{5}{2} = x \Rightarrow$ x-intercept: $(-\frac{5}{2}, 0)$

$g(x) = 2x + 5$
$g(0) = 2(0) + 5$
$g(0) = 5 \Rightarrow$ y-intercept: (0, 5)

21.
$f(x) = \frac{2}{3}x - 6$
$0 = \frac{2}{3}x - 6$
$6 = \frac{2}{3}x$
$9 = x \Rightarrow$ x-intercept: (9, 0)

$f(x) = \frac{2}{3}x - 6$
$f(0) = \frac{2}{3}(0) - 6$
$f(0) = -6 \Rightarrow$ y-intercept: (0, – 6)

23.
$f(x) = \frac{3}{5}x + 10$
$0 = \frac{3}{5}x + 10$
$-10 = \frac{3}{5}x$
$-\frac{50}{3} = x \Rightarrow$ x-intercept: $(-\frac{50}{3}, 0)$

$f(x) = \frac{3}{5}x + 10$
$f(0) = \frac{3}{5}(0) + 10$
$f(0) = 10 \Rightarrow$ y-intercept: (0, 10)

25. $\frac{7-5}{-2-4} = \frac{2}{-6} = -\frac{1}{3}$

27. $\frac{-1-(-5)}{5-(-3)} = \frac{4}{8} = \frac{1}{2}$

29. $\frac{-3-0}{0-5} = \frac{3}{5}$

31. $\frac{-3-0}{0-(-2)} = -\frac{3}{2}$

33. $\frac{0-(-2)}{3-3} = \frac{2}{0}$; undefined

35. $\frac{-4-(-4)}{4-(-2)} = \frac{0}{6} = 0$

37. $\frac{-4.2-5.3}{2.6-1.5} = \frac{-9.5}{1.1} = -\frac{95}{11}$

39. $\frac{-1\frac{1}{4}-4\frac{1}{2}}{2\frac{1}{2}-(-1\frac{3}{4})} = -\frac{5.75}{4.25} = -\frac{575}{425} = -\frac{23}{17}$

41. $\frac{2\frac{1}{3}-3\frac{2}{3}}{1\frac{2}{3}-(-1\frac{1}{3})} = \frac{-\frac{4}{3}}{3} = -\frac{4}{3} \cdot \frac{1}{3} = -\frac{4}{9}$

43. a. $\Delta x = \$10$ sales, $\Delta y = \$0.60$ tax; slope $= \frac{\$0.60}{\$10} = \$0.06$ tax/\$ sales

b. Working backward in the table, x-intercept = (0, 0); \$0 sales means \$0 tax.

c. y-intercept is also (0, 0), there is 0 sales tax if there is 0 sales.

45. a. $\Delta x = 1$ trip, $\Delta y = -\$0.75$ value; slope = –\$0.75 value/trip

b. Working forward on the table, x-intercept = $(26\frac{2}{3}, 0)$; maximum number of trips is 26.

c. y-intercept is in the table (0, 20); original value of mass transit ticket is \$20.

47. $\Delta x = 0.5$ sec, Δy is not constant; function is not linear.

49. From (4, 5), move –2 units in y and 3 units in x; $(4 + 3, 5 - 2) = (7, 3)$.

51. From (–4, 1), move 3 units in y and 5 units in x; $(-4 + 5, 1 + 3) = (1, 4)$.

53. x-intercept $(a, 0)$, y-intercept $(0, b)$; slope is $\frac{b-0}{0-a} = -\frac{b}{a}$.

55. The slope of a vertical line is undefined.

57. A vertical line has more than one output for each input, which is contrary to the definition of a function.

59. a.
$y = 3x - 8$
$y - y + 8 = 3x - 8 + 8 - y$
$8 = 3x - y$
$3x - y = 8 \Rightarrow$ Linear in two variables

b.
$x = 5 + y$
$x - y = 5 + y - y$
$x - y = 5 \Rightarrow$ Linear in two variables

c.
$C = x + 0.15x + 0.06x$
$C - 1.21x = 1.21x - 1.21x$
$C - 1.21x = 0 \Rightarrow$ Linear in two variables

d.
$3x = 4$
$3x - 4 = 4 - 4$
$3x - 4 = 0 \Rightarrow$ Linear in one variable

e.
$x^2 = 0$
Non-linear

f.
$4n - 5 = 10n + 7$
$4n - 5 + 5 - 4n = 10n + 7 + 5 - 4n$
$0 = 6n + 12$
$6n + 12 = 0$
$n + 2 = 0 \Rightarrow$ Linear in one variable

g.
$2\pi r = 11$
$2\pi r - 11 = 11 - 11$
$2\pi r - 11 = 0 \Rightarrow$ Linear in one variable

h.
$6x^2 - 2x(3x - 4) = 7$
$6x^2 - 6x^2 + 8x = 7$
$6x^2 - 6x^2 + 8x - 7 = 7 - 7$
$8x - 7 = 0 \Rightarrow$ Linear in one variable

i.
$t - 6n = 4n - 2$
$t - 6n + 6n + 2 = 4n - 2 + 2 + 6n$
$t + 2 - t = 10n - t$
$2 = 10n - t$
$10n - t = 2 \Rightarrow$ Linear in two variables

j.
$P = 4L$
$P - P = 4L - P$
$0 = 4L - P$
$4L - P = 0 \Rightarrow$ Linear in two variables

Exercises 1.3: Modeling with a Linear Function

1. $y = 0.055x$; slope = \$0.055/\$; y-intercept = (0, \$0)

3. $y = 3.00x + 10$; slope = \$3.00/person; y-intercept = (0, \$10)

5. $C = 2\pi r$; slope = 2π; vertical axis intercept = (0, 0)

7. $F = \mu N$; slope = μ, vertical axis intercept = (0, 0)

9. $C = a + bY$; slope = b, vertical axis intercept = (0, a)

11. Slope = 8; y-intercept = (0, –4): $y = 8x - 4$

13. Slope = $\frac{1}{2}$; y-intercept = (0, –8): $y = \frac{1}{2}x - 8$

15. Slope = –2; y-intercept = (0, 0): $y = -2x$

17. (3, 6) and (0, –2) $\Rightarrow$ y-intercept ; slope = $\frac{-2-6}{0-3} = \frac{8}{3}$; $y = \frac{8}{3}x - 2$

19. (–2, 4) and (5, –3); slope = $\frac{-3-4}{5-(-2)} = \frac{-7}{7} = -1$; $b = 4 - (-1)(-2)$, $b = 2$

$y = -1x + 2$ or $y = -x + 2$

21. a. Pulse rate is a function of age.

b. Answers will vary.

c. Max. pulse rate is 220 – age. Let x = age and P = pulse rate. $P = 0.5(220 - x)$

d. $P = 0.7(220 - x)$

e. $P = 0.5(220 - 50)$
$P = 0.5(170)$
$P = 85$

$P = 0.7(220 - 50)$
$P = 0.7(170)$
$P = 119$

f. $95 = 0.5(220 - x)$
$190 = 220 - x$
$x = 30$

$133 = 0.7(220 - x)$
$190 = 220 - x$
$x = 30$

23. (2, 5) and (5, 4); slope = $\frac{4-5}{5-2} = -\frac{1}{3}$; $b = 5 - (-\frac{1}{3})(2)$, $b = \frac{17}{3}$; $y = -\frac{1}{3}x + \frac{17}{3}$

25. (5, 4) and (4, 1); slope = $\frac{1-4}{4-5} = \frac{-3}{-1} = 3$; $b = 4 - 3(5)$, $b = -11$; $y = 3x - 11$

27. (4, 1) and (2, 5); slope = $\frac{5-1}{2-4} = \frac{4}{-2} = -2$; $b = 1 - (-2)(4)$, $b = 9$; $y = -2x + 9$

29. slope $= \dfrac{212-32}{100-0} = \dfrac{180}{100} = \dfrac{9}{5}$; b = 32; F = $\frac{9}{5}$C + 32

31. Let y = cost in \$ and x = size in inches; points are (4, 0.99) and (6, 1.99);

slope $= \dfrac{1.99-0.99}{6-4} = \dfrac{1}{2} = 0.5$;

$y-0.99 = 0.5(x-4)$

$y-0.99 = 0.5x-2$

$y = 0.5x-1.01$

If x = 8, y = 0.5(8) – 1.01= \$2.99.

33. Let y = cost in \$ and x = pounds; points are (7, 5.99) and (16, 8.99);

slope $= \dfrac{8.99-5.99}{16-7} = \dfrac{3}{9} = \dfrac{1}{3} \approx 0.333$;

$y-5.99 = 0.333(x-7)$

$y-5.99 = 0.333x-2.331$

$y = 0.333x+3.659$

If x = 10, y = 0.333(10) + 3.66 = \$6.99.

35. Let y = cost in \$ and x = area of print; points are (17.5, 0.39) and (80, 4.49);

slope $= \dfrac{4.49-0.39}{80-17.5} = \dfrac{4.1}{62.5} = 0.0656$;

$y-0.39 = 0.0656(x-17.5)$

$y-0.39 = 0.0656x-1.148$

$y = 0.0656x-0.758$

If x = 35, y = 0.0656(35) – 0.758 = \$1.54

The actual current cost is \$1.19.

37.

Δx	x (cups)	y (\$)	Δy	Slope
	1	3.19		
1			0.80	0.80
	2	3.99		
2			1.00	0.50
	4	4.99		
4			2.50	0.625
	8	7.49		

Average slope ≈ 0.64; working backwards in table y-intercept = 2.55; y = 0.64x + 2.55

39.

Δx	x (#)	y ($)	Δy	slope
	10	7.99		
20			2.00	0.10
	30	9.99		
50			5.00	0.10
	80	14.99		

Average slope = 0.10; working backwards in table y-intercept = 6.99; $y = 0.10x + 6.99$

41.

Δx	x	y	Δy
	1	−8	
1			5
	2	−3	
1			5
	3	2	
1			5
	4	7	
1			5
	5	12	

slope = 5; $b = -8 - 5(1) = -13$

$y = 5x - 13$

43.

Δx	x	y	Δy
	1	9	
1			2
	2	11	
1			2
	3	13	
1			2
	4	15	
1			2
	5	17	

slope = 2; $b = 9 - 2(1) = 7$

$y = 2x + 7$

45.

Δx	x	y	Δy
	1	2	
1			6
	2	8	
1			6
	3	14	
1			6
	4	20	
1			6
	5	26	

slope = 6; $b = 2 - 6(1) = -4$

$y = 6x - 4$

47.

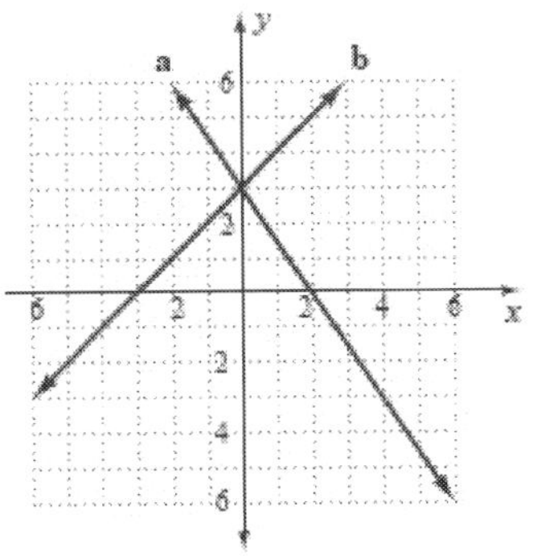

a: $y = -\frac{3}{2}x + 3$

b: $y = x + 3$

49.

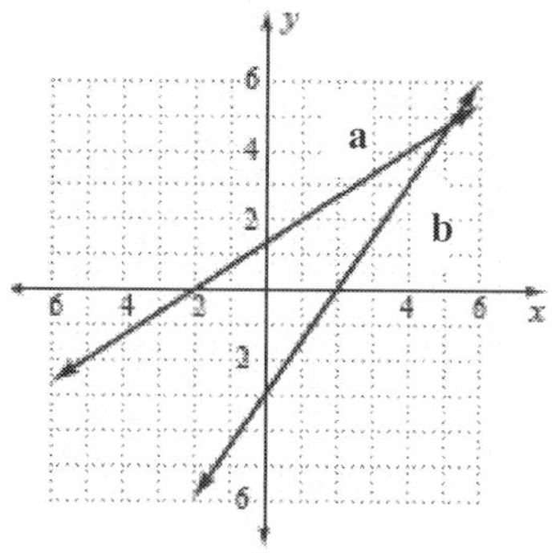

a: $y = \frac{2}{3}x + \frac{4}{3}$

b: $y = \frac{3}{2}x - 3$

51. Answers will vary, but should include plotting the ordered pair (c, d) and moving a units vertically and b units horizontally to find a second ordered pair.

53. Start at the intercept, move vertically up (or down if slope is negative) the number of units in the numerator of the slope, and then move right horizontally the number of units in the denominator.

55. a. See part c for graph.

b. Yes the data looks linear.

c.

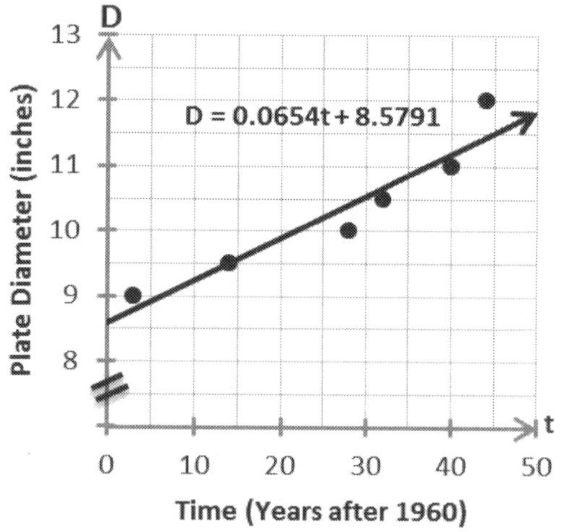

Answers may vary. The regression line, from section 1.4, is $D = 0.0654t + 8.5791$

d. Answers may vary. From the regression line, (0, 8.6) or about 8.6 inches is the plate size. It may be reasonable if plate sizes were increasing before 1960 but plate sizes are more likely to have been uniform for a long period of time.

e. The regression line suggests about 9.9 inches in 1980.

Exercises 1.4: Modeling Special Lines and Functions

1. a. $2x = y - 2(3 - x)$
$2x = y - 6 + 2x$
$2x - 2x + 6 = y - 6 + 2x - 2x + 6$
$6 = y$ or $y = 6$

b. $y - 4 = 2x + y$
$y - 4 + 4 - y = 2x + y + 4 - y$
$0 = 2x$
$x = 0$

Lines are perpendicular; zero slope vs. undefined slope.

3. a. $x = -6y$
$x \div -6 = -6y \div -6$
$-\frac{x}{6} = y$ or $y = -\frac{x}{6}$

b. $3x = y - 3x - 4$
$3x + 3x + 4 = y - 3x - 4 + 3x + 4$
$6x + 4 = y$ or $y = 6x + 4$

Lines are perpendicular; slopes are negative reciprocals.

5. a. $x = 4$

b. $y = x + y - 5$
$y - y = x + y - 5 - y$
$0 = x - 5$ or $x = 5$

Lines are parallel; same slope - both undefined.

7. a. $y + 2x = 2$
$y + 2x - 2x = 2 - 2x$
$y = -2x + 2$

b. $y = 2(x - 1)$
$y = 2x - 2$

Lines are neither parallel nor perpendicular.

9. $C = 78x$, $C = 98x$, $C = 108x$; not parallel, different slopes

11. $V = 5.00 - 0.05x$, $V = 10.00 - 0.05x$, $V = 20.00 - 0.05x$; parallel lines, same slope

13. If postage cost is \$0.44 per stamp, $C = 100(0.44)x = 44x$; $C = 50(0.44)x = 22x$; $C = 20(0.44)x = 8.80x$; not parallel, different slopes

In exercises 15 to 23, change to $y = mx + b$ form (where necessary) to find the slope of the original equation before solving the problem.

15. $2x + 3y = 6$
$3y = -2x + 6$
$y = -\frac{2}{3}x + 2$
Parallel line has same slope.
y-intercept is (0, 0)
$y = -\frac{2}{3}x$

17. $y = -\frac{1}{2}x + 3$
Perpendicular line has negative reciprocal slope.
slope = 2
y-intercept is (0, 0)
$y = 2x$

19. $y = \frac{5}{8}x - 3$

Perpendicular line has negative reciprocal slope.

slope $= -\frac{8}{5}$

$b = 3 - (-\frac{8}{5})(2) = \frac{31}{5}$

$y = -\frac{8}{5}x + \frac{31}{5}$

21. $4x - 3y = 12$

$4x - 12 = 3y$

$y = \frac{4}{3}x - 4$

Parallel lines have same slope.

$b = 1 - (\frac{4}{3})(-2) = \frac{11}{3}$

$y = \frac{4}{3}x + \frac{11}{3}$

23. $5x - 2y = 8$

$5x - 8 = 2y$

$y = \frac{5}{2}x - 4$

Perpendicular lines have negative

reciprocal slope.

slope $= -\frac{2}{5}$

$b = -1 - (-\frac{2}{5})(2) = -\frac{1}{5}$

$y = -\frac{2}{5}x - \frac{1}{5}$

25. Starting at (1, 3) and moving clockwise around the figure:

$\frac{4-3}{3-1} = \frac{1}{2}$, $\frac{2-4}{4-3} = -2$, $\frac{1-2}{2-4} = \frac{1}{2}$,

$\frac{3-1}{1-2} = -2$

Opposite lines are parallel (same slope) and adjacent lines are perpendicular (negative reciprocal slopes).

27. Constant function; output is always \$35.

29. Constant function; output always \$100

31. Constant function; output is always 1.

33. Identity function; output equals input.

35. a. Yes the data looks linear.

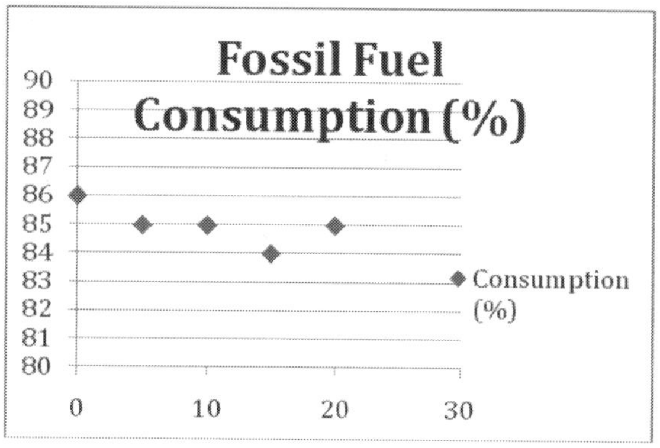

b. $y = -0.06x + 85.6$

c. The slope indicates that our percent consumption of fossil fuels is falling at about 0.06 percent per year. The y-intercept is (0, 85.6), the percent consumption in 1985.

d. $y = -0.06(2020 - 1985) + 85.6 = -0.06(35) + 85.6 = 83.5\%$

37. a. Yes the data looks linear.

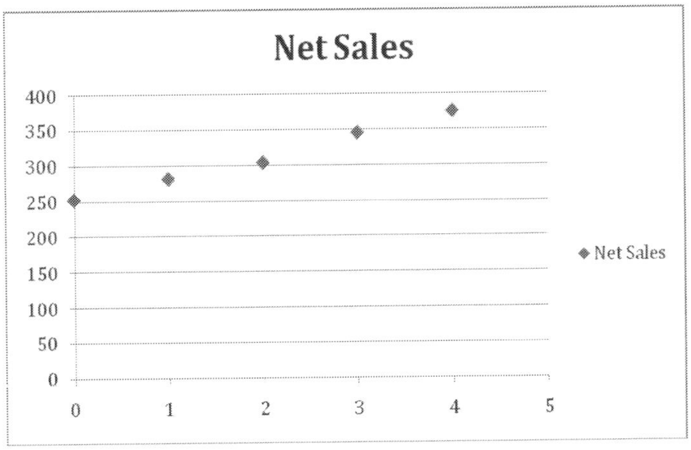

b. y = 30.7x + 250.4

c. The slope is 30.7, an increase in billions per year for Wal-Mart net sales. The *y*-intercept is (0, \$250.4), the net sales in billions in 2004.

d. $y = 30.7(6) + 250.4 \approx \$434.6$ billion

39. a. Yes, although it does have some curve as though the number were leveling off.

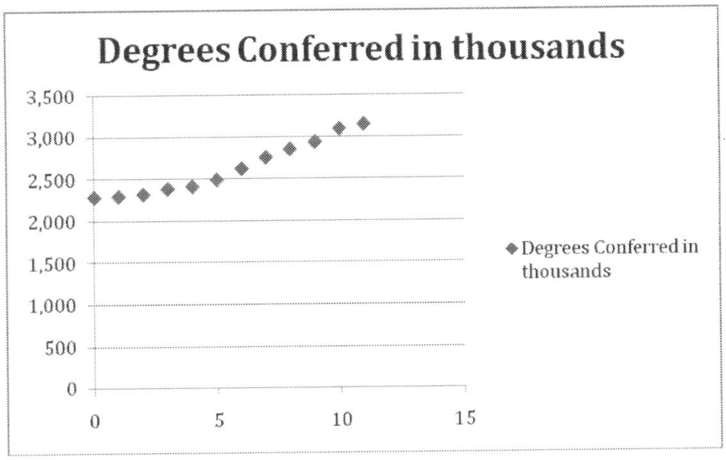

b. $y = 85.07x + 2165.68$

c. The slope is 85.07, the increase in thousands of college degrees every year. The intercept is (0, 2166), the number of degrees, in thousands, in 1997, which is close to the actual number of degrees in 1997.

d. $y = 85.07(2015 - 1997) + 2165.68 = 85.07(18) + 2165.68 \approx 3{,}697$ thousand degrees awarded

41. Dozen lg. eggs = 12(50 g) = 600 g, dozen extra lg. eggs = 12(56 g) = 672 g, dozen jumbo eggs = 12(63 g) = 756 g; $y \approx 0.0013x + 0.225$, with x in grams and y in \$

43. $y = 0.4x - 0.21$

45. a. Dot graph, pages are integers

b.

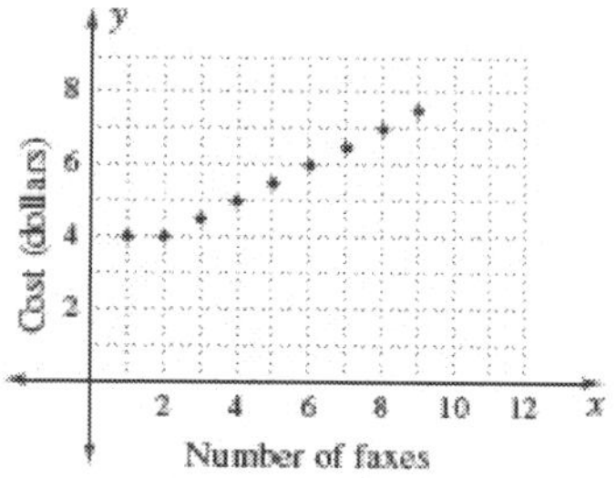

47. a. Step graph, round *x* up to whole hour

b.

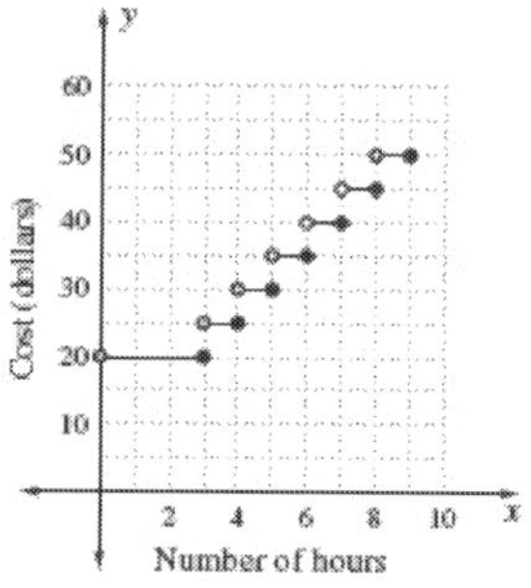

49. a. Dot graph, skaters are integers

b. Note: Dots in graph appear as a solid line due to selection scale on x-axis.

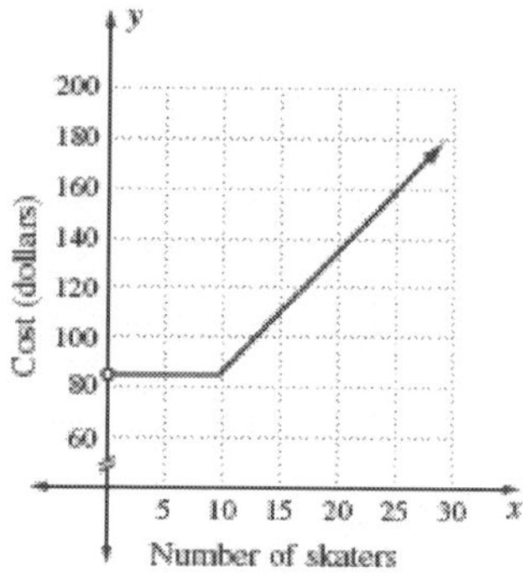

51. a. Step graph, round *x* up to whole minute

b.

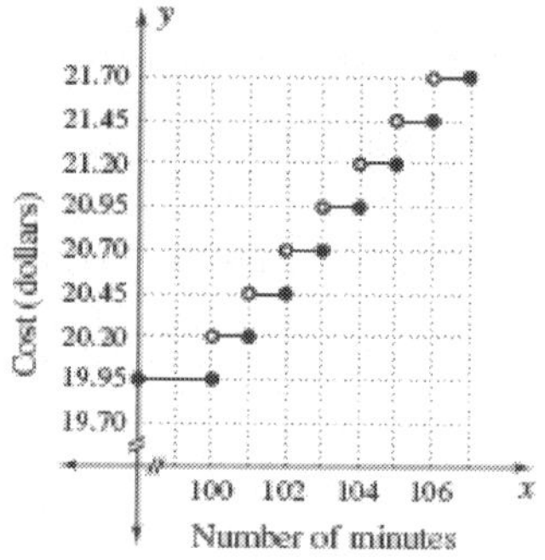

53. part of an hour, portion of a minute

55. Opposite sides should have the same slopes and adjacent sides should have negative reciprocal slopes.

$\frac{5}{2}, -\frac{5}{2}, \frac{3}{2}, -\frac{3}{2}$; not a rectangle;

$\frac{-2-(-1)}{1-(-1)} = -\frac{1}{2}, \frac{2-(-2)}{3-1} = 2, \frac{3-2}{1-3} = -\frac{1}{2}, \frac{-1-3}{-1-1} = 2$; rectangle

Exercises 1.5: Solving Linear Inequalities and Absolute Value Equations and Inequalities

1. The graph of $y = -2$ is below the graph of $y = -2x + 3$ for $x < 2.5$.

$-2 < -2x + 3 \quad -5 < -2x$

$2.5 > x$ or $x < 2.5$

3. The graph of $y = -2x + 3$ is above the graph of $y = 0$ for $x < 1.5$.

$-2x + 3 > 0$

$-2x > -3 \Rightarrow x < 1.5$

5. The graph of $y = x + 4$ is above the graph of $y = -\frac{1}{2}x + 1$ for $x > -2$.

$x + 4 > -\frac{1}{2}x + 1$

$1.5x > -3 \Rightarrow x > -2$

7. The graph of $y = 0$ is below the graph of $y = x + 4$ for $x > -4$.

$0 < x + 4$

$-4 < x$ or $x > -4$

9. The graph of $y = -\frac{1}{2}x + 1$ is below or the same as the graph of $y = 0$ for $x \geq 2$.

$-\frac{1}{2}x + 1 \leq 0$

$1 \leq \frac{1}{2}x \Rightarrow 2 \leq x$ or $x \geq 2$

11. The graph of $y = -2x + 3$ is above or the same as the graph of $y = 3$ for $x \leq 0$.

$-2x + 3 \geq 3$

$-2x \geq 0 \Rightarrow x \leq 0$

13. The graph of $y = x + 4$ is below or the same as the graph of $y = 4$ for $x \leq 0$.

$x + 4 \leq 4 \Rightarrow x \leq 0$

15. $1 - 2x < x$

$1 < 3x$

$\frac{1}{3} < x$ or $x > \frac{1}{3}$

17. $-\frac{1}{2}x + 1 < 4 - x$

$\frac{1}{2}x < 3 \Rightarrow x < 6$

19. $0.4x - 6 \leq 4$

$0.4x \leq 10 \Rightarrow x \leq 25$

21. $8 - 0.75x > 11$

$-0.75x > 3$

$x < -4$

23. $3(1 - x) - 1 < 2.5x - 9$

$3 - 3x - 1 < 2.5x - 9$

$11 < 5.5x \Rightarrow 2 < x$ or $x > 2$

25. $5 - 3.5(x + 1) \geq 4 - 1.5x$

$5 - 3.5x - 3.5 \geq 4 - 1.5x$

$-2.5 \geq 2x \Rightarrow -1.25 \geq x$ or $x \leq -1.25$

27. $10 < 5x < 20$

$10 \div 5 < 5x \div 5 < 20 \div 5$

$2 < x < 4$

29. $10 < -5x < 20$

$10 \div -5 > -5x \div -5 > 20 \div -5$

$-2 > x > -4$

31. $-3 < 2 - x < 4$

$-3 - 2 < 2 - x - 2 < 4 - 2$

$-5 < -x < 2$

$-5 \div -1 > -x \div -1 > 2 \div -1$

$5 > x > -2$

33. $-16 \le 4 - 5x \le 29$
$-16 - 4 \le 4 - 5x - 4 \le 29 - 4$
$-20 \le -5x \le 25$
$-20 \div -5 \ge -5x \div -5 \ge 25 \div -5$
$4 \ge x \ge -5$

35. a. $\{-2, 6\}$
b. $\{2\}$
c. no solution
d. $\{-3, 7\}$

Exercise 37 is solved using the graph provided in the text.

37. a. $x = -1$ or $x = 4$ **b.** $x = \frac{3}{2}$ **c.** $-2 < x < 5$ **d.** All real numbers
e. No solution **f.** $x = 0$ or $x = 3$ **g.** $x < 0$ or $x > 3$ **h.** $-\frac{1}{2} < x < \frac{7}{2}$

39. $|x| = 4$
$x = 4$ or $x = -4 \Rightarrow \{\pm 4\}$

41. $|x + 2| = 3$
$x + 2 = 3$ or $x + 2 = -3$
$x = 1$ or $x = -5 \Rightarrow \{-5, 1\}$

43. $|x - 5| = 2$
$x - 5 = 2$ or $x - 5 = -2$
$x = 7$ or $x = 3 \Rightarrow \{3, 7\}$

45. $|x - 4| = 2$
$x - 4 = 2$ or $x - 4 = -2$
$x = 6$ or $x = 2 \Rightarrow \{2, 6\}$

47. $|2x - 3| > 7$
$2x - 3 > 7$ or $-(2x - 3) > 7$
$2x > 10$ or $-2x + 3 > 7$
$x > 5$ or $-2x > 4$
$x < -2$
$x < -2$ or $x > 5$

49. $|5 - 4x| < 11$
$5 - 4x < 11$ and $-(5 - 4x) < 11$
$-4x < 6$ and $-5 + 4x < 11$
$x > -1.5$ and $4x < 16$
$x < 4$
$-1.5 < x < 4$

51. $|3 - \frac{1}{2}x| > 4.5$
$3 - \frac{1}{2}x > 4.5$ or $-(3 - \frac{1}{2}x) > 4.5$
$-\frac{1}{2}x > 1.5$ or $-3 + \frac{1}{2}x > 4.5$
$x < -3$ or $\frac{1}{2}x > 7.5$
$x > 15$
$x < -3$ or $x > 15$

53. $|0.5 - 0.25x| \le 2$
$0.5 - 0.25x \le 2$ or $-(0.5 - 0.25x) \le 2$
$-0.25x \le 1.5$ or $-0.5 + 0.25x \le 2$
$x \ge -6$ or $0.25x \le 2.5$
$x \le 10$
$-6 \le x \le 10$

55. a. 2, $|x + 2|$; abs(x + 2)
b. 3, $|x| + 2$; abs(x) + 2
c. 4, $\frac{1}{|x+2|}$; 1 ÷ (abs(x+2))
d. 1, $\frac{1}{|x| + 2}$; 1 ÷ (abs(x) + 2)

57. $0.04x \le 5$
$x \le \$125$

59. $5 \le 0.02x \le 25$
$\$250 \le x \le \$1250$

61. In $4 < x + 5 < 6$, look for where $y = x + 5$ is above $y = 4$ and below $y = 6$.

63. Add the opposite of the negative x term to both sides first.

Chapter 1: Review Exercises

1. The ***vertical-line test*** is used to find out if a graph is a function. The ***two-output test*** is used on a table to see if it is a function.

2. The set of all inputs is the ***domain***; the set of all outputs is the ***range.***

3. A ***dot graph*** has only integer inputs.

4. A ***step graph*** has inputs rounded off before outputs are found.

5. A ***linear function*** is a set of data with a constant slope.

6. ***Linear regression is*** used to find a linear equation on a graphing calculator.

7. A function for which the output exactly matches the input is an ***identity function.***

8. A function with only one output (regardless of input) is a ***constant function.***

9. A function with a zero or positive output for any real-number input is the ***absolute value function.*** (Note: squaring function is not in the list.)

10. Numbers that are zero or positive are ***non-negative numbers.***

11. The 3 ways to describe a set of numbers are ***inequality, compound inequality, interval.***

12. The rate of change of a graph is the ***slope.***

13. The ways to find a linear equation are ***point-slope, slope-intercept, arithmetic sequence, table, linear regression.***

14. Viewed from left to right, rising functions are ***increasing functions***; falling functions are ***decreasing functions***, and horizontal functions are ***constant functions.***

15. a. $-6.2 \le x \le 2.2$ **b.** $-3 \le y \le 3$ **c.** not a function
Note: the values in part a are estimated.

16. a. $\{0, 1, 2, 3, 4, 5, 6\}$ **b.** $\{0, 2, 4, 6, 8, 10\}$ **c.** Function

17. a. $f(0) = 1$ **b.** $f(2) = 4$ **c.** $f(-1) = \frac{1}{2}$
d. $f(3) = 8$ **e.** $f(1) = 2$ **f.** $x \ge 1$
g. none **h.** $x \le 2$ **i.** 1
j. $\Re$ **k.** $y > 0$

18. a. $f(x) = 1$ $\{0\}$ **b.** $f(x) = 3$ $\{\approx 1.5\}$ **c.** $f(x) = \frac{1}{2}$ $\{-1\}$

d. $f(x) = 4$ $\{2\}$ **e.** $f(x) = -2$ $\{\ \}$

19. $f(1) = 2(1)^2 - 3(1) + 1$
$f(1) = 2 - 3 + 1$
$f(1) = 0$

20. $f(2) = 2(2)^2 - 3(2) + 1$
$f(2) = 8 - 6 + 1$
$f(2) = 3$

21. $f(0.5) = 2(0.5)^2 - 3(0.5) + 1$
$f(0.5) = 0.5 - 1.5 + 1$
$f(0.5) = 0$

22. $f(1.5) = 2(1.5)^2 - 3(1.5) + 1$
$f(1.5) = 4.5 - 4.5 + 1$
$f(1.5) = 1$

23. $f(-2) = 2(-2)^2 - 3(-2) + 1$
$f(-2) = 8 + 6 + 1$
$f(-2) = 15$

24. $f(-0.5) = 2(-0.5)^2 - 3(-0.5) + 1$
$f(-0.5) = 0.5 + 1.5 + 1$
$f(-0.5) = 3$

25. $f(\square) = 2(\square)^2 - 3(\square) + 1$

26. $f(-a) = 2(-a)^2 - 3(-a) + 1$
$f(-a) = 2a^2 + 3a + 1$

	Ordered Pairs	**Slope**	**Equation**	Hor/Ver	**x-intercept**	**y-intercept**
27.	(1, 3) (2, 2)	$\frac{3-2}{1-2} = -1$	$b = 3 - (-1)(1)$, $b = 4$ $y = -x + 4$	neither	$0 = -x + 4$ $x = 4$ (4, 0)	$y = 4$ (0, 4)
28.	(2, 3) (2, 2)	$\frac{3-2}{2-2} \Rightarrow$ Undefined	$x = 2$	vertical	(2, 0)	no y-intercept
29.	(−3, 3) (0, 0)	$\frac{3-0}{-3-0} = -1$	$b = 0$ $y = -x$	neither	$x = 0$ (0, 0)	$y = 0$ (0, 0)
30.	(−3, 3) (3, 3)	$\frac{3-3}{3-(-3)} = 0$	$y = 3$	horizontal	no x-intercept	$y = 3$ (0, 3)
31.	(−1, −1) (1, 3)	$\frac{-1-3}{-1-1} = 2$	$b = 3 - 2(1)$ $b = 1$ $y = 2x + 1$	neither	$0 = 2x + 1$ $x = -\frac{1}{2}$ $\left(-\frac{1}{2}, 0\right)$	$y = 1$ (0, 1)

32.	(2, 3) (−3, 4)	$\frac{4-3}{-3-2} = -\frac{1}{5}$	$b = 3 - \left(-\frac{1}{5}\right)(2)$ $b = \frac{17}{5}$ $y = -\frac{1}{5}x + \frac{17}{5}$	neither	$0 = -\frac{1}{5}x + \frac{17}{5}$ $x = 17$ $(17, 0)$	$y = \frac{17}{5}$ $\left(0, \frac{17}{5}\right)$
33.	(−3, 3) (2, 3)	$\frac{3-3}{-3-2} = 0$	y = 3	horizontal	no x-intercept	y = 3 (0, 3)
34.	(0, 4) (−1, −1)	$\frac{-1-4}{-1-0} = 5$	b = 4 y = 5x + 4	neither	0 = 5x + 4 $x = -\frac{4}{5}$ $\left(-\frac{4}{5}, 0\right)$	y = 4 (0, 4)

35. Ex. 27 and 29 are parallel.

36. Ex. 28 & 30, 32 & 34 are perpendicular.

37. slope is negative reciprocal, $-\frac{1}{3}$

a. y-intercept is (0, 0); $y = -\frac{1}{3}x$

b. $b = 4 - (-\frac{1}{3})(3)$, $b = 5$; $y = -\frac{1}{3}x + 5$

38. slope is same, $\frac{2}{3}$

a. y-intercept is (0, 0); $y = \frac{2}{3}x$

b. $b = 3 - (\frac{2}{3})(4)$, $b = \frac{1}{3}$; $y = \frac{2}{3}x + \frac{1}{3}$

39. a. y = 0.065x

b. Slope = \$0.065 tax/\$1 purchased; y-intercept = (\$0, \$0), no tax on \$0 purchases

40. a. y = 18.95x

b. Slope = \$18.95/person; y-intercept = (0, \$0), no cost if no one attends

41. a. y = 45x + 500

b. Slope = \$45/hour of repair; y-intercept = (0, \$500), basic inspection cost

42. a. C = 1.259x + 8.90

b. Slope = \$1.259/kgal; y-intercept = (0, \$8.90), basic monthly charge

Exercises 43 and 44 used LinReg on a graphing calculator to find the equation. The solution is given for reference only.

43. $y \approx 11{,}528 - 42x$

44. $y \approx 11{,}244 - 45x$

45. Δx (ft) = 11 & 6; Δy ($) = 16.50 & 9; slopes are $16.50 \div 11 = 1.50$ & $9 \div 6 = 1.50$ using the first data set: $b = 45.50 - 1.50(25)$, $b = 8$; $y = 1.50x + 8$

46. $\Delta x = 1$, $\Delta y = -2$; working backwards, when $x = 0$, $y = 18 + 2 = 20$; $y = -2x + 20$

47. $\Delta x = 1$, $\Delta y = -4$; working backwards when $x = 0$, $y = 31 + 4 = 35$; $y = -4x + 35$

48. $\Delta x = 1$, $\Delta y = 3$; working backwards, when $x = 0$, $y = -5 - 3 = -8$; $y = 3x - 8$

49. Let x = # of people, y = total cost; $y = 85$ for $0 < x \le 10$; $y = 85 + 4.75(x - 10)$ for $x > 10$, inputs are positive integers only, dot graph

50. Let x = # of children; y = total cost; $y = 40$ for $0 < x \le 10$; $y = 40 + 4.5(x - 10)$ for $x > 10$, inputs are positive integers only, dot graph

51. Let x = # of hrs; y = cost; $y = 26$ for $0 < x \le 2$; $y = 26 + 19(x - 2)$ for $x > 2$; inputs may be any non-negative number, step graph

52. Let x = # of days; y = cost; $y = 108$ for $0 < x \le 2$; $y = 108 + 63(x - 2)$ for $x > 2$; inputs may be any non-negative number, step graph

53. a.

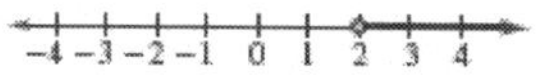

$-\frac{3}{4}x - \frac{1}{2} < -2$

$4(-\frac{3}{4}x - \frac{1}{2}) < 4(-2)$

$-3x - 2 < -8$

$-3x < -6$

$x > 2$

b.

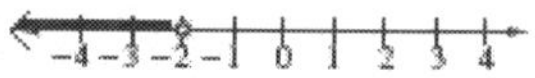

$\frac{1}{3}x + \frac{5}{3} < -\frac{3}{4}x - \frac{1}{2}$

$12(\frac{1}{3}x + \frac{5}{3}) < 12(-\frac{3}{4}x - \frac{1}{2})$

$4x + 20 < -9x - 6$

$13x < -26$

$x < -2$

c.

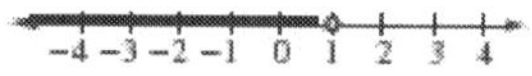

$\frac{1}{3}x + \frac{5}{3} < 2$

$3(\frac{1}{3}x + \frac{5}{3}) < 3(2)$

$x + 5 < 6$

$x < 1$

54.

$3x + 1 > 4$

$3x > 3$

$x > 1$

55.

$-2x + 1 \le 3$

$-2x \le 2$

$x \ge -1$

56. $\frac{1}{2}x - \frac{3}{2} \geq 0$

$2(\frac{1}{2}x - \frac{3}{2}) \geq 2(0)$

$x - 3 \geq 0$

$x \geq 3$

57. $\frac{1}{2}x - \frac{3}{2} < -1$

$2(\frac{1}{2}x - \frac{3}{2}) < 2(-1)$

$x - 3 < -2$

$x < 1$

58. $3x + 1 > -2x + 1$

$5x > 0$

$x > 0$

59. $\frac{1}{2}x - \frac{3}{2} \leq -2x + 1$

$2(\frac{1}{2}x - \frac{3}{2}) \leq 2(-2x + 1)$

$x - 3 \leq -4x + 2$

$5x \leq 5$

$x \leq 1$

60. $\frac{1}{2}x - \frac{3}{2} \geq 3x + 1$

$2(\frac{1}{2}x - \frac{3}{2}) \geq 2(3x + 1)$

$x - 3 \geq 6x + 2 \quad -5 \geq 5x$

$-1 \geq x$ or $x \leq -1$

61. a. $19 < x\left(\frac{704.5}{74^2}\right) < 24$

$19(74^2) < 704.5x < 24(74^2)$

$147.7 < x < 186.5$

b. $19 < x\left(\frac{704.5}{62^2}\right) < 24$

$19(62^2) < 704.5x < 24(62^2)$

$103.7 < x < 131.0$

62. a. $1 < 5 + \frac{2}{3}x < 13$

$-4 < \frac{2}{3}x < 8$

$-6 < x < 12$

b. $-6 < 4 + 2x < 18$

$-10 < 2x < 14$

$-5 < x < 7$

c. $-9 < 3 - 4x < 11$

$-12 < -4x < 8$

$3 > x > -2$ or $-2 < x < 3$

d. $0 < 2 - \frac{1}{2}x < 6$

$-2 < -\frac{1}{2}x < 4$

$4 > x > -8$ or $-8 < x < 4$

63. $8 < 0.05x < 9$

$160 < x < 180$

64. $4.50 < 0.15x < 12$

$30 < x < 80$

65. a. $|x - 1| = 4$

$x - 1 = 4$ or $x - 1 = -4$

$x = 5$ or $x = -3$

$\{-3, 5\}$

b. $|x - 1| = 2$

$x - 1 = 2$ or $x - 1 = -2$

$x = 3$ or $x = -1$

$\{-1, 3\}$

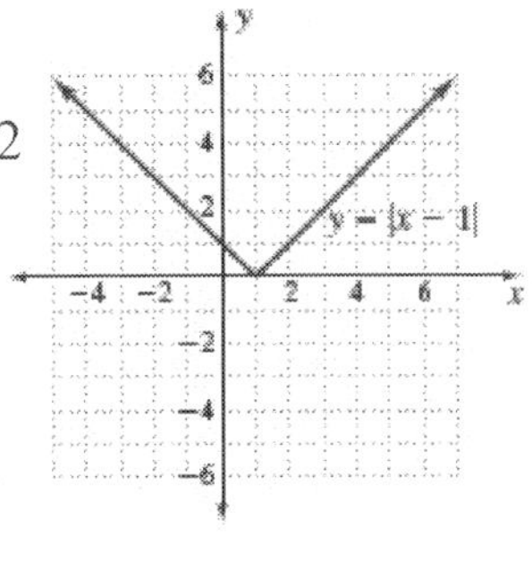

c. $|x - 1| = 0$

$x - 1 = 0$

$x = 1$

$\{1\}$

d. $|x - 1| = -2$

absolute value is always positive{ } or ∅

66. a. $|x - 3| = 4$

$x - 3 = 4$ or $x - 3 = -4$

$x = 7$ or $x = -1$

$\{-1, 7\}$

b. $|5 - x| = 4$

$5 - x = 4$ or $5 - x = -4$

$x = 1$ or $x = 9$

$\{1, 9\}$

67. domain = $\Re$; range $y \geq 0$

68. domain = $\Re$; range $y \geq -1$

69. domain = $\Re$; range = $\Re$

70. domain = $\Re$; range $y = 1$

71. $|4x - 3| < 6$

$4x - 3 < 6$ and $-(4x - 3) < 6$

$4x < 9$ and $-4x + 3 < 6$

$x < 2.25$ and $-4x < 3$

$x > -0.75$

$-0.75 < x < 2.25$

72. $|2x - 1| > 5$

$2x - 1 > 5$ or $-(2x - 1) > 5$

$2x > 6$ or $-2x + 1 > 5$

$x > 3$ or $-2x > 4$

$x < -2$

$x < -2$ or $x > 3$

73. $|0.75x - 5| > 1$

$0.75x - 5 > 1$ or $-(0.75x - 5) > 1$

$0.75x > 6$ or $-0.75x + 5 > 1$

$x > 8$ or $-0.75 > -4$

$x < 5\frac{1}{3}$

$x < 5\frac{1}{3}$ or $x > 8$

74. $|0.5x + 2| < 4$

$0.5x + 2 < 4$ and $-(0.5x + 2) < 4$

$0.5x < 2$ and $-0.5x - 2 < 4$

$x < 4$ and $-0.5x < 6$

and $x > -12$

$-12 < x < 4$

75.

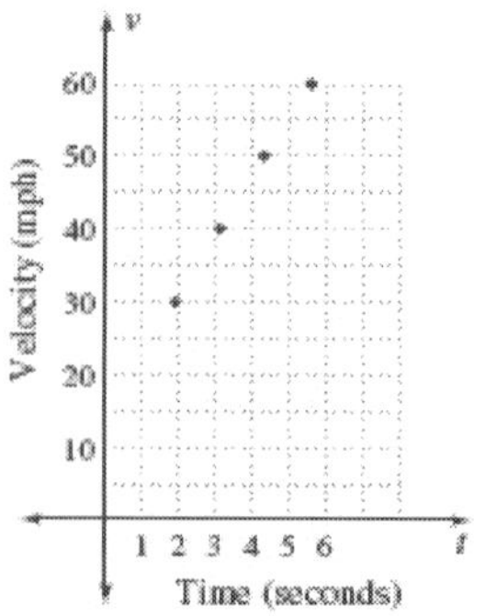

Answers may vary. The data seems to be linear. In order for the car to go faster, more time is needed. This speed would end at some point, since a car can only go so fast. Also, it would seem that the car would start at about 15 mph at 0 seconds which does not make sense. Perhaps not linear; graph should start at (0, 0) to fit 0 mph.

$y = 8.13x + 14.73$

76.

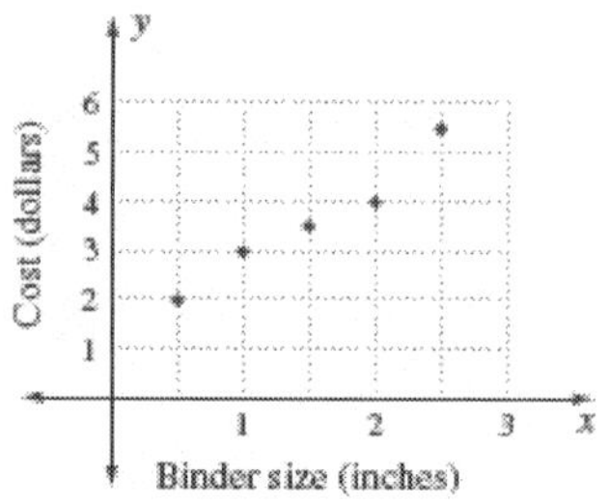

Answers may vary. The 2½-inch binder is relatively expensive, as costs increase less rapidly for the 1½-inch and 2-inch binders than for smaller ones.

$y = 1.604x + 1.178$

77. One solution: Let L = Length of truck rounded down to the nearest multiple of 10 feet.
Let F = fare in dollars.
F = \$16.25 for L < 20 feet;

$$F = \$16.25 + \$32.50 + \$16.25\left(\frac{L-20}{10}\right) \text{ for } 20 \text{ ft } < L < 80 \text{ ft}$$

$$F = \$16.25 + \$32.50 + \$16.25\left(\frac{L-20}{10}\right) + \$1.70(L-79) \text{ for } L \geq 80 \text{ ft}$$

Check: F = \$130, L = 75 ft rounds down to 70 ft;

$$F = \$16.25 + \$32.50 + \$16.25\left(\frac{L-20}{10}\right)$$

$$\$130 = \$16.25 + \$32.50 + \$16.25\left(\frac{70-20}{10}\right)$$

$$\$130 = \$130$$

78.

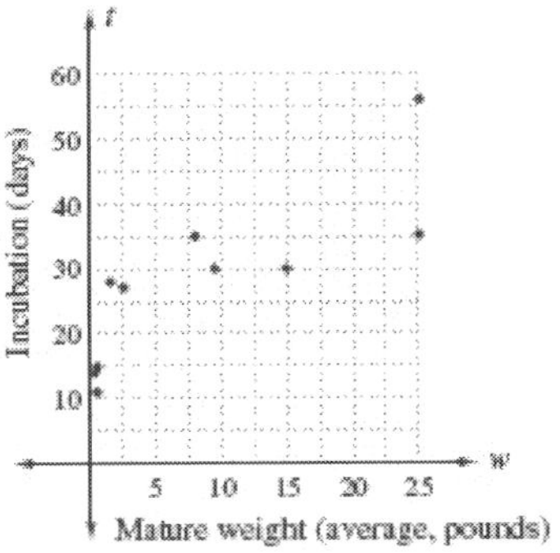

Answers may vary. There seems to be somewhat of a linear relationship. The heavier the bird's weight is related to the longer incubation time.

79. a. $\frac{4-5}{3-0} = -\frac{1}{3}$, $\frac{-5-4}{0-3} = 3$, $\frac{4-(-5)}{3-0} = 3$, $\frac{5-4}{0-3} = -\frac{1}{3}$; negative reciprocals

b. $\frac{3-5}{-4-0} = \frac{1}{2}$, $\frac{-5-3}{0-(-4)} = -2$, $\frac{3-(-5)}{-4-0} = -2$, $\frac{5-3}{0-(-4)} = \frac{1}{2}$; negative reciprocals

c. slope between (a, 0) and (0, b) = $-\frac{b}{a}$; slopes are 1, –1, –1, 1; negative reciprocals

d. BAF, BCF and BDF are all right angles.

e. $\frac{-4-5}{-3-0} = 3$, $\frac{-4-(-5)}{-3-0} = -\frac{1}{3}$; negative reciprocals

f. A right angle is formed by a point on the circumference connected with the end points on the diameter.

Chapter 1: Test

1. function

2. not a function, one input has two outputs

3. **a.** $f(-2) = 3(-2)^2 - 2(-2) - 4$
$f(-2) = 12 + 4 - 4$
$f(-2) = 12$

b. $f(0) = 3(0)^2 - 2(0) - 4$
$f(0) = 0 - 0 - 4$
$f(0) = -4$

c. $f(2) = 3(2)^2 - 2(2) - 4$
$f(2) = 12 - 4 - 4$
$f(2) = 4$

4. **a.** slope $= \dfrac{4-2}{-2-5} = -\dfrac{2}{7}$

b. $b = 4 - (-\frac{2}{7})(-2)$, $b = 3\frac{3}{7}$; $y = -\frac{2}{7}x + 3\frac{3}{7}$

c. parallel line = same slope: $-\frac{2}{7}$

d. perpendicular line = negative reciprocal slope: $\frac{7}{2}$; $b = -1 - (\frac{7}{2})2$, $b = -8$; $y = \frac{7}{2}x - 8$

5. **a.** The slope of a horizontal line is ***zero.***

b. A line that falls from left to right has a ***negative*** slope and is said to be a ***decreasing*** function.

c. If the slope of a graph between all pairs of points is constant, the graph is a ***linear*** function.

d. A horizontal linear graph is also called a ***constant*** function.

e. Linear equations have a ***constant*** slope.

f. The set of inputs to a number pattern is the ***positive integers or natural numbers.***

6. **a.** $y = 7x + 2.50$, y in \$

b. Slope is \$7 per mile.

7. **a.** Reasonable inputs and output would be non-negative numbers; x = number of batteries, y = cost in dollars.

b. (4, 3.29), (16, 8.99)

c. slope $= \dfrac{8.99-3.29}{16-4} = 0.475$, $b = 3.29 - 0.475(4)$, $b = 1.39$; $y = 0.475x + 1.39$

d. If $x = 8$, $y = 0.475(8) + 1.39 = 5.19$. Would recommend \$5.19.

e. 8 is not half way between the given amount of batteries (4 and 16).

8. From LinReg on graphing calculator: $y \approx 10.1x - 13.8$, y in \$

9. **a.** $\Delta y = 8$, next number is $42 + 8 = 50$; when $x = 0$, $y = 10 - 8 = 2$; $y = 8x + 2$.

b. $\Delta y = 7$, next number is $12 + 7 = 19$; when $x = 0$, $y = -16 - 7 = -23$; $y = 7x - 23$.

10. $y = |x| - 3$

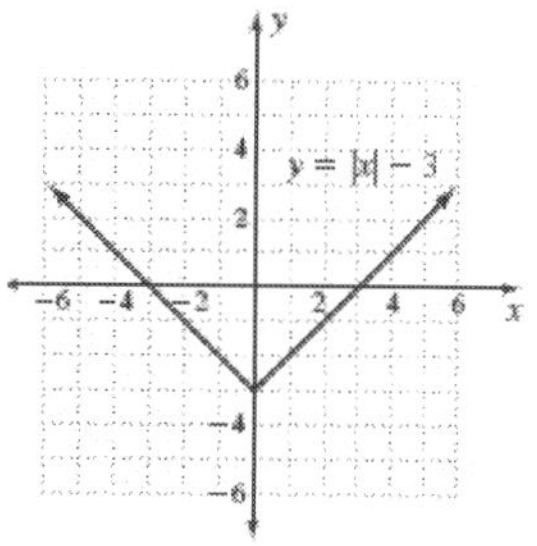

11. $y = |x - (-2)|$

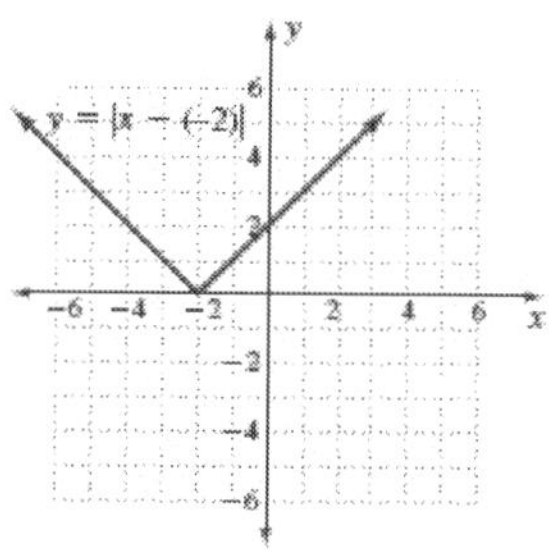

12. $-0.5 < 0.5 - 0.25x < 2.5$

$-1 < -0.25x < 2$

$4 > x > -8;\ -8 < x < 4$

13. $|0.5x + 2| = 5$

$0.5x + 2 = 5$ or $0.5x + 2 = -5$

$0.5x = 3$ or $0.5x = -7$

$x = 6$ or $x = -14$

$\{-14, 6\}$

14. $19 \le \dfrac{704.5}{68^2}x \le 24$

$\dfrac{68^2 \cdot 19}{704.5} \le x \le \dfrac{68^2 \cdot 24}{704.5}$

$124.7 \text{ lb} \le x \le 157.5 \text{ lb}$

15. $|2 - 0.5x| < 4$

$2 - 0.5x < 4$ and $-(2 - 0.5x) < 4$

$-0.5x < 2$ and $-2 + 0.5x < 4$

$x > -4$ and $0.5x < 6$

$x < 12$

$-4 < x < 12$

16.

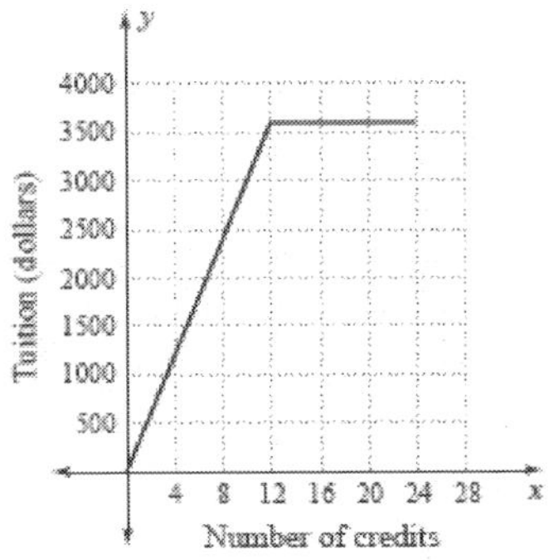

17.

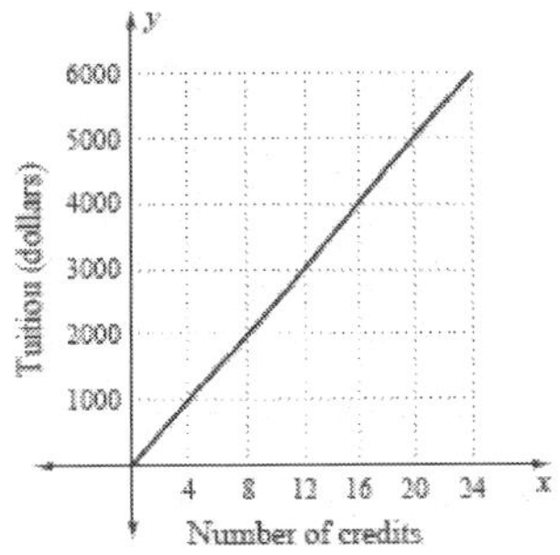

18.

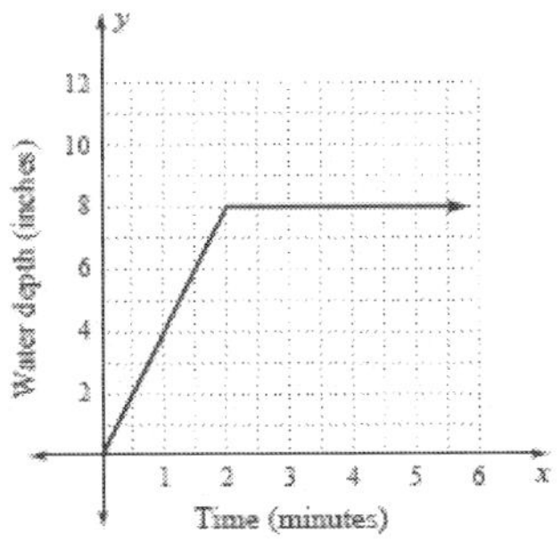

Water overflows bucket when y = 8.

19.

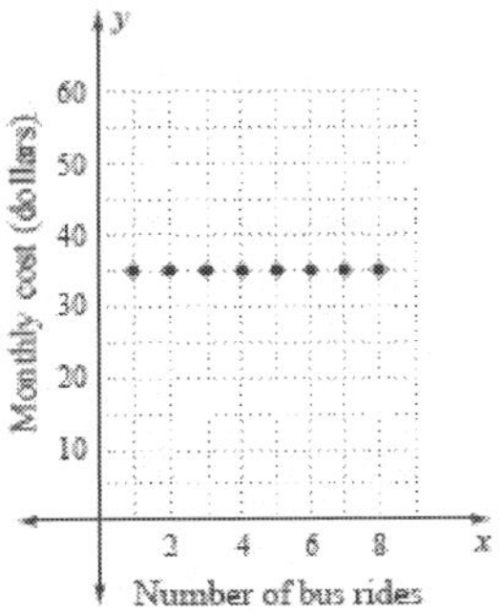

20.

Transcript Copies	1	2	3	4	5	6
Cost $	5	5	7	9	11	13

Points should not be connected; only whole copies are reasonable.

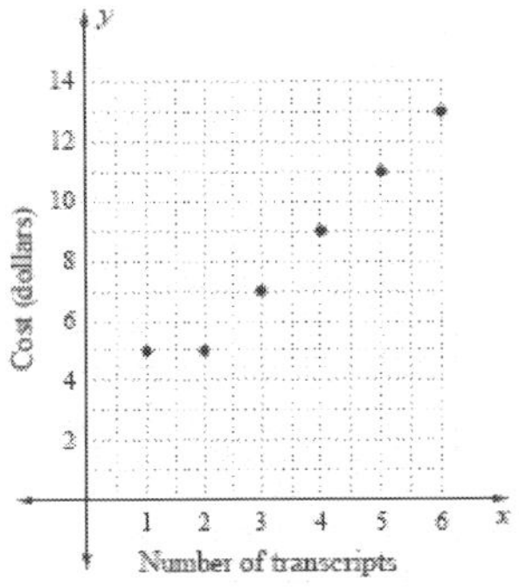

Cumulative Review: Chapters 0 to 1

1. a. $3(x^2+3x-4)-x(x^2+3x-4)=3x^2+9x-12-x^3-3x^2+4x=-x^3+13x-12$

b. $a(a^2-2ab+b^2)-b(a^2+2ab-b^2)=a^3-2a^2b+ab^2-a^2b-2ab^2+b^3$
$=a^3-3a^2b-ab^2+b^3$

2. $3[4-2(5-8)-6]=3[4-2(-3)-6]=3[4+6-6]=3[4]=12$

3. $a(b+c)-b(a+c)+c(a-b)=ab+ac-ab-bc+ac-bc=2ac-2bc$

4. $\pi(7.5)^2=56.25\pi\ ft^2$

5. $\frac{1}{4}\pi(15\text{ ft})^2=56.25\pi\text{ ft}^2$

6. $\dfrac{ac-bc}{c}=\dfrac{c(a-b)}{c}=a-b$

7. $\dfrac{16+21x}{6}$ does not simplify.

8. a. $3x+12=5x+3$
$12-3=5x-3x$
$9=2x$
$x=4.5$

b. $15-4x=5(6-x)$
$15-4x=30-5x$
$-4x+5x=30-15$
$x=15$

9. a. $9-4(x-3)=45$
$9-4x+12=45$
$21-4x=45$
$-4x=24$
$x=-6$

b. $(9-4)(x-3)=45$
$5(x-3)=45$
$5x-15=45$
$5x=60$
$x=12$

10. a. $\frac{2}{3}x+\frac{1}{4}x+x=46$
$\frac{23}{12}x=46$
$x=24$

b. $\frac{9}{5}x+32=212$
$\frac{9}{5}x=180$
$x=100$

11. a. $\frac{x}{8}+5=2$
$8\left(\frac{x}{8}+5\right)=8(2)$
$x+40=16$
$x=-24$

b. $3x=x+15$
$3x-x=15$
$2x=15$
$x=7.5$

12. a. $f(x)=-|x|$; $f(-1)=-|-1|=-1$; $f(1)=-|1|=-1$; $f(2)=-|2|=-2$

b. $f(x)=1+|x|$; $f(-1)=1+|-1|=2$; $f(1)=1+|1|=2$; $f(2)=1+|2|=3$

c. $f(x)=-x^2$; $f(-1)=-(-1)^2=-1$; $f(1)=-(1)^2=-1$; $f(2)=-(2)^2=-4$

d. $f(x)=-x$; $f(-1)=-(-1)=1$; $f(1)=-(1)=-1$; $f(2)=-(2)=-2$

13. a. $3 - 4x = 2(x + 1)$
$3 - 4x = 2x + 2$
$1 = 6x$
$x = \frac{1}{6}$

b. Graph $f(x)$ and $g(x)$ on the same coordinate system and look for the x-coordinate of their intersection.

14. Choose substitution because variable is isolated.
$y = 2.5x + 4$
$2y - 5x = 8$
$2(2.5x + 4) - 5x = 8$
$5x + 8 - 5x = 8$
$8 = 8$; identity
Coincident lines

15. Choose substitution because variable is isolated.
$2y - 3x = 8$
$y = 1.5x - 3$
$2(1.5x - 3) - 3x = 8$
$3x - 6 - 3x = 8$
$-6 = 8$; contradiction
Parallel lines

16. Choose elimination because standard form is easily obtained.
$2x - 12 = -3y$ (a); $2x + 3y = 12$ (a std form)
$6 - 4y = 3x$ (b); $3x + 4y = 6$ (b std form)
$3(2x + 3y) = 3(12)$ (c); $6x + 9y = 36$ (c)
$-2(3x + 4y) = -2(6)$ (d); $-6x - 8y = -12$ (d)
(c) + (d): $\mathbf{y = 24}$ $\quad 2x - 12 = -3(24)$
$2x = -60$; $\mathbf{x = -30}$

17. Choose substitution because variables are isolated.
$y = -0.8x - 1.8$
$y = -1.2x + 3$
$-0.8x - 1.8 = -1.2x + 3$
$0.4x = 4.8$
$\mathbf{x = 12}$
$y = -0.8(12) - 1.8$
$\mathbf{y = -11.4}$ $(12, -11.4)$

18. $x + y = 1$ pt
$0.03x + 0y = 0.02(1)$
$x = \frac{2}{3}$ pt 3% hydrogen peroxide
$y = 1 - \frac{2}{3} \Rightarrow y = \frac{1}{3}$ pt wáter

19. $\text{slope} = \dfrac{1-(-3)}{-6-4} = -\dfrac{2}{5}$; $b = 1 - (-\frac{2}{5})(-6) = -\frac{7}{5}$; $y = -\frac{2}{5}x - \frac{7}{5}$

20. $\text{slope} = \dfrac{-1-(-7)}{2-(-2)} = \dfrac{3}{2}$; $b = -1 - \left(\frac{3}{2}\right)(2) = -4$; $y = \frac{3}{2}x - 4$

21. Slope is negative reciprocal or $-\frac{1}{3}$; $b = 0$; $y = -\frac{1}{3}x$.

22. Next pair is (4, 16); $f(x) = x^2$.

23. Next pair is (4, 4); $f(x) = x$.

24. Next pair is (4, 8); $f(x) = 2x$.

25. Next pair is (4, 4); $f(x) = 4$.

26. Let x = miles and y = cost in \$; $y = 0.57x + 60$.

27. Let x = number of workers and y = cost in \$; $y = 65x + 500$.

28. If $a = b$, then $a + c = b + c$. If $a = b$, then $ac = bc$.
The multiplication on each side (or division on each side) by a negative number will change the sign of the inequality: if $a < b$ and c is negative, then $ac > bc$.

29. a. A and a are not the same variable.

b. Possible equations: $a = \frac{2A}{h} - b$; $a = \frac{2A - hb}{h}$; $a = \frac{2}{h}\left(A - \frac{bh}{2}\right)$
Different orders of operations were used to solve the equation.

30. From LinReg on a graphing calculator: $y \approx 17.2 - 5.2x$.

31. a. Let B = Bounty and x = number of fish caught.
$B = 4x$ for $0 \le x \le 100$; $B = 4(100) + 5(x - 100)$ for $100 < x \le 400$;
$B = 4(100) + 5(300) + 8(x - 400)$ for $400 > x$

b. $19084 = 400 + 1500 + 8(x - 400) + 1000$
$19084 = 1900 + 8x - 3200 + 1000$
$19084 = 8x - 300$
$19384 = 8x$
$x = 2423$ ⇒ A total of 2423 fish were caught.

c. The amount of money earned is a function of the number of fish caught.

d. For example: x in [0, 3000] with scale of 300, y in [0, 24000], with scale of 2000.

32. a.

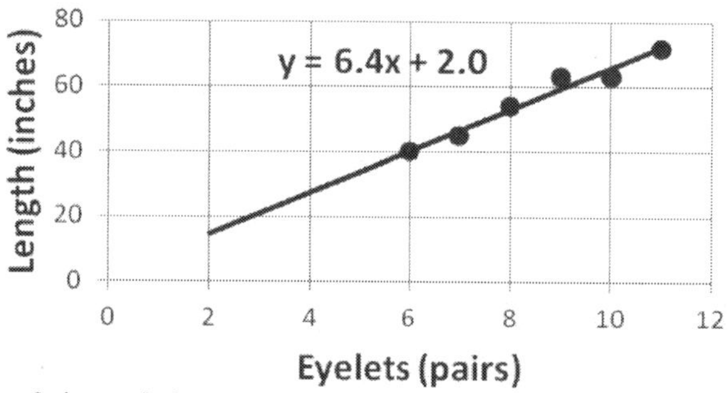

b. $y = 6.4x + 2.0$

c. see part a.

d. No, but the line from parts b and c passes close to the points.

e. Other factors that might control length of the lace are the width and length of the shoe, the make of the shoe, or the width of the lace.

CHAPTER 2

Quadratic and Polynomial Functions

Exercises 2.1: Quadratics Functions

1. a.

Outputs	Differences	
1	**1st**	**2nd**
	2	
3		4
	6	
9		12
	18	
27		36
	54	
81		108
	162	
243		

Each number is 3 times the previous number; next number is 3(243) = 729. Sequence is neither linear nor quadratic.

b.

Outputs	Differences
11	**1st**
	7
18	
	7
25	
	7
32	
	7
39	

Each number is 7 more than the previous number; next number is 39 + 7 = 46. Sequence is linear.

c.

Outputs	Differences
56	**1st**
	–9
47	
	–9
38	
	–9
29	
	–9
20	

Each number is 9 less than the previous number; next number is 20 – 9 = 11; sequence is linear.

d.

Outputs	Differences	
–1	**1st**	**2nd**
	2	
1		4
	6	
7		4
	10	
17		4
	14	
31		

Using 2nd difference, next 1st difference is 18, next number is 49. Sequence is quadratic.

3. a. Equation has degree 3; neither linear nor quadratic.

b. Equation has degree 5; neither linear nor quadratic.

c. Equation has degree 2; quadratic.

d. Equation has degree 1; linear.

5. a. In $f(r) = \pi r^2 + 2\pi r(3 \text{ meters})$, the variable is **r**; $a = \pi$, $b = 2\pi(3)$, $c = 0$.

b. In $f(T) = \frac{g}{4\pi^2}T^2$, the variable is **T**; $a = \frac{g}{4\pi^2}$, $b = 0$, $c = 0$.

c. $y = \frac{1}{2}x(x + 1) = \frac{1}{2}x^2 + \frac{1}{2}x$, the variable is **x**, $a = \frac{1}{2}$, $b = \frac{1}{2}$, $c = 0$

7.

x	$y = 2x^2 + x + 1$
−3	$2(-3)^2 + (-3) + 1 = 16$
−2	$2(-2)^2 + (-2) + 1 = 7$
−1	$2(-1)^2 + (-1) + 1 = 2$
0	$2(0)^2 + (0) + 1 = 1$
1	$2(1)^2 + (1) + 1 = 4$
2	$2(2)^2 + (2) + 1 = 11$
3	$2(3)^2 + (3) + 1 = 22$

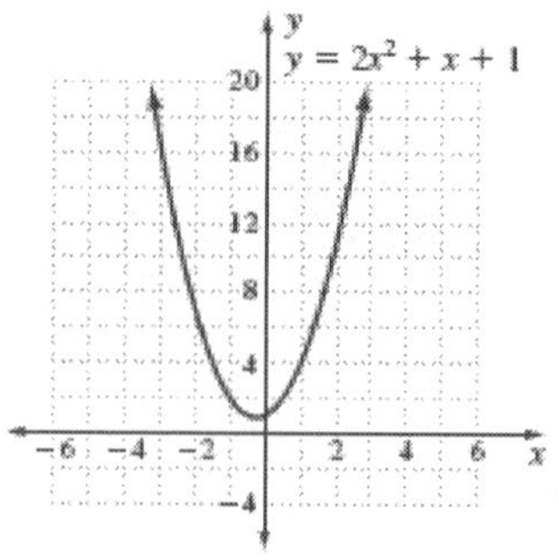

9.

x	$y = -x^2 - 2$
−3	$-(-3)^2 - 2 = -11$
−2	$-(-2)^2 - 2 = -6$
−1	$-(-1)^2 - 2 = -3$
0	$-(0)^2 - 2 = -2$
1	$-(1)^2 - 2 = -3$
2	$-(2)^2 - 2 = -6$
3	$-(3)^2 - 2 = -11$

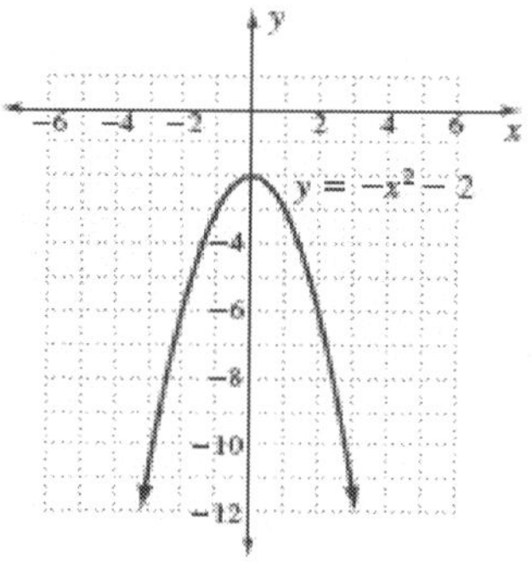

11. The x-intercepts are where $f(x) = 0$; $(1, 0)$ and $(3, 0)$. The y-intercept is where $x = 0$; $(0, 3)$. Noting symmetry in table, the vertex is $(2, -1)$.

13. From the table: $f(3) = 0$; $f(-1) = 8$

$f(6) = 6^2 - 4(6) + 3$

$f(6) = 36 - 24 + 3$

$f(6) = 15$

15. The function is increasing when $x > 2$. The function is decreasing when $x < 2$.

17. a. From the graph, x-intercepts are at $(-2, 0)$ and $(4, 0)$.

b. From the graph, y-intercept is $(0, -8)$.

c. Axis of symmetry goes through the vertex at $x = 1$.

d. Vertex is at $(1, -9)$.

d. Range of $f(x)$ is $y \geq -9$.

e. $f(x)$ is increasing when $x > 1$. $f(x)$ is decreasing when $x < 1$.

19.

x	$f(x) = -x^2 - 2x + 8$
−4	$-(-4)^2 - 2(-4) + 8 = 0$
−3	$-(-3)^2 - 2(-3) + 8 = 5$
−2	$-(-2)^2 - 2(-2) + 8 = 8$
−1	$-(-1)^2 - 2(-1) + 8 = 9$
0	$-(0)^2 - 2(0) + 8 = 8$
1	$-(1)^2 - 2(1) + 8 = 5$
2	$-(2)^2 - 2(2) + 8 = 0$

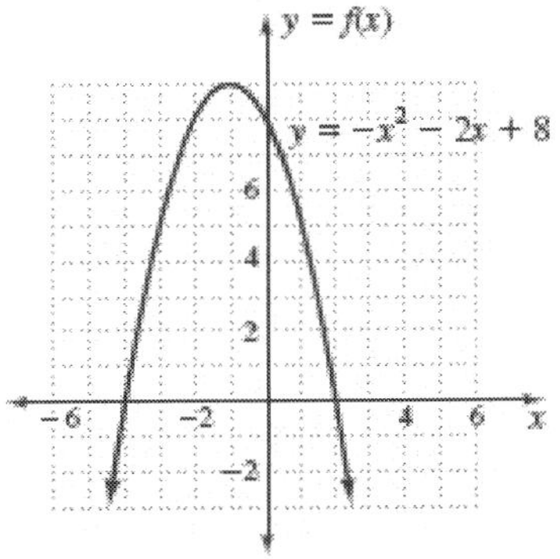

a. x-intercepts are (−4, 0) and (2, 0). **b.** y-intercept is (0, 8).
c. Axis of symmetry is $x = -1$. **d.** Vertex is (−1, 9). **e.** Range is $y \le 9$.
e. $f(x)$ is increasing when $x < -1$. $f(x)$ is decreasing when $x > -1$.

21.

r	$A = 2\pi r^2 + 2\pi r(6)$
0	$2\pi(0)^2 + 12\pi(0) = 0$
1	$2\pi(1)^2 + 12\pi(1) \approx 44.0$
2	$2\pi(2)^2 + 12\pi(2) \approx 100.5$
3	$2\pi(3)^2 + 12\pi(3) \approx 169.6$
4	$2\pi(4)^2 + 12\pi(4) \approx 251.3$
5	$2\pi(5)^2 + 12\pi(5) \approx 345.6$
6	$2\pi(6)^2 + 12\pi(6) \approx 452.4$

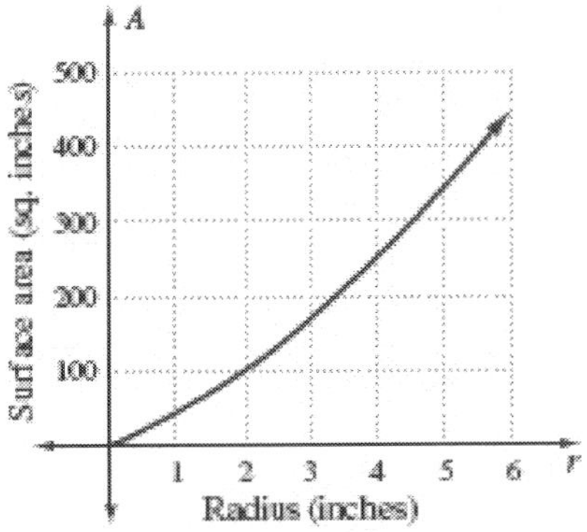

a. From the table & graph, ≈ 3.4 in **b.** From the table & graph, ≈ 5.5 in.
c. The top and bottom surface areas vary with the square of the radius.

23. a. zero **b.** zero

c. Refer to the function $f(x) = x^3 - 3x + 5$ and to the table at the right. Six is the constant third difference.

d. In part a, leading coefficient is 1, second difference is 2.
In part b, leading coefficient is −1, second difference is -2.
In part c, leading coefficient is 3, second difference is 6.
The second difference is 2a.

Outputs	Differences		
	1st	2nd	3rd
5			
	−2		
3		6	
	4		6
7		12	
	16		6
23		18	
	34		
57			

25. a. A quadratic equation $ax^2 + bx + c = 0$ can be solved so that a is positive.
b. Neither the function nor $y = ax^2 + bx + c$ can be solved for a positive a without introducing a negative sign on the $f(x)$ or y.

27. Possible functions include $f(x) = |x|$ or the constant function $f(x) = c$. A function cannot have a horizontal axis of symmetry because the vertical-line test would fail if there were two outputs for some inputs. The graph of $x = |y|$ has a horizontal axis of symmetry but is not a function.

29. In $y = ax^2 + bx + c$, if a is positive the graph opens up. If a is negative the graph opens down.

Exercises 2.2: Modeling Quadratic Functions

1.

Input	Output	Differences 1st	Differences 2nd
1	7		
		9	
2	16		2
		11	
3	27		2
		13	
4	40		2
		15	
5	55		

Quadratic; $2a = 2$; $a = 1$

Working backwards, when $x = 0$, $y = 0$;

$c = 0$; $7 = 1(1)^2 + b(1) + 0$; $b = 6$.

$y = x^2 + 6x$

3.

Input	Output	Differences 1st	Differences 2nd
1	4		
		4	
2	8		0
		4	
3	12		0
		4	
4	16		0
		4	
5	20		

Linear; $m = 4$; when $x = 0$, $y = 0$

$b = 0$; $y = 4x$.

5.

Input	Output	Differences 1st	Differences 2nd
1	8		
		19	
2	27		4
		23	
3	50		4
		27	
4	77		4
		31	
5	108		

Quadratic; $2a = 4$, $a = 2$. When $x = 0$

$y = -7$, $c = -7$; $8 = 2(1)^2 + b(1) - 7$;

$b = 13$; $y = 2x^2 + 13x - 7$

7.

Input	Output	Differences 1st	Differences 2nd
1	5		
		7	
2	12		2
		9	
3	21		2
		11	
4	32		2
		13	
5	45		

Quadratic; $2a = 2$, $a = 1$. When $x = 0$

$y = 0$; $c = 0$; $5 = 1(1)^2 + b(1) + 0$

$b = 4$; $y = x^2 + 4x$

9.

Input	Output	Differences	
1	13	**1st**	**2nd**
		12	
2	25		0
		12	
3	37		0
		12	
4	49		0
		12	
5	61		

Linear; m = 12. When x = 0, y = 1; b = 1
$y = 12x + 1$

11.

Input	Output	Differences	
1	1	**1st**	**2nd**
		0	
2	1		1
		1	
3	2		0
		1	
4	3		1
		2	
5	5		

Neither linear nor quadratic

13.

Input	Output	Differences	
1	9	**1st**	**2nd**
		7	
2	16		–2
		5	
3	21		–2
		3	
4	24		–2
		1	
5	25		

Quadratic; $2a = -2$, $a = -1$. When $x = 0$
$y = 0$, $c = 0$; $9 = -1(1)^2 + b(1) + 0$
$b = 10$; $y = -x^2 + 10x$

For exercises 15 to 19, equations are found on a graphing calculator. Results are given for reference only.

15. a. Quadratic; a linear equation would suggest that posts of 2 or 4 inches are free.

b. $y = 0.5x - 2.16$, $r = 0.9911$

c. $0.0375x^2 - 0.175x + 0.26$

d. \$5.84 (linear); \$7.49 (quadratic)

17. a. B(–50, 0), C(50, 0), D(0, 30); $y = -0.012x^2 + 30$

b. for B(0, 0): C (100, 0), D(50, 30); $y = -0.012x^2 + 1.2x$
for C(0, 0): B(–100, 0), D(–50, 30); $y = -0.012 - 1.2x$
for D(0, 0): B(–50, –30), C(50, –30), $y = -0.012x^2$

19. for A(0,0): B(200, 150), C(400,0) $y = -0.00375x^2 + 1.5x$
for B(0, 0): A(–200, –150), C(200, –150), $y = -0.00375x^2$
for C(0, 0): A(–400, 0), B(–200, 150), $y = -0.00375x^2 - 1.5x$
for D(0, 0): A(–200, 0), B(0, 150), C(200, 0), $y = -0.00375x^2 + 150$

21. a. If price is a function of area, then the plot of the graph is slightly curved, $y = -0.00167x^2 + 1.087x + 3.300$.

b. The 150-square-inch boxes should cost \$129 every six months. The size of the box is within the existing data.

c. Using the current equation, this price would be \$171 which is a lower price than that for a smaller box. We have passed the vertex of the parabola and prices for area beyond the vertex are irrelevant.

d. There seems to be a discount. If one divides the cost by the area, the largest box is \$0.65 per square inch, compared with the second (and relatively more expensive) box at \$1.24 per square inch.

23. $y \approx 1.533x^2 - 2.632x + 3.429$

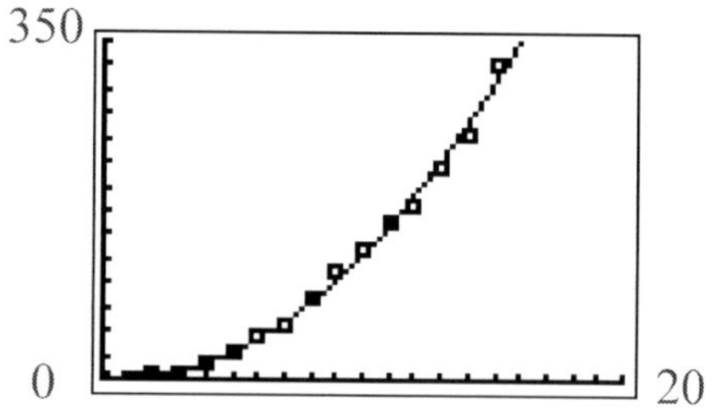

25. a. $h \approx -5.14t^2 - 25.89t + 10264$

b. $$\frac{-24.6\text{ mi}}{\text{hr}} \cdot \frac{5280\text{ ft}}{1\text{ mi}} \cdot \frac{1\text{ hr}}{60\text{ min}} \cdot \frac{1\text{ min}}{60\text{ sec}} = -36.08\text{ feet per second} \approx -36.1\text{feet per second}$$

$h \approx -16.1t^2 - 36.1t + 10079$

c. elapsed time in seconds.

d. height in feet

e. The graph of Pat's fall because her fall is slowed by her jump position and her clothing.

27. ***Answers will vary depending on which ordered pairs are selected. Given are from suggested ordered pairs in the text.***

a. (0, 1), (1, 2), (4, 14)

b. $f(0) = 1,\ f(1) = 2,\ f(4) = 14$

c. $1 = a(0)^2 + b(0) + c \Rightarrow 1 = c$

$2 = a(1)^2 + b(1) + c \Rightarrow 2 = a + b + c$

$14 = a(4)^2 + b(4) + c \Rightarrow 14 = 16a + 4b + c$

d. $a + b + 1 = 2 \Rightarrow a + b = 1$

$16a + 4b + 1 = 14 \Rightarrow 16a + 4b = 13$ (1)

$-4(a + b) = -4(1) \Rightarrow -4a - 4b = -4$ (2) *add (1) and (2)*

$12a = 9 \Rightarrow a = \frac{3}{4};\quad \frac{3}{4} + b = 1 \Rightarrow b = \frac{1}{4}$

Quadratic function is: $y = \frac{3}{4}x^2 + \frac{1}{4}x + 1$

Exercises 2.3: Polynomial Functions and Operations

1. **a.** $\sqrt{y}+x\sqrt{y}$ is not a polynomial. **b.** x^2+y^2 is a binomial.

3. **a.** $2\pi r$ is a monomial. **b.** $-\frac{1}{2}gt^2+vt+h$ is a trinomial.

5. **a.** $(x^3+2x^2+4x)-(2x^2+4x+8)=x^3+2x^2+4x-2x^2-4x-8=x^3-8$
b. $(x^3-6x^2+9x)-(3x^2-18x+27)=x^3-6x^2+9x-3x^2+18x-27=x^3-9x^2+27x-27$

7. **a.** $(x-6)+(2x+5)=x-6+2x+5=3x-1$
b. $(3x-1)+(4x-5)=3x-1+4x-5=7x-6$

9. **a.** $x(x-2)+2(x-2)=x^2-2x+2x-4=x^2-4$
b. $4(3-x)+x(3-x)=12-4x+3x-x^2=12-x-x^2$

11. **a.** $(x^3+2x^2+x)+(x^2+2x+1)=x^3+2x^2+x+x^2+2x+1=x^3+3x^2+3x+1$
b. $(a^3-a^2b+ab^2)+(a^2b-ab^2+b^3)=a^3-a^2b+ab^2+a^2b-ab^2+b^3=a^3+b^3$

13. $x(x^2-2xy+y^2)-y(x^2-2xy+y^2)=x^3-2x^2y+xy^2-x^2y+2xy^2-y^3=$
$x^3-3x^2y+3xy^2-y^3$

15.

Multiply	2x	+3
3x	**(3x)(2x) = $6x^2$**	**(3)(3x) = 9x**
−1	**(−1)(2x) = −2x**	**(−1)(3) = −3**

$(2x+3)(3x-1)=6x^2+7x-3$

17.

Multiply	3x	+1
3x	**(3x)(3x) = $9x^2$**	**(1)(3x) = 3x**
+1	**(1)(3x) = 3x**	**(1)(1) = 1**

$(3x+1)(3x+1)=9x^2+6x+1$

19.

Multiply	x^2	−2x	+4
x	**(x)(x^2) = x^3**	**(x)(−2x) = $-2x^2$**	**(4)(x) = 4x**
+2	**(2)(x^2) = $2x^2$**	**(2)(−2x) = −4x**	**(2)(4) = 8**

$(x+2)(x^2-2x+4)=x^3+8$

21. $(x + 6)(x - 3) =$
$x^2 - 3x + 6x - 18 =$
$x^2 + 3x - 18$

23. $(x - 6)(x + 3) =$
$x^2 + 3x - 6x - 18 =$
$x^2 - 3x - 18$

25. $(x - 9)(x - 2) =$
$x^2 - 2x - 9x + 18 =$
$x^2 - 11x + 18$

27. $(x + 4)(x - 4) =$
$x^2 - 4x + 4x - 16 =$
$x^2 - 16$

29. $(x - 5)(x - 5) =$
$x^2 - 5x - 5x + 25 =$
$x^2 - 10x + 25$

31. $(2x - 3)(x + 4) =$
$2x^2 + 8x - 3x - 12 =$
$2x^2 + 5x - 12$

33. $(2x + 3)(x - 4) =$
$2x^2 - 8x + 3x - 12 =$
$2x^2 - 5x - 12$

35. $(3x - 1)(3x + 2) =$
$9x^2 + 6x - 3x - 2 =$
$9x^2 + 3x - 2$

37. $(x + 2)(x + 2) =$
$x^2 + 2x + 2x + 4 =$
$x^2 + 4x + 4$

39. $(2x - 1)(2x + 1) =$
$4x^2 + 2x - 2x - 1 =$
$4x^2 - 1$

41. $(x - 2)(x^2 + 2x + 4) =$
$x^3 + 2x^2 + 4x - 2x^2 - 4x - 8 =$
$x^3 - 8$

43.

Multiply	x^2	$-3x$	-2
x^2	$\mathbf{(x^2)(x^2) = x^4}$	$\mathbf{(x^2)(-3x) = -3x^3}$	$\mathbf{(x^2)(-2) = -2x^2}$
$+3x$	$\mathbf{(3x)(x^2) = 3x^3}$	$\mathbf{(3x)(-3x) = -9x^2}$	$\mathbf{(3x)(-2) = -6x}$
-2	$\mathbf{(-2)(x^2) = -2x^2}$	$\mathbf{(-2)(-3x) = 6x}$	$\mathbf{(-2)(-2) = 4}$

$(x^2 - 3x - 2)(x^2 + 3x - 2) = x^4 - 13x^2 + 4$

45.

Factor	**6x**	**+1**
2x	$12x^2$	$+2x$
–3	$-18x$	-3

$12x^2 - 16x - 3 = (6x + 1)(2x - 3)$

47.

Factor	**x**	**–3**
3x	$3x^2$	$-9x$
–4	$-4x$	$+12$

$3x^2 - 13x + 12 = (x - 3)(3x - 4)$

49. $x^2 + 7x + 12 =$
$x^2 + 4x + 3x + 12 =$
$x(x + 4) + 3(x + 4) =$
$(x + 4)(x + 3)$

51. $x^2 + x - 12 =$
$x^2 + 4x - 3x - 12 =$
$x(x + 4) - 3(x + 4) =$
$(x + 4)(x - 3)$

53. $x^2 - 7x + 12 =$
$x^2 -3x - 4x + 12 =$
$x(x - 3) - 4(x - 3) =$
$(x - 3)(x - 4)$

55. $x^2 - 11x - 12 =$
$x^2 + x - 12x - 12 =$
$x(x + 1) - 12(x + 1) =$
$(x + 1)(x - 12)$

57. $x^2 + 3x - 28 =$
$x^2 + 7x - 4x - 28 =$
$x(x + 7) - 4(x + 7) =$
$(x + 7)(x - 4)$

59.

Factor	**2x**	**+5**
3x	$6x^2$	$+15x$
+2	$+4x$	$+10$

Diagonal sum = 19x
Diagonal product = $60x^2$
Factors are 4x and 15x.
$(2x + 5)(3x + 2)$

61.

Factor	**2x**	**+5**
3x	$6x^2$	$+15x$
–2	$-4x$	-10

Diagonal sum = 11x
Diagonal product = $-60x^2$
Factors are –4x and 15x.
$(2x + 5)(3x - 2)$

63.

Factor	**6x**	**–5**
x	$6x^2$	$-5x$
–2	$-12x$	$+10$

Diagonal sum = -17x
Diagonal product = $60x^2$
Factors are -12x and –5x.
$(6x - 5)(x - 2)$

65.

Factor	**3x**	**–2**
5x	$15x^2$	$-10x$
+3	$+9x$	-6

Diagonal sum = –1x
Diagonal product = $-90x^2$
Factors are –10x and 9x.
$(3x - 2)(5x + 3)$

67.

Factor	**3x**	**+5**
2x	$6x^2$	$+10x$
+1	$+3x$	$+5$

Diagonal sum = +13x
Diagonal product = $30x^2$
Factors are 10x and 3x.
$(3x + 5)(2x + 1)$

69.

Factor	x	+6
10x	$10x^2$	+60x
+1	+x	+6

Diagonal sum = 61x
Diagonal product = $60x^2$
Factors are 60x and x.
$(x + 6)(10x + 1)$

71. $6x^2 - 32x + 10 = 2(3x^2 - 16x + 5)$

Factor	x	–5
3x	$3x^2$	–15x
–1	–x	+5

Diagonal sum = –16x
Diagonal product = $15x^2$
Factors are –x and –15x.
$2(x - 5)(3x - 1)$

73. $x^3 + 2x^2 + 4x = x(x^2 + 2x + 4)$

75. $x^3 - 6x^2 + 9x = x(x^2 - 6x + 9)$
$x^2 - 6x + 9 =$
$x^2 - 3x - 3x + 9$
$x(x - 3) - 3(x - 3) =$
$(x - 3)(x - 3)$
$x^3 - 6x^2 + 9x = x(x - 3)^2$

77. $a^3 - a^2b + ab^2 = a(a^2 - ab + b^2)$

79. $a^3 + a^2b + ab^2 = a(a^2 + ab + b^2)$

81. $3x^2 - 48 = 3(x^2 - 16)$
$x^2 - 16 = x^2 + 4x - 4x - 16 =$
$x(x + 4) - 4(x + 4) =$
$(x + 4)(x - 4)$
$3x^2 - 48 = 3(x + 4)(x - 4)$

83. $12x^2 - 27 = 3(4x^2 - 9)$

Factor	2x	–3
2x	$4x^2$	–6x
+3	6x	–9

Diagonal sum is 0.
Diagonal product is $-36x^2$.
Factors are 6x and –6x.
$12x^2 - 27 = 3(2x - 3)(2x + 3)$

85. $20x^2 - 45 = 5(4x^2 - 9) =$
$5(2x - 3)(2x + 3)$ *see ex. 83*

The solution for example 87 is based on observed patterns in previous examples.

87. $28 - 63x^2 = 7(4 - 9x^2) = 7(2 + 3x)(2 - 3x)$

89. a.

Multiply	x^2	$+x$	$+1$
x^2	$(x^2)(x^2) = x^4$	$(x^2)(x) = x^3$	$(x^2)(1) = x^2$
$-x$	$(-x)(x^2) = -x^3$	$(-x)(x) = -x^2$	$(-x)(1) = -x$
-1	$(-1)(x^2) = -x^2$	$(-1)(x) = -x$	$(-1)(1) = -1$

$x^4 - x^2 - 2x - 1$

b.

Multiply	x^2	$-x$	$+1$
x^2	$(x^2)(x^2) = x^4$	$(x^2)(-x) = -x^3$	$(x^2)(1) = x^2$
$+x$	$(x)(x^2) = x^3$	$(x)(-x) = -x^2$	$(x)(1) = x$
-1	$(-1)(x^2) = -x^2$	$(-1)(-x) = x$	$(-1)(1) = -1$

$x^4 - x^2 + 2x - 1$

c.

Multiply	x^2	$+x$	$+1$
x^2	$(x^2)(x^2) = x^4$	$(x^2)(x) = x^3$	$(x^2)(1) = x^2$
$+x$	$(x)(x^2) = x^3$	$(x)(x) = x^2$	$(x)(1) = x$
-1	$(-1)(x^2) = -x^2$	$(-1)(x) = -x$	$(-1)(1) = -1$

$x^4 + 2x^3 + x^2 - 1$

d.

Multiply	x^2	$-x$	$+1$
x^2	$(x^2)(x^2) = x^4$	$(x^2)(-x) = -x^3$	$(x^2)(1) = x^2$
$-x$	$(-x)(x^2) = -x^3$	$(-x)(-x) = x^2$	$(-x)(1) = -x$
-1	$(-1)(x^2) = -x^2$	$(-1)(-x) = x$	$(-1)(1) = -1$

$x^4 - 2x^3 + x^2 - 1$

e.

Multiply	x^2	$+x$	$+1$
x^2	$(x^2)(x^2) = x^4$	$(x^2)(x) = x^3$	$(x^2)(1) = x^2$
$+x$	$(x)(x^2) = x^3$	$(x)(x) = x^2$	$(x)(1) = x$
$+1$	$(1)(x^2) = x^2$	$(1)(x) = x$	$(1)(1) = 1$

$x^4 + 2x^3 + 3x^2 + 2x + 1$

91. From the first equation, the graph turns down, $a < 0$, and $y = 4$ is the y-intercept. From the factored form, the graph has x-intercepts $x = 2$ and $x = -2$.

93. Answers may vary. Note that, $x(x + 1)(x + 2) = x^3 + 3x^2 + 2x$;
Some things that are always true about the product: divisible by 3, always even, always a whole number, always divisible by 6. Consider the following cases: If x is even, then is $x^3 + 3x^2 + 2x$ even? If x is odd, then is $x^3 + 3x^2 + 2x$ even? For more ideas, try other ways of writing the consecutive numbers; that is, $3k$, $3k + 1$, $3k + 2$ or $(x - 1)x(x + 1)$ or $(x - 2)(x - 1)x$.
The product of any four consecutive whole numbers will always be even, always a whole number, and always divisible by 2, 3, 4, 6, 8, 12 and 24.

Exercises 2.4: Special Products and Completing the Square

1. $(x - 3)(x + 4) =$
$x^2 + 4x - 3x - 12 =$
$x^2 + x - 12$

3. $(2x + 3)(x - 4) =$
$2x^2 - 8x + 3x - 12 =$
$2x^2 - 5x - 12$

5. $(x + 3)(x + 3) =$
$x^2 + 2(3x) + 9 =$
$x^2 + 6x + 9$ (perfect sq. trinomial)

7. $(x + 6)(x - 6) =$
$x^2 - 6x + 6x - 36 =$
$x^2 - 36$ (difference of squares)

9. $(2x - 5)^2 =$
$(2x)^2 + 2(2x)(-5) + (-5)^2 =$
$4x^2 - 20x + 25$ (perfect sq. trinomial)

11. $(x - 6)^2 =$
$x^2 + 2(-6x) + (-6)^2 =$
$x^2 - 12x + 36$ (perfect sq. trinomial)

13. $(x - \frac{1}{2})^2 =$
$x^2 + 2(-\frac{1}{2}x) + (-\frac{1}{2})^2 =$
$x^2 - x + \frac{1}{4}$ (perfect sq. trinomial)

15. $(2x - 3)^2 =$
$(2x)^2 + 2(2x)(-3) + (-3)^2 =$
$4x^2 - 12x + 9$ (perfect sq. trinomial)

17.

	a	**+7**
a	a^2	7a
+7	7a	**49**

$(a + 7)^2 = a^2 + 14a + 49$

Missing entries: +7, +7, 49

19.

	x	$-\frac{7}{2}$
x	x^2	$-\frac{7}{2}x$
$-\frac{7}{2}$	$-\frac{7}{2}x$	$\frac{49}{4}$

$(x - \frac{7}{2})^2 = x^2 - 7x + \frac{49}{4}$

Missing entries: $-\frac{7}{2}, -\frac{7}{2}, \frac{49}{4}$

21. $x^2 + 4x + 4 =$
$x^2 + 2(2x) + 2^2 =$
$(x + 2)^2$

23. $n^2 + 2n + 1 =$
$n^2 + 2(n) + 1^2 =$
$(n + 1)^2$

25. $49x^2 + 14x + 1 =$
$(7x)^2 + 2(7x)(1) + 1^2$
$(7x + 1)^2$

27. $9x^2 - 6xy + y^2 =$
$(3x)^2 + 2(3x)(-y) + (-y)^2 =$
$(3x - y)^2$

29. $x^2 + 5x + 6\frac{1}{4} =$
$x^2 + 2(x)(\frac{5}{2}) + (\frac{5}{2})^2 =$
$(x + 2\frac{1}{2})^2$

31. $x^2 + \underline{2(4)}x + (4)^2 = (x + 4)^2$
$x^2 + \underline{\mathbf{8}}x + 16 = (x + 4)^2$

33. $x^2 - \underline{2(7)}x + (7)^2 = (x - 7)^2$
$x^2 - \underline{\mathbf{14}}x + 49 = (x - \underline{\mathbf{7}})^2$

35. $x^2 - 5x + \underline{(-2.5)^2} = (x - 2.5)$
$x^2 - 5x + \underline{\mathbf{6.25}} = (x - 2.5)^2$

37. $x^2 + 7x + \underline{(\frac{7}{2})^2} = (x + \frac{7}{2})^2$

$x^2 + 7x + \underline{\mathbf{12.25}} = (x + \underline{\mathbf{3.5}})^2$

39. $x^2 - 24x + \underline{(\frac{24}{2})^2} = (x - \underline{\frac{24}{2}})^2$

$x^2 - 24 + \underline{\mathbf{144}} = (x - \underline{\mathbf{12}})^2$

41. $x^2 - 4 =$

$x^2 - 2^2 = (x - 2)(x + 2)$

43. $n^2 + 16$

Cannot be factored.

45. $49x^2 - 1 =$

$(7x)^2 - 1^2 = (7x - 1)(7x + 1)$

47. $y^2 - 36z^2 =$

$y^2 - (6z)^2 = (y + 6z)(y - 6z)$

49. $9x^2 - y^2 =$

$(3x)^2 - y^2 = (3x + y)(3x - y)$

51. $x^2 - 144 =$

$x^2 - (12)^2 = (x - 12)(x + 12)$

53. $0.25x^2 - 0.01 =$

$(0.5x)^2 - 0.1^2 =$

$(0.5x + 0.1)(0.5x - 0.1)$

55. $(x + 1)(x^2 - x + 1) =$

$x^3 + 1$

sum of cubes

57. $(x - 3)(x^2 - 6x + 9) =$

$x^3 - 6x^2 + 9x - 3x^2 + 18x - 27$

$x^3 - 9x^2 + 27x - 27$

59. $(x + 1)(x^2 - 2x + 1) =$

$x^3 - 2x^2 + x + x^2 - 2x + 1$

$x^3 - x^2 - x + 1$

61.

Factor	$\mathbf{x^2}$	$\mathbf{-4x}$	$\mathbf{+16}$
x	x^3	$\mathbf{-4x^2}$	$\mathbf{+16x}$
$+4$	$\mathbf{+4x^2}$	$\mathbf{-16x}$	$+64$

$x^3 + 64 = (x + 4)\mathbf{(x^2 - 4x + 16)}$

63. a. $x^3 + 27 = x^3 + 3^3 =$

$(x + 3)(x^2 - 3x + 3^2) =$

$(x + 3)(x^2 - 3x + 9)$

b. $x^3 + 8 = x^3 + 2^3 =$

$(x + 2)(x^2 - 2x + 2^2) =$

$(x + 2)(x^2 - 2x + 4)$

c. $x^3 - 64 = x^3 - 4^3 =$

$(x - 4)(x^2 + 4x + 4^2) =$

$(x - 4)(x^2 + 4x + 16)$

65. a. $x^3 - 0.001 = x^3 - 0.1^3 =$

$(x - 0.1)(x^2 + 0.1x + 0.1^2) =$

$(x - 0.1)(x^2 + 0.1x + 0.01)$

b. $x^3 - \frac{1}{27} = x^3 - (\frac{1}{3})^3 =$

$(x - \frac{1}{3})(x^2 + \frac{1}{3}x + (\frac{1}{3})^2) =$

$(x - \frac{1}{3})(x^2 + \frac{1}{3}x + \frac{1}{9})$

67.

Multiply	**n**	
n	n^2	**−4n**
	+4n	+16

Diagonals must both add to zero and multiply to $+16n^2$. Both results are not possible.

69. a. $A = \pi(R^2 - r^2)$

$A = \pi(R + r)(R - r)$

b. R − r is the distance between the 2 radii.

71. a.

x	$y = x^3$	**differences**		
1	$(1)^3 = 1$	**1st**	**2nd**	**3rd**
		7		
			12	
2	$(2)^3 = 8$			6
		19		
			18	
3	$(3)^3 = 27$			6
		37		
			24	
4	$(4)^3 = 64$			
		61		
5	$(5)^3 = 125$			

x	$y = 2x^3 + x$	**differences**		
1	$2(1)^3 + 1 = 3$	**1st**	**2nd**	**3rd**
		15		
			24	
2	$2(2)^3 + 2 = 18$			12
		39		
			36	
3	$2(3)^3 + 3 = 57$			12
		75		
			48	
4	$2(4)^3 + 4 = 132$			
		123		
5	$2(5)^3 + 5 = 255$			

x	$y = 0.5x^3 + x^2$	**differences**		
1	$0.5(1)^3 + (1)^2 = 1.5$	**1st**	**2nd**	**3rd**
		6.5		
			8	
2	$0.5(2)^3 + (2)^2 = 8$			3
		14.5		
			11	
3	$0.5(3)^3 + (3)^2 = 22.5$			3
		25.5		
			14	
4	$0.5(4)^3 + (4)^2 = 48$			
		39.5		
5	$0.5(5)^3 + (5)^2 = 87.5$			

b. 3rd difference is constant; predict that n^{th} row of differences would be constant for an nth degree polynomial.

c.

sequence	differences		
	1st	**2nd**	**3rd**
3			
	7		
		12	
10			6
	19		
		18	
29			6
	37		
		24	
66			
	61		
127			

Cubic polynomial

sequence	Differences		
	1st	**2nd**	**3rd**
3			
	6		
		12	
9			24
	18		
		36	
27			72
	54		
		108	
81			
	162		
243			

Not a cubic polynomial

sequence	differences		
	1st	**2nd**	**3rd**
–4			
	7		
		12	
3			6
	19		
		18	
22			6
	37		
		24	
59			
	61		
120			

Cubic polynomial

d.

x	$f(x) = ax^3 + bx^2 + cx + d$	differences		
		1st	**2nd**	**3rd**
1	1a + 1b + 1c + d			
		7a + 3b + c		
			12a + 2b	
2	8a + 4b + 2c + d			6a
		19a + 5b + c		
			18a + 2b	
3	27a + 9b + 3c + d			6a
		37a + 7b + c		
			24a + 2b	
4	64a + 16b + 4c + d			
		61a + 9b + c		
5	125a + 25b + 5c + d			

3rd difference can be used to find a, 2nd difference and a can be used to find b, 1st difference with a and b can be used to find c.

Exercises 2.5: Solving Quadratic Equations with Tables, Graphs and Factors

1. The axis of symmetry is $x = 2$. Factors: $(x - 1)(x - 3)$

3. **a.** From the table: $\{-1, 5\}$
b. From the table: $\{1, 3\}$
c. From the table and ex. 17: $\{-2, 6\}$
d. From the table: no real number solution

5. **a.** From the graph: $x^2 - 2x - 8 = -9$, $\{1\}$
b. From the graph: $x^2 - 2x - 8 = -5$, $\{-1, 3\}$
c. From the graph: $x^2 - 2x - 8 = 0$, $\{-2, 4\}$
Symmetric to $x = 1$, which is the line of symmetry.

For exercises 7 through 29, results of factoring are given but details are not. Please refer to sections 2.3 and 2.4 for factoring details.

7.
$x^2 - x - 6 = 0$
$(x + 2)(x - 3) = 0$
$x + 2 = 0$ or $x - 3 = 0$
$x = -2$ or $x = 3$
$\{-2, 3\}$

9.
$2x^2 + x - 1 = 0$
$(2x - 1)(x + 1) = 0$
$2x - 1 = 0$ or $x + 1 = 0$
$x = \frac{1}{2}$ or $x = -1$
$\{-1, \frac{1}{2}\}$

11.
$x^2 - 121 = 0$
$(x + 11)(x - 11) = 0$
$x + 11 = 0$ or $x - 11 = 0$
$x = -11$ or $x = 11$
$\{\pm 11\}$ (ds)

13.
$3x^2 - 48 = 0$
$3(x + 4)(x - 4) = 0$
$x + 4 = 0$ or $x - 4 = 0$
$x = -4$ or $x = 4$
$\{\pm 4\}$ (ds)

15.
$4x^2 - 2x = 0$
$2x(2x - 1) = 0$
$2x = 0$ or $2x - 1 = 0$
$x = 0$ or $x = \frac{1}{2}$
$\{0, \frac{1}{2}\}$

17.
$2x^2 = 6x$
$2x^2 - 6x = 0$
$2x(x - 3) = 0$
$2x = 0$ or $x - 3 = 0$
$x = 0$ or $x = 3$
$\{0, 3\}$

19.
$2x^2 - 6x + 4 = 0$
$2(x^2 - 3x + 2) = 0$
$2(x - 2)(x - 1) = 0$
$x - 2 = 0$ or $x - 1 = 0$
$x = 2$ or $x = 1$
$\{1, 2\}$

21.
$2x^2 - 11x + 12 = x - 6$
$2x^2 - 12x + 18 = 0$
$2(x^2 - 6x + 9) = 0$
$2(x - 3)^2 = 0$
$x - 3 = 0$
$x = 3 \Rightarrow \{3\}$ (pst) double root

23. $\frac{1}{2}x^2 - \frac{3}{2}x = 5$
$x^2 - 3x = 10$
$x^2 - 3x - 10 = 0$
$(x + 2)(x - 5) = 0$
$x + 2 = 0$ or $x - 5 = 0$
$x = -2$ or $x = 5 \Rightarrow \{-2, 5\}$

25. $0.25x^2 + 0.50x - 6 = 0$
$x^2 + 2x - 24 = 0$
$(x - 4)(x + 6) = 0$
$x - 4 = 0$ or $x + 6 = 0$
$x = 4$ or $x = -6$
$\{-6, 4\}$

27. $0.4x^2 + 0.2x = 0.2$
$4x^2 + 2x = 2$
$4x^2 + 2x - 2 = 0$
$2(2x + x - 1) = 0$
$2(2x - 1)(x + 1) = 0$
$2x - 1 = 0$ or $x + 1 = 0$
$x = 0.5$ or $x = -1 \Rightarrow \{-1, 0.5\}$

29. $x^2 - \frac{5}{3}x = \frac{2}{3}$
$3x^2 - 5x = 2$
$3x^2 - 5x - 2 = 0$
$(3x + 1)(x - 2) = 0$
$3x + 1 = 0$ or $x - 2 = 0$
$x = -\frac{1}{3}$ or $x = 2$
$\{-\frac{1}{3}, 2\}$

31. $x - 4 = 0$ or $x + 3 = 0$
$x = 4$ or $x = -3$
x-intercepts: $(-3, 0)$, $(4, 0)$

33. $2x - 5 = 0$ or $x + 3 = 0$
$x = 2.5$ or $x = -3$
x-intercepts: $(-3, 0)$, $(2.5, 0)$

35. $3x - 4 = 0$ or $2x + 1 = 0$
$x = 1\frac{1}{3}$ or $x = -0.5$
x-intercepts: $(-0.5, 0)$, $(1\frac{1}{3}, 0)$

37. $3 = a(-4 + 1)(-4 + 5)$
$3 = a(-3)(1)$
$3 = -3a \Rightarrow a = -1$

39. $7.5 = a(0 + 1)(0 + 5)$
$7.5 = a(1)(5)$
$7.5 = 5a$
$a = 1.5$

41. $y = a(x - 3)(x + 2)$
$12 = a(0 - 3)(0 + 2)$
$12 = a(-3)(2)$
$12 = -6a \Rightarrow a = -2$
$y = -2(x - 3)(x + 2)$
$y = -2(x^2 - x - 6)$
$y = -2x^2 + 2x + 12$

43. $y = a(x + 3)(x - 5)$
$6 = a(3 + 3)(3 - 5)$
$6 = a(6)(-2)$
$6 = -12a$
$a = -0.5$
$y = -0.5(x + 3)(x - 5)$
$y = -0.5(x^2 - 2x - 15)$
$y = -0.5x^2 + x + 7.5$

45. $y = a(x - 10)(x - 30)$
$12 = a(20 - 10)(20 - 30)$
$12 = a(10)(-10)$
$12 = -100a$
$a = -0.12$
$y = -0.12(x - 10)(x - 30)$
$y = -0.12(x^2 - 40x + 300)$
$y = -0.12x^2 + 4.8x - 36$

47. $y = -\frac{3}{40}(x - 10)(x - 40)$. To find the vertex, average x-intercept values and substitute into original equation to find y.

49. To find x-intercepts, use the zero product rule, set each factor equal to zero, and solve each equation for x.

51. a. $(8.5 - 2x)(11 - 2x)$

b. $8.5 - 2x = 0$ or $11 - 2x = 0$

$x = 4.25$ or $x = 5.5$

c. $x = 4.25$ uses the entire width of the paper. It is impossible to make squares with $x = 5.5$.

d. $(8.5 - 2x)(11 - 2x) = 51$

$93.5 - 39x + 4x^2 = 51$

$4x^2 - 39x + 42.5 = 0$

$(4x - 5)(x - 8.5) = 0$

$4x - 5 = 0$ or $x - 8.5 = 0$

$x = 1.25$ or $x = 8.5$

$x = 1.25$ is a reasonable solution.

$x = 8.5$ is impossible in this problem setting.

53. a. $h_0 = 320$ ft; $v_0 = 16$ ft/sec. To find elapsed time to ground: $0 = -16t^2 + 16t + 320$. Divide both sides by -16: $0 = t^2 - t - 20$. Factor: $0 = (t - 5)(t + 4)$. Solve: $t - 5 = 0$, $t = 5$ sec, or $t + 4 = 0$, $t = -4$ sec (discard negative).

Time to reach the ground = 5 sec.

To find speed when it reaches the ground, use $v = -32t + 16$ and substitute $t = 5$ sec:

$v = -32(5) + 16$; $\mathbf{v = -144}$ **ft/sec** or ≈ 98 mph.

The worker's friend should have protection similar to that of a major league baseball catcher. Thus, the baseball catcher's mitt, protective face mask and helmet are suggested.

b. $v_0 = -16$ ft/sec: $0 = -16t^2 - 16t + 320$; $0 = t^2 + t - 20$; $0 = (t + 5)(t - 4)$;

$t + 5 = 0$, $t = -5$ (discard negative), or $t - 4 = 0$, $t = 4$

Time to reach the ground = 4 sec.

$v = -32(4) - 16$; $\mathbf{v = -144}$ **ft/sec**

Speed is the same; same precautions are recommended.

Exercises 2.6: The Roles of a, b, c and Binomial Squares in Graphing Quadratic Functions

1.

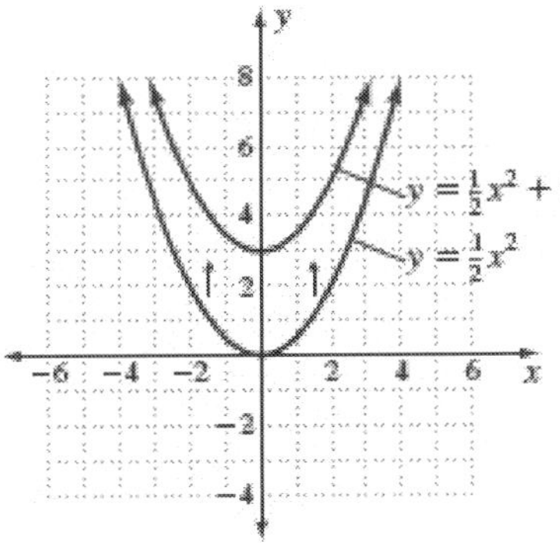

3.

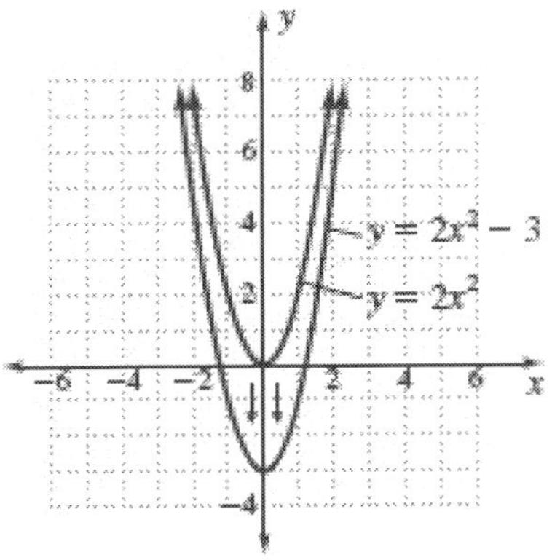

5. $f(x) = 2x^2$, $g(x) = \pi x^2$

$\pi > 2$; $g(x)$ is steeper.

7. $j(x) = 4x^2$, $g(x) = \pi x^2$

$4 > \pi$; $j(x)$ is steeper.

9. $f(x) = \frac{2}{3}x^2$, $h(x) = \frac{3}{4}x^2$

$\frac{3}{4} > \frac{2}{3}$; $h(x)$ is steeper.

11. $p(x) = 2.41x^2$, $q(x) = 2.288x^2$

$2.41 > 2.288$; $p(x)$ is steeper.

13. Exercises 16, 17, 18, 19, 21, 24, 25, 26, 27, 28, 29, 31, 32, 34, 35, 36

15. & 17.

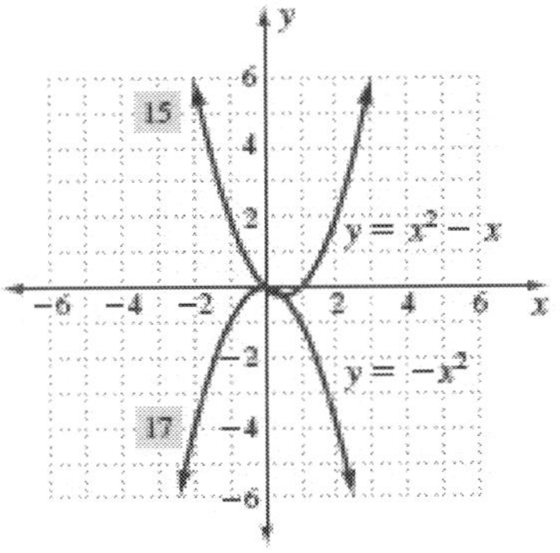

19. & 21.

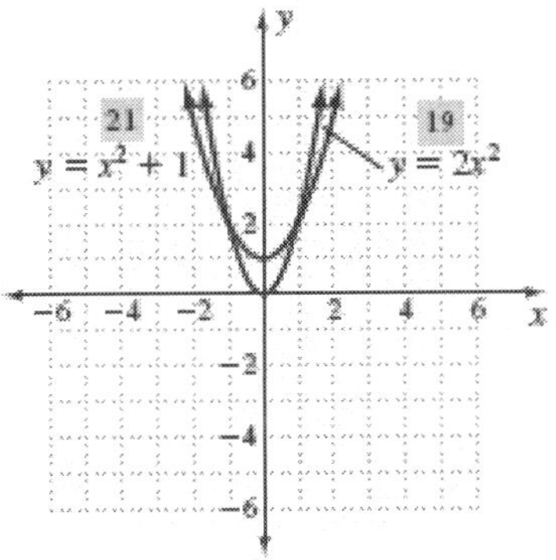

23. & 25.

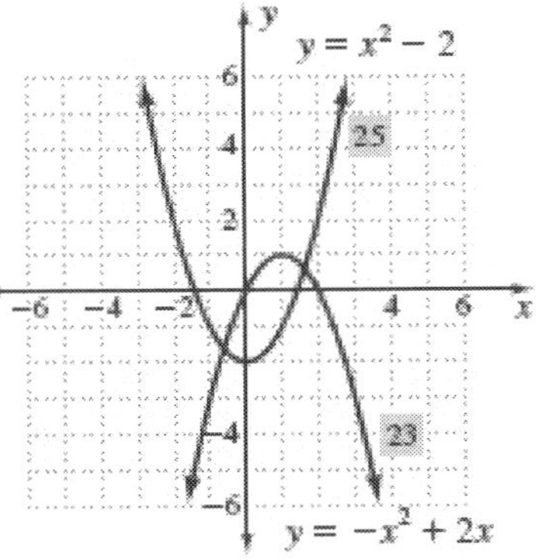

27. & 29.

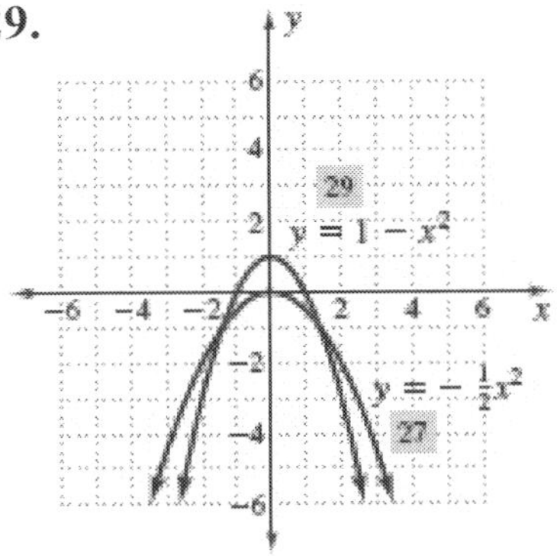

31. & 33.

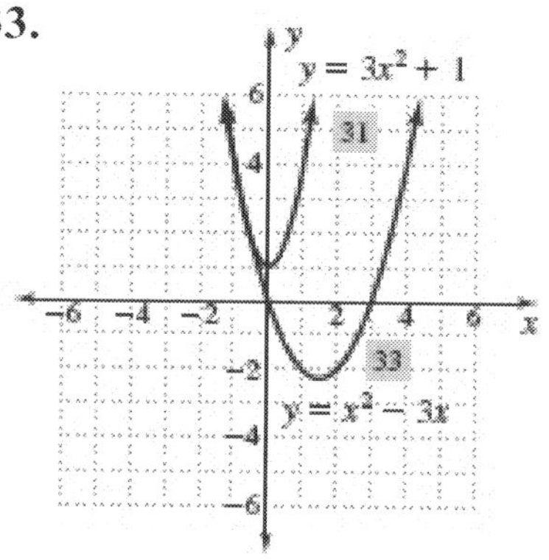

35.

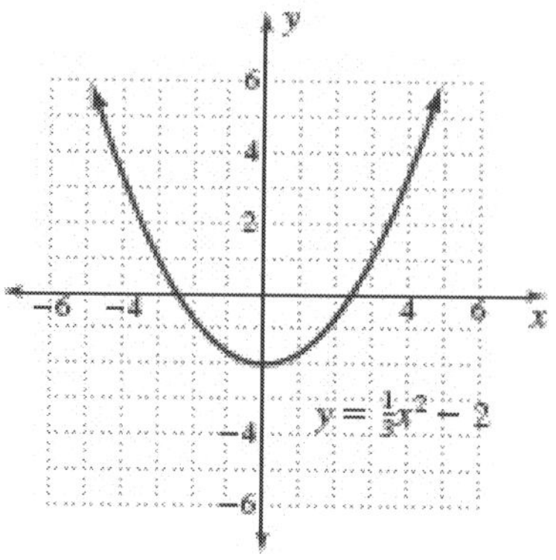

37. The equation of a parabolic cable supporting a bridge would have a positive coefficient on x^2; $y = \frac{4}{125}x^2$.

39. The coefficient on x^2 in $y = -0.00375x^2 + 1.5x + 0$ is negative, indicating that the graph will turn down.

41. The absolute value of a is 0.00375, which is less than the absolute value of −1. The graph will be flatter.

43. a. If the origin is 20 feet directly below the cannon, all the points on the water-spray graph shift up 20. The cannon is at (0, 20), the vertex is at (200, 170), and the fire is at (400, 20).

b. The new equation has a y-intercept of 20 instead of 0. The equation will be $y = -0.00375x^2 + 1.5x + 20$.

45. $y = (x - 8)(x + 8)$ is the same as $y = x^2 - 64$; shift $y = x^2$ down 64 units.

47. The two possibilities are a graph with the same shape opening up or opening down. The coefficient on x^2 is the same, the y-intercept is −3; $y = x^2 - 3$ or $y = -x^2 - 3$.

49. All data sets appear somewhat parabolic and open downward. Each year tends to shift upward from prior year Ages 65 and over showed the least increase.
Equations that fit $y = -40.6x^2 + 3670x - 39500$, $y = -53.7x^2 + 4840x - 50100$ and $y = -56.8x^2 + 5130x - 49800$.
Prices have risen, so money in 1993 might actually be worth more than money in 2003.

Chapter 2 Review

1. $y = 3(x - 3)^2 - 4$
$y = 3(x^2 + 2(x)(-3) + (-3)^2) - 4$
$y = 3(x^2 - 6x + 9) - 4$
$y = 3x^2 - 18x + 27 - 4$
$y = 3x^2 - 18x + 23$
$a = 3, b = -18, c = 23$

2. $y = -4(x + 2)^2 + 5$
$y = -4(x^2 + 2(x)(2) + (2)^2) + 5$
$y = -4(x^2 + 4x + 4) + 5$
$y = -4x^2 - 16x - 16 + 5$
$y = -4x^2 - 16x - 11$
$a = -4, b = -16, c = -11$

3. $x(x - 3) = x + 7$
$x^2 - 3x = x + 7$
$x^2 - 4x - 7 = 0$
$a = 1, b = -4, c = -7$

4. $x - 4x(2 - x) = x - x^2$
$x - 8x + 4x^2 = x - x^2$
$5x^2 - 8x = 0$
$a = 5, b = -8, c = 0$

5. The vertex of the graph of a quadratic function is the location of the highest or lowest y-value. ***true***

6. The x-coordinate of the vertex of the graph of a quadratic function is halfway between the x-intercepts of the graph. ***true***

7. The vertex of the parabola lies on the axis of symmetry of the graph. ***true***

8. The first differences in a quadratic sequence are constant. ***false***

9. a.

x	$f(x) = x^2 - x - 6$
−2	$(-2)^2 - (-2) - 6 = 0$
−1	$(-1)^2 - (-1) - 6 = -4$
0	$(0)^2 - (0) - 6 = -6$
1	$(1)^2 - (1) - 6 = -6$
2	$(2)^2 - (2) - 6 = -4$

b. Extend table or use symmetry: (−2, 0), (3, 0)

c. Average x-values of intercepts: $(-2 + 3) \div 2 = 0.5$; $x = 0.5$

d. $(0.5)^2 - (0.5) - 6 = -6.25$; vertex is at (0.5, – 6.25).

e.

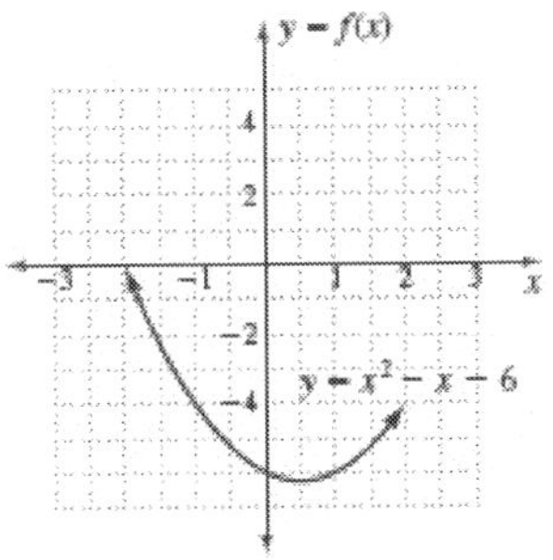

f. $x^2 - x - 6 = 6$
$x^2 - x - 12 = 0$
$(x - 4)(x + 3) = 0$
$x - 4 = 0$ or $x + 3 = 0$
$x = 4$ or $x = -3$
$\{-3, 4\}$

g. $x^2 - x - 6 = 0$

Note: x-intercepts $\{-2, 3\}$

h. $x^2 - x - 6 = -7$

$x^2 - x + 1 = 0$

Cannot be factored.

No real number solution.

i. Range is $y \geq -6.25$.

j. $f(x)$ is decreasing when $x < 0.5$.

10. a.

x	$f(x) = x^2 + x - 12$
−2	$(-2)^2 + (-2) - 12 = -10$
−1	$(-1)^2 + (-1) - 12 = -12$
0	$(0)^2 + (0) - 12 = -12$
1	$(1)^2 + (1) - 12 = -10$
2	$(2)^2 + (2) - 12 = -6$

b. Extend table or use symmetry: $(-4, 0)$, $(3, 0)$

c. Average x-values of intercepts: $(-4 + 3) \div 2 = -0.5$, $x = -0.5$

d. $(-0.5)^2 + (-0.5) - 12 = -12.25$; vertex is at $(-0.5, -12.25)$.

e.

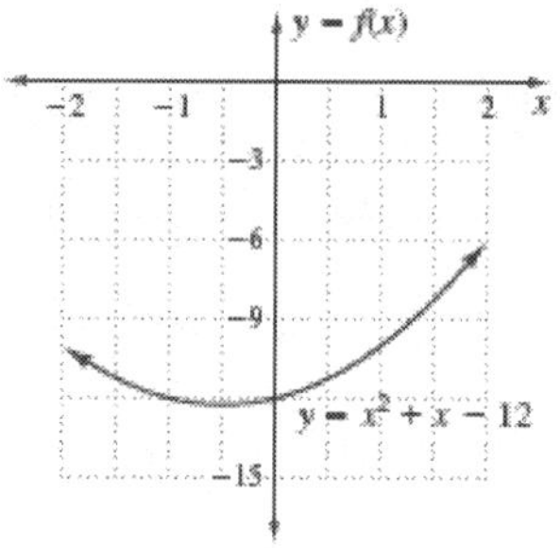

f. $x^2 + x - 12 = -6$

$x^2 + x - 6 = 0$

$(x + 3)(x - 2) = 0$

$x + 3 = 0$ or $x - 2 = 0$

$x = -3$ or $x = 2$

$\{-3, 2\}$

g. $x^2 + x - 12 = 0$

Note: x-intercepts $\{-4, 3\}$

h. $x^2 + x - 12 = -10$

$x^2 + x - 2 = 0$

$(x + 2)(x - 1) = 0$

$x + 2 = 0$ or $x - 1 = 0$

$x = -2$ or $x = 1$

$\{-2, 1\}$

i. Range is $y \geq -12.25$.

j. $f(x)$ is increasing when $x > -0.5$.

11. Variable has 2 as highest exponent.

r	$A = 4\pi r^2$
0	$4\pi(0)^2 = 0$
1	$4\pi(1)^2 = 4\pi$
2	$4\pi(2)^2 = 16\pi$
3	$4\pi(3)^2 = 36\pi$
4	$4\pi(4)^2 = 64\pi$
5	$4\pi(5)^2 = 100\pi$
6	$4\pi(6)^2 = 144\pi$
7	$4\pi(7)^2 = 196\pi$
8	$4\pi(8)^2 = 256\pi$
9	$4\pi(9)^2 = 320\pi$
10	$4\pi(10)^2 = 400\pi$

12. Variable has 2 as highest exponent.

r	$A = 20{,}000(1 - r)^2$
0	$20{,}000(1 - 0)^2 = 20{,}000$
0.01	$20{,}000(1 - 0.01)^2 = 19{,}602$
0.02	$20{,}000(1 - 0.02)^2 = 19{,}208$
0.03	$20{,}000(1 - 0.03)^2 = 18{,}818$
0.04	$20{,}000(1 - 0.04)^2 = 18{,}432$
0.05	$20{,}000(1 - 0.05)^2 = 18{,}050$
0.06	$20{,}000(1 - 0.06)^2 = 17{,}672$
0.07	$20{,}000(1 - 0.07)^2 = 17{,}298$
0.08	$20{,}000(1 - 0.08)^2 = 16{,}928$

13.

x	y	differences	
		1st	**2nd**
1	6		
		9	
2	15		4
		13	
3	28		4
		17	
4	45		4
		21	
5	66		

Quadratic: $2a = 4$, $a = 2$; when $x = 0$, $y = 1$, $c = 1$; $9 = 3(2) + b$, $b = 3$
$y = 2x^2 + 3x + 1$.

14.

x	y	differences	
		1st	**2nd**
1	−1		
		4	
2	3		6
		10	
3	13		6
		16	
4	29		6
		22	
5	51		

Quadratic: $2a = 6$, $a = 3$; when $x = 0$, $y = 1$, $c = 1$; $4 = 3(3) + b$, $b = -5$
$y = 3x^2 - 5x + 1$.

15.

x	y	differences	
1	−1	**1st**	**2nd**
		3	
2	2		0
		3	
3	5		0
		3	
4	8		0
		3	
5	11		

Linear: $m = 3$; when $x = 0$, $y = -4$, $b = -4$; $y = 3x - 4$.

16.

x	y	differences	
1	8	**1st**	**2nd**
		−6	
2	2		0
		−6	
3	−4		0
		−6	
4	−10		0
		−6	
5	−16		

Linear: $m = -6$; when $x = 0$, $y = 14$, $b = 14$; $y = -6x + 14$.

17.

x	y	differences	
1	15	**1st**	**2nd**
		−3	
2	12		−2
		−5	
3	7		−2
		−7	
4	0		−2
		−9	
5	−9		

Quadratic: $2a = -2$, $a = -1$; when $x = 0$, $y = 16$, $c = 16$; $3(-1) + b = -3$, $b = 0$
$y = -x^2 + 16$.

18.

x	y	differences	
1	16	**1st**	**2nd**
		−1	
2	15		−6
		−7	
3	8		−12
		−19	
4	−11		−18
		−37	
5	−48		

Neither quadratic nor linear.

19.

x	y	differences	
		1st	**2nd**
1	7		
		2	
2	9		−6
		−4	
3	5		−12
		−16	
4	−11		−18
		−34	
5	−45		

Neither quadratic nor linear.

20.

x	y	differences	
		1st	**2nd**
1	24		
		−3	
2	21		−2
		−5	
3	16		−2
		−7	
4	9		−2
		−9	
5	0		

Quadratic: $2a = -2$, $a = -1$; when $x = 0$, $y = 25$, $c = 25$; $3(-1) + b = -3$, $b = 0$; $y = -x^2 + 25$.

21. There are two points given: starting point, (0, 0), and vertex, (30, 100). End point is found from symmetry, (60, 0). Using the x-intercepts we get: $y = a(x - 0)(x - 60)$. Substitute the vertex to find a: $100 = a(30)(30 - 60)$; $100 = -900a$; $a = -\frac{1}{9}$, or $a \approx -0.111$.
$y = -0.111(x)(x - 60)$; $y = -0.111(x^2 - 60x)$; $\mathbf{y = -0.111x^2 + 6.667x}$

22. There are two points given: starting point, (0, 0), and vertex, (25, 75). End point is found from symmetry, (50, 0). Using the x-intercepts we get: $y = a(x - 0)(x - 50)$. Substitute the vertex to find a: $75 = a(25)(25 - 50)$; $75 = -625a$; $a = -0.12$
$y = -0.12(x)(x - 50)$; $y = -0.12(x^2 - 50x)$; $\mathbf{y = -0.12x^2 + 6x}$

23. a. $x^3 - y^3$ is a ***binomial***.

b. $\sqrt{x} - \sqrt{y}$ is not a polynomial.

c. $x^2 + x^{-1}$ is not a polynomial.

d. $3x + 3$ is a ***binomial***.

e. $3x^3 - 3x^2 + 3x$ is a ***trinomial***.

f. $|x + 1|$ is not a polynomial.

24. a. $(x^2 + 3x) + (x^3 - 4x^2 - 5x) =$
$x^2 + 3x + x^3 - 4x^2 - 5x =$
$x^3 - 3x^2 - 2x$

b. $(x^2 + 3x) - (x^3 - 4x^2 - 5x) =$
$x^2 + 3x - x^3 + 4x^2 + 5x =$
$-x^3 + 5x^2 + 8x$

c. $x^2 - x + 1 - x^2 + x + 1 = 2$

d. $x^2 - x + 1 - (x^2 + x + 1) =$
$x^2 - x + 1 - x^2 - x - 1 = -2x$

e. $14 - 6(x + 3) =$
$14 - 6x - 18 = -6x - 4$

f. $12 - 7(x - 2) =$
$12 - 7x + 14 = -7x + 26$

25. a. $(x - 3)(x + 3) = x^2 - 9$

b. $(2x - 5)(2x - 5) =$
$(2x)^2 + 2(2x)(-5) + (-5)^2 =$
$4x^2 - 20x + 25$

c. $(x - 1)(x^2 + x + 1) = x^3 - 1$

d. $(n + 4)(n + 4) =$
$n^2 + 2(n)(4) + 4^2 =$
$n^2 + 8n + 16$

e. $(2x - 3)(3x + 4) =$
$6x^2 + 8x - 9x - 12 =$
$6x^2 - x - 12$

f. $(x^2 + x - 1)(x^2 - x - 1) =$
$x^4 - x^3 - x^2 + x^3 - x^2 - x - x^2 + x + 1 =$
$x^4 - 3x^2 + 1$

26. a. $(x - 5)(x - 5) =$
$x^2 + 2(x)(-5) + (-5)^2 =$
$x^2 - 10x + 25$

b. $(x + 2)(x^2 - 2x + 4) = x^3 + 8$

c. $(n - 2)(n + 2) = n^2 - 4$

d. $(3x + 4)(3x + 4) =$
$9x^2 + 2(3x)(4) + 16 =$
$9x^2 + 24x + 16$

e. $(3x - 2)(4x + 3) =$
$12x^2 + 9x - 8x - 6 =$
$12x^2 + x - 6$

f. $(x^2 - 3x + 3)(x^2 + 3x - 3) =$
$x^4 + 3x^3 - 3x^2 - 3x^3 - 9x^2 + 9x + 3x^2 + 9x - 9 =$
$x^4 - 9x^2 + 18x - 9$

27. a. $x^2 + 3x - 4 =$
$x^2 - x + 4x - 4 =$
$x(x - 1) + 4(x - 1) =$
$(x - 1)(x + 4)$

b. $2x^2 - 3x =$
$x(2x - 3)$

c. $2x^2 + x - 3 =$
$2x^2 - 2x + 3x - 3 =$
$2x(x - 1) + 3(x - 1) =$
$(x - 1)(2x + 3)$

d. $9x^2 + 12x + 4 =$
$9x^2 + 6x + 6x + 4 =$
$3x(3x + 2) + 2(3x + 2) =$
$(3x + 2)(3x + 2) = (3x + 2)^2$

e. $x^2 - x =$
$x(x - 1)$

f. $3x^2 + 6x + 3 =$
$3(x^2 + 2x + 1) = 3(x + 1)^2$

28. a. $x^2 + x =$
$x(x + 1)$

b. $3x^2 - x - 4 =$
$3x^2 + 3x - 4x - 4 =$
$3x(x + 1) - 4(x + 1)$
$(x + 1)(3x - 4)$

c. $2x^2 - 9x + 10 =$
$2x^2 - 4x - 5x + 10 =$
$2x(x - 2) - 5(x - 2) =$
$(x - 2)(2x - 5)$

d. $3x^2 + 10x + 8 =$
$3x^2 + 6x + 4x + 8 =$
$3x(x + 2) + 4(x + 2) =$
$(x + 2)(3x + 4)$

e. $x^2 + x - 6 =$
$x^2 + 3x - 2x - 6 =$
$x(x + 3) - 2(x + 3) =$
$(x + 3)(x - 2)$

f. $4x^2 - 16x + 16 =$
$4(x^2 - 4x + 4) =$
$4(x - 2)^2$

29. a. $(2x + 3)(2x - 3) = 4x^2 - 9$
Difference of squares

b. $(2x - 3)(2x - 3) = 4x^2 - 12x + 9$
Perfect square trinomial

c. $(2x + 3)(2x + 3) =$
$4x^2 + 12x + 9$
Perfect square trinomial

d. $(3x - 2)(2x + 3) =$
$6x^2 + 9x - 4x - 6 =$
$6x^2 + 5x - 6$

30. a. $x^2 - \underline{2\sqrt{225}}\,x + 225 = (x - \underline{\sqrt{225}})^2$
$x^2 - \underline{\mathbf{30}}x + 225 = (x - \underline{\mathbf{15}})^2$

b. $x^2 + 20x + \underline{(\tfrac{20}{2})}^2 = (x + \underline{\tfrac{20}{2}})^2$
$x^2 + 20x + \underline{\mathbf{100}} = (x + \underline{\mathbf{10}})^2$

c. $x^2 + \underline{2\sqrt{64}}\,x + 64 = (x + \underline{\sqrt{64}})^2$
$x^2 + \underline{\mathbf{16}}x + 64 = (x + \underline{\mathbf{8}})^2$

31. a. $x^2 + 10x + \underline{\left(\tfrac{10}{2}\right)^2} = (x + \underline{\tfrac{10}{2}})^2$
$x^2 + 10x + \underline{\mathbf{25}} = (x + \underline{\mathbf{5}})^2$

b. $x^2 + \underline{2\sqrt{169}}\,x + 169 = (x + \underline{\sqrt{169}})^2$
$x^2 + \underline{\mathbf{26}}x + 169 = (x + \underline{\mathbf{13}})^2$

c. $9x^2 + \underline{2(3)}x + 1 = (3x + \underline{1})^2$
$9x^2 + \underline{\mathbf{6}}x + 1 = (3x + \underline{1})^2$

32. a. $x^2 - 8x +$ ________ $= x^2 - 2(4)x + 4^2 = x^2 - 8x + \underline{\mathbf{16}}$

b. $x^2 + 3x +$ ________ $= x^2 + 2(\tfrac{3}{2})x + (\tfrac{3}{2})^2 = x^2 + 3x + \underline{\tfrac{9}{4}}$

c. $x^2 - x +$ ________ $= x^2 - 2(\tfrac{1}{2})x + (\tfrac{1}{2})^2 = x^2 - x + \underline{\tfrac{1}{4}}$

33.

	x	$\frac{5}{12}$
x	x^2	$\frac{5}{12}x$
$\frac{5}{12}$	$\frac{5}{12}x$	$\frac{25}{144}$

$(x + \tfrac{5}{12})^2 = x^2 + \tfrac{5}{6}x + \underline{\tfrac{25}{144}}$

34.

	x	$\frac{3}{8}$
x	x^2	$\frac{3}{8}x$
$\frac{3}{8}$	$\frac{3}{8}x$	$\frac{9}{64}$

$(x + \tfrac{3}{8})^2 = x^2 + \tfrac{3}{4}x + \underline{\tfrac{9}{64}}$

35. $x^2 - 49 = (x + 7)(x - 7)$

36. $4x^2 - 9 = (2x + 3)(2x - 3)$

37. $4x^2 - 4x + 1 = (2x - 1)^2$

38. $x^2 + 24x + 144 = (x + 12)^2$

39. $x^2 + 16$; cannot be factored.

40. $x^2 - 225 = (x + 15)(x - 15)$

41. $4x^2 + 12x + 9 = (2x + 3)^2$

42. $x^2 + 36$; cannot be factored.

43. $(a - b)(a^2 - 2ab + b^2) =$
$a^3 - 2a^2b + ab^2 - a^2b + 2ab^2 - b^3 =$
$a^3 - 3a^2b + 3ab^2 - b^3$

44. $(a - b)(a^2 + ab + b^2) =$
$a^3 - b^3$

45. $(a + b)(a^2 - ab + b^2) =$
$a^3 + b^3$

46. $(a + b)(a^2 + 2ab + b^2) =$
$a^3 + 2a^2b + ab^2 + a^2b + 2ab^2 + b^3 =$
$a^3 + 3a^2b + 3ab^2 + b^3$

47. $a^3 - 8 =$
$(a - 2)(a^2 + 2a + 4)$

48. $x^3 - 125 =$
$(x - 5)(x^2 + 5x + 25)$

49. $x^3 + 27 =$
$(x + 3)(x^2 - 3x + 9)$

50. $a^3 + 1 =$
$(a + 1)(a^2 - a + 1)$

51. $x^3 - 1000 =$
$(x - 10)(x^2 + 10x + 100)$

52. $a^3 + 64 =$
$(a + 4)(a^2 - 4a + 16)$

53. The area of the shaded section is the difference between the area of the large triangle and the area of the small triangle, A – a. First find the two areas, then subtract.

$A = \frac{\sqrt{3}}{4}X^2, \quad a = \frac{\sqrt{3}}{4}x^2$; Shaded area $= \frac{\sqrt{3}}{4}X^2 - \frac{\sqrt{3}}{4}x^2 = \frac{\sqrt{3}}{4}(X + x)(X - x)$

54. The area of the shaded section is the difference between the area of the large triangle and the area of the small triangle, A – a. First find the two areas, then subtract.

$A = \frac{1}{2}BH, \quad a = \frac{1}{2}bh$

Shaded area $= \frac{1}{2}BH - \frac{1}{2}bh = \frac{1}{2}(BH - bh)$

Note: As instructed, answers to exercise 55 are estimated to the nearest 0.5.

55. a. From the graph: $2x^2 - 8x + 4 = 0$
$\approx \{0.5, 3.5\}$

b. From the graph: $2x^2 - 8x + 4 = 4$
$\approx \{0, 4\}$

Answers to exercise 56b could vary. One possible solution is given for reference.

56. a. $2x^2 - 8x + 4 = -4$

b. $2x^2 - 8x + 4 < -4$

c. 3 solutions are not possible.

57.
$x^2 + 3x - 4 = 0$
$(x + 4)(x - 1) = 0$
$x + 4 = 0$ or $x - 1 = 0$
$x = -4$ or $x = 1$
$\{-4, 1\}$

58.
$x^2 - 5x + 6 = 0$
$(x - 2)(x - 3) = 0$
$x - 2 = 0$ or $x - 3 = 0$
$x = 2$ or $x = 3$
$\{2, 3\}$

59.
$x^2 + 16 = 8x$
$x^2 - 8x + 16 = 0$
$(x - 4)^2 = 0$
$x - 4 = 0$
$x = 4$ (double root)
$\{4\}$

60.
$x^2 = 6x - 9$
$x^2 - 6x + 9 = 0$
$(x - 3)^2 = 0$
$x - 3 = 0$
$x = 3$ (double root)
$\{3\}$

61.
$9x^2 + 12x + 4 = 0$
$(3x + 2)^2 = 0$
$3x + 2 = 0$
$x = -\frac{2}{3}$
$\{-\frac{2}{3}\}$

62.
$3x^2 + x - 4 = 0$
$3x^2 - 3x + 4x - 4 = 0$
$3x(x - 1) + 4(x - 1) = 0$
$(x - 1)(3x + 4) = 0$
$x - 1 = 0$ or $3x + 4 = 0$
$x = 1$ or $x = -1\frac{1}{3}$
$\{-1\frac{1}{3}, 1\}$

63.
$2x^2 + x = 3$
$2x^2 + x - 3 = 0$
$2x^2 - 2x + 3x - 3 = 0$
$2x(x - 1) + 3(x - 1) = 0$
$(x - 1)(2x + 3) = 0$
$x - 1 = 0$ or $2x + 3 = 0$
$x = 1$ or $x = -1.5$
$\{-1.5, 1\}$

64.
$9x^2 = 4$
$9x^2 - 4 = 0$
$(3x - 2)(3x + 2) = 0$
$3x - 2 = 0$ or $3x + 2 = 0$
$x = \frac{2}{3}$ or $x = -\frac{2}{3}$
$\{-\frac{2}{3}, \frac{2}{3}\}$

65. $6x^2 + 5x + 2 = 8$
$6x^2 + 5x - 6 = 0$
$6x^2 + 9x - 4x - 6 = 0$
$3x(2x + 3) - 2(2x + 3) = 0$
$(2x + 3)(3x - 2) = 0$
$2x + 3 = 0$ or $3x - 2 = 0$
$x = -1.5$ or $x = \frac{2}{3}$
$\{-1.5, \frac{2}{3}\}$

66. $4x(x - 5) = 15 - 3x$
$4x^2 - 20x + 3x - 15 = 0$
$4x^2 + 3x - 20x - 15 = 0$
$x(4x + 3) - 5(4x + 3) = 0$
$(4x + 3)(x - 5) = 0$
$4x + 3 = 0$ or $x - 5 = 0$
$x = -0.75$ or $x = 5$
$\{-0.75, 5\}$

67. $24(x + 6) = -x^2$
$x^2 + 24x + 144 = 0$
$(x + 12)^2 = 0$
$x + 12 = 0$
$x = -12$ (double root)
$\{-12\}$

68. $9x^2 + 1 = 6x$
$9x^2 - 6x + 1 = 0$
$(3x - 1)^2 = 0$
$3x - 1 = 0$
$x = \frac{1}{3}$ (double root)
$\{\frac{1}{3}\}$

69. $4x(x - 1) = 9 - 4x$
$4x^2 - 4x + 4x - 9 = 0$
$4x^2 - 9 = 0$
$(2x + 3)(2x - 3) = 0$
$2x + 3 = 0$ or $2x - 3 = 0$
$x = -\frac{3}{2}$ or $x = \frac{3}{2}$
$\{\pm\frac{3}{2}\}$

70. $5x(5x + 2) = 2(8 + 5x)$
$25x^2 + 10x - 10x - 16 = 0$
$25x^2 - 16 = 0$
$(5x + 4)(5x - 4) = 0$
$5x + 4 = 0$ or $5x - 4 = 0$
$x = -\frac{4}{5}$ or $x = \frac{4}{5}$
$\{\pm\frac{4}{5}\}$

71. $y = a(x - 4)(x + 2)$
For any real number a.

72. $y = a(x + 3)(x - 1)$
For any real number a.

73. $3 = a(1 - 4)(1 + 2)$
$3 = a(-3)(3)$
$3 = -9a$
$a = -\frac{1}{3}$
$y = -\frac{1}{3}(x - 4)(x + 2)$
$y = -\frac{1}{3}(x^2 + 2x - 4x - 8)$
$y = -\frac{1}{3}(x^2 - 2x - 8)$
$y = -\frac{1}{3}x^2 + \frac{2}{3}x + \frac{8}{3}$

74. $4 = a(-1 + 3)(-1 - 1)$
$4 = a(2)(-2)$
$4 = -4a$
$a = -1$
$y = -1(x + 3)(x - 1)$
$y = -1(x^2 - x + 3x - 3)$
$y = -1(x^2 + 2x - 3)$
$y = -x^2 - 2x + 3$

75. $0 = a(x + 2)(x - 4)$
a is any real number.

76. $0 = a(x + 3)(x - 5)$
a is any real number.

77. $x = -\frac{2}{5}$ or $x = \frac{1}{5}$

$5x = -2$ or $5x = 1$

$5x + 2 = 0$ or $5x - 1 = 0$

$0 = a(5x + 2)(5x - 1)$

a is any real number.

78. $x = -\frac{3}{4}$ or $x = \frac{1}{4}$

$4x = -3$ or $4x = 1$

$4x + 3 = 0$ or $4x - 1 = 0$

$0 = a(4x + 3)(4x - 1)$

a is any real number.

79., 81., & 83.

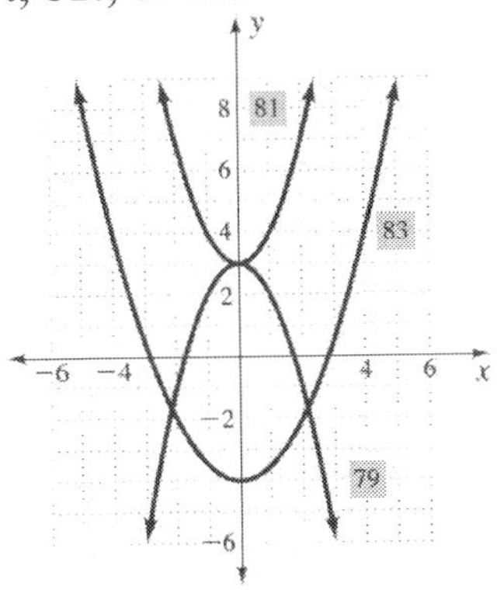

80., 82., & 84.

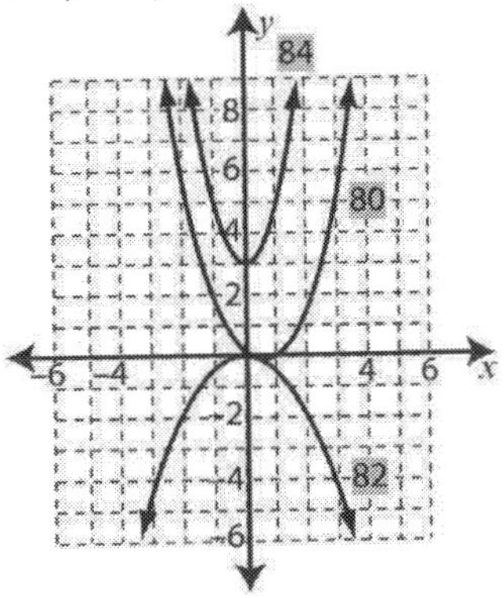

85. a. Olympic diver: input variable t indicates vertical motion equation, a = –4.905 is a steep downward curve, c = 10 meters is diver's initial height, and b = 5 is a small initial velocity.

b. Fireworks: input variable t indicates vertical motion equation, a = –4.905 is a steep downward curve, c = 0 indicates the ground as a starting point, and b = 115 is a large initial velocity.

c. Bridge cable: a = 0.03125 is a shallow upward curve and c = 60 feet indicates starting point above the ground.

d. Golf ball: a = –0.0005812 is a very shallow downward curve.

86. The vertex of the graph of $y = (x + 4)^2$ is on the x-axis.

87. The two factors of $(x + 4)^2$ are the same; this is called a ***double root.***

88. $C_1 = 2\pi r_1$

$C_2 = 2\pi r_2$

$C_1 + 1 \text{ foot} = C_2$

$2\pi r_1 + 1 \text{ foot} = 2\pi r_2$

$1 \text{ foot} = 2\pi r_2 - 2\pi r_1$

$1 \text{ foot} = 2\pi(r_2 - r_1)$

$1 \text{ foot} \div 2\pi = r_2 - r_1$

On the right side is the difference between the radii, or the clearance between the earth and the string. On the left is 1 foot (or 12 inches) divided by 2π. 12 in. ÷ 2π = 1.9 in., plenty of room for a mouse to run under the string ... for earth or basketball or medicine ball!

Chapter 2 Test

1. a. One solution is shown in the table: $y = -10$ when $x = -4$. We can tell that the other solution will be between $x = -2$ and $x = 0$. Add $x = -1$ to table: $y = (-1)^2 + 5(-1) - 6$, $y = -10$. The other solution is $x = -1$. **{–4, –1}**

b. One solution is shown in the table: $y = 8$ when $x = 2$. We can tell that the other solution will be between $x = -6$ and $x = -8$. Add $x = -7$ to table: $y = (-7)^2 + 5(-7) - 6$, $y = 8$. The other solution is $x = -7$. **{–7, 2}**

c. One solution is shown in the table: $y = 0$ when $x = -6$. We can tell that the other solution is between $x = 0$ and $x = 2$. Add $x = 1$ to the table: $y = (1)^2 + 5(1) - 6$, $y = 0$. The other solution is $x = 1$. **{–6, 1}**

d. It appears from the symmetry of the table that the lowest point will be between $x = -2$ and $x = -4$. Add $x = -3$ to the table: $y = (-3)^2 + 5(-3) - 6$, $y = -12$. Add $x = -2.5$ to the table: $y = (-2.5)^2 + 5(-2.5) - 6$, $y = -12.25$. This point is halfway between two equal y values; it is the lowest point on the graph. Therefore, $y = -20$ is not possible. **No real number solution.**

e. From part d above, the axis of symmetry is the x-value of the vertex. $\mathbf{x = -2.5}$

f. The vertex is given in d above. **(–2.5, –12.25)**

g. The domain is $\Re$. The range is $y \geq -12.25$.

h. The function is increasing when $x > -2.5$.

2. a.

Value	Differences 1st	2nd
–6		
	8	
2		2
	10	
12		2
	12	
24		2
	14	
38		

Next number is $38 + (14 + 2) = 54$.

Quadratic; $2a = 2$, $a = 1$;

$3(1) + b = 8$, $b = 5$;

$-6 = 1(1)^2 + 5(1) + c$, $c = -12$

$y = x^2 + 5x - 12$

b.

Value	Differences 1st	2nd
23		
	–8	
15		0
	–8	
7		0
	–8	
–1		0
	–8	
–9		

Next number is $-9 - 8 = -17$.

Linear; $m = -8$; $23 = -8(1) + b$,

$b = 31$; $y = -8x + 31$

3. $(x - 6)(x - 2) =$

$x^2 - 2x - 6x + 12 =$

$x^2 - 8x + 12$; other

4. $(x - 12)(x + 1) =$

$x^2 + x - 12x - 12 =$

$x^2 - 11x - 12$; other

5. $(2x-3)(x+4) =$
$2x^2 + 8x - 3x - 12 =$
$2x^2 + 5x - 12$; other

6. $(x-2)^2 =$
$x^2 + 2(x)(-2) + (-2)^2 =$
$x^2 - 4x + 4$; perfect square trinomial

7. $(2x+1)(2x-1) =$
$4x^2 - 1$; difference of squares

8. $(x-2)(x^2+2x+4) =$
$x^3 - 8$; other

9. $x^2 - 7x + \underline{\quad\quad} =$
$x^2 - 2(\frac{7}{2})x + \underline{\frac{49}{4}} =$
$\underline{\left(x-\frac{7}{2}\right)^2}$

10. $2x^2 - 7x + 6 =$
$2x^2 - 4x - 3x + 6 =$
$2x(x-2) - 3(x-2) =$
$(x-2)(2x-3)$

11. $x^2 - 6x + 9 =$
$(x-3)^2$

12. $x^2 - 49 =$
$(x-7)(x+7)$

13. $12x^2 + 30x - 18 =$
$6(2x^2 + 5x - 3) =$
$6(2x^2 + 6x - x - 3) =$
$6[2x(x+3) - 1(x+3)] =$
$6(x+3)(2x-1)$

14. $x^3 - 27 =$
$(x-3)(x^2+3x+9)$

15. $x^3 + 125 =$
$(x+5)(x^2-5x+25)$

16. $2x^2 - 7x + 6 = 0$
$2x^2 - 4x - 3x + 6 = 0$
$2x(x-2) - 3(x-2) = 0$
$(x-2)(2x-3) = 0$
$x - 2 = 0$ or $2x - 3 = 0$
$x = 2$ or $x = 1.5$
$\{1.5, 2\}$

17. $x^2 = 4(3x+7)$
$x^2 - 12x - 28 = 0$
$(x+2)(x-14) = 0$
$x + 2 = 0$ or $x - 14 = 0$
$x = -2$ or $x = 14$
$\{-2, 14\}$

18.

$4x^2 = 2x + 30$

$4x^2 - 2x - 30 = 0$

$2x^2 - x - 15 = 0$

$2x^2 - 6x + 5x - 15 = 0$

$2x(x - 3) + 5(x - 3)$

$(x - 3)(2x + 5) = 0$

$x - 3 = 0 \text{ or } 2x + 5 = 0$

$x = 3 \text{ or } x = -2.5$

$\{-2.5, 3\}$

19.

$y = a(x + 3)(x - 4)$

$2 = a(0 + 3)(0 - 4)$

$2 = a(3)(-4)$

$2 = -12a$

$a = -\frac{1}{6}$

$y = -\frac{1}{6}(x + 3)(x - 4)$

$y = -\frac{1}{6}(x^2 - 4x + 3x - 12)$

$y = -\frac{1}{6}x^2 + \frac{1}{6}x + 2$

20. a. From calculator linear regression: $y = -0.167x + 59.2$; $-0.167(52) + 59.2 = 50.5$. This suggests that the voter turnout will decline to 50.5% in 2012.

b. From calculator quadratic regression: $y = 0.0141x^2 - 0.846x + 64.2$: $0.0141(52)^2 - 0.846(52) + 64.2 = 58.3$. This suggests that voter turnout will have bottomed out in 1996 and will gradually rise to 58.3% in 2012.

21. Answers may vary, but should resemble:

For the path of water in a drinking fountain or the path of a thrown ball, the x-intercepts are the starting and ending points and the vertex is the highest point.

For the cable in a suspension bridge, the x-intercepts are where the cable attaches to the end supports (or where the end supports are attached to the ground) and the vertex is the lowest point.

The y-intercepts would represent the initial height of the person throwing the ball or the nozzle on the water fountain. On the bridge the y-intercept might be the height where the cable attaches to the end support.

22. $y = x^2 + 2$ is shifted up 2 units from $y = x^2$. It has no x-intercepts; the vertex moves from (0, 0) to (0, 2).

23. The negative sign in $y = -x^2$ turns the graph of $y = x^2$ upside down.

24. The first equation, $y = x - \frac{32.2}{2500}x^2$, could describe the path of a ball because the graph opens downward.

25. The first graph has the same vertex, axis of symmetry, and orientation but is steeper because every output of x^2 is multiplied by 3 in the second equation.

Cumulative Review: Chapters 0 to 2

1. $8 - 3(x - 2) - (x - 3) =$

$8 - 3x + 6 - x + 3 =$

$17 - 4x$

2. $3x - 3(2x - 1) = 27$

$3x - 6x + 3 = 27$

$-3x = 24$

$x = -8$

3. $2y = 3x + 4 \Rightarrow y = 1.5x + 2;\ m = \frac{3}{2}$; if function passes through the origin, b = 0,

$y = \frac{3}{2}x$

4. a. $5.50x = 40(8.00)$

$5.5x = 320$

$x = 58.2$ hours

b. $6.00(x) = 3(35.00)$

$6x = 105$

$x = 17.5$ hours

5. a. $c = nt^2$

$n = \dfrac{c}{t^2}$

b. $s = 2a + d(n - 1)$

$s - 2a = d(n - 1)$

$\dfrac{s - 2a}{d} = n - 1$

$n = \dfrac{s - 2a}{d} + 1 \quad or \quad n = \dfrac{s - 2a + d}{d}$

6. $\dfrac{x + 7}{2} > x + 6$

$x + 7 > 2x + 12$

$7 - 12 > 2x - x$

$-5 > x$ or $x < -5$

7. a. $f(-1) = 3(-1)^2 - 5(-1) + 2$

$f(-1) = 3 + 5 + 2 = 10$

b. $f(0) = 3(0)^2 - 5(0) + 2$

$f(0) = 2$

c. $f(2) = 3(2)^2 - 5(2) + 2$

$f(2) = 3(4) - 10 + 2$

$f(2) = 12 - 8$

$f(2) = 4$

d. $f(x + 1) = 3(x + 1)^2 - 5(x + 1) + 2$

$f(x + 1) = 3(x^2 + 2x + 1) - 5x - 5 + 2$

$f(x + 1) = 3x^2 + 6x + 3 - 5x - 5 + 2$

$f(x + 1) = 3x^2 + x$

8. a.

Sequence	differences	
5	**1st**	**2nd**
	7	
12		2
	9	
21		2
	11	
32		2
	13	
45		

Next number is 45 + (13 + 2) = 60.
Quadratic: 2a = 2, a = 1; 3(1) + b = 7, b = 4; $5 = (1)^2 + 4(1) + c$, c = 0
$y = x^2 + 4x$

b.

Sequence	differences	
8	**1st**	**2nd**
	7	
15		0
	7	
22		0
	7	
29		0
	7	
36		

Next number is 36 + 7 = 43.
Linear: m = 7; 8 = 7(1) + b, b = 1;
y = 7x + 1

c.

Sequence	differences	
6	**1st**	**2nd**
	6	
12		0
	6	
18		0
	6	
24		0
	6	
30		

Next number is 30 + 6 = 36.
Linear: m = 6; 6 = 6(1) + b, b = 0;
y = 6x

d.

Sequence	differences	
4	**1st**	**2nd**
	11	
15		4
	15	
30		4
	19	
49		4
	23	
72		

Next number is 72 + (23 + 4) = 99.
Quadratic: 2a = 4, a = 2; 3(2) + b = 11, b = 5; $4 = 2(1)^2 + 5(1) + c$, c = –3;
$y = 2x^2 + 5x - 3$

9. $y = a(x - 3)(x + 4)$

$-4 = a(2 - 3)(2 + 4)$

$-4 = a(-1)(6)$

$-4 = -6a$

$a = \frac{2}{3}$

$y = \frac{2}{3}(x - 3)(x + 4)$

$y = \frac{2}{3}(x^2 + 4x - 3x - 12)$

$y = \frac{2}{3}(x^2 + x - 12)$

$y = \frac{2}{3}x^2 + \frac{2}{3}x - 8$

10. a. $6(5\frac{2}{3}) = 6(5) + 6(\frac{2}{3}) = 30 + 4 = 34$

Distributive property of multiplication over addition

b. $\frac{2}{7} + \frac{3}{5} + \frac{5}{7} = \frac{2}{7} + \frac{5}{7} + \frac{3}{5} = 1\frac{3}{5}$

Commutative property of addition

11.

x	4	2	7	**1**	10	**8**
f(x)	**19**	7	52	4	103	67

$f(x) = x^2 + 3$; most helpful were $f(10)$ and $f(2)$.

12. $5 > 4 - x > 2$

$5 - 4 > -x > 2 - 4$

$1 > -x > -2$

$-1 < x < 2$

13. $|2x - 1| > 6$

$2x - 1 > 6$ or $-(2x - 1) > 6$

$2x > 7$ or $-2x > 5$

$x > 3.5$ or $x < -2.5$

14. $2x - 3y = 16$ (1)

$3x = 10 + y \Rightarrow y = 3x - 10$ (2)

Substitute (2) into (1).

$2x - 3(3x - 10) = 16$

$2x - 9x + 30 = 16$

$-7x + 30 = 16$

$-7x = -14$

$\mathbf{x = 2}$

$3(2) = 10 + y$

$6 - 10 = y$

$\mathbf{y = -4}$

15. $3x + 4y = 1$ (1)

$2x + 6y = 9$ (2)

Multiply (1) by -2 and (2) by 3 then add

$$\begin{array}{r} -6x - 8y = -2 \\ +\ \underline{6x + 18y = 27} \\ 10y = 25 \end{array}$$

$\mathbf{y = 2.5}$

$3x + 4(2.5) = 1$

$3x + 10 = 1$

$3x = -9$

$\mathbf{x = -3}$

16. The points form a square, with alternate sides being perpendicular. The sides have the same length as observed from their positions, slopes of opposite sides are equal, slopes of adjacent sides are negative reciprocals. Slopes are -5 and $\frac{1}{5}$.

17. a. $x^2 - 2x - 3 = 0$ $\{-1, 3\}$

b. $x^2 - 2x - 3 = -3$ $\{0, 2\}$

18. a.
$x^2 + 6x - 16 = 0$
$(x + 8)(x - 2) = 0$
$x + 8 = 0$ or $x - 2 = 0$
$x = -8$ or $x = 2$
$\{-8, 2\}$

b.
$5x^2 + x - 4 = 0$
$(5x - 4)(x + 1) = 0$
$5x - 4 = 0$ or $x + 1 = 0$
$x = \frac{4}{5}$ or $x = -1$
$\{-1, \frac{4}{5}\}$

c.
$7x^2 + 4x = 3$
$7x^2 + 4x - 3 = 0$
$(7x - 3)(x + 1) = 0$
$x = \frac{3}{7}$ or $x = -1$
$\{-1, \frac{3}{7}\}$

d.
$6x^2 + 3 = 11x$
$6x^2 - 11x + 3 = 0$
$(3x - 1)(2x - 3) = 0$
$x = \frac{1}{3}$ or $x = \frac{3}{2}$
$\{\frac{1}{3}, \frac{3}{2}\}$

19. a.
$(5x - 3)(3x - 5) =$
$15x^2 - 25x - 9x + 15 =$
$15x^2 - 34x + 15$

b.
$(3 - x)(3 + x) =$
$9 + 3x - 3x - x^2 =$
$9 - x^2$

20. a. $16 - x^2 = (4)^2 - x^2 = (4 - x)(4 + x)$

b. $x^3 - 64 = x^3 - (4)^3 = (x - 4)(x^2 + 4x + 16)$

21.
$y = \frac{2}{3}x + 4$
$3y = 2x - 6$
$3(\frac{2}{3}x + 4) = 2x - 6$
$2x + 12 = 2x - 6$
$12 = -6$
Substitution leads to a contradiction, which means there is no solution.

Solving the second equation for y leads to $y = \frac{2}{3}x - 2$.
The two lines have the same slope, which means they are parallel lines.

22.
Let x = amount of hours needed to raise GPA.
Let y = total number of credit hours that leads to 3.0 GPA.
$48 + x = y$
$4x + 2.75(48) = 3y$
$4x + 132 = 3(48 + x)$
$4x + 132 = 144 + 3x$
$x = 12$ hours

CHAPTER 3

Square Roots; Quadratic Equations and Inequalities

Exercises 3.1: The Square Root Function and Pythagorean Theorem

1. $x^2 = 225$
$x = \pm\sqrt{225}$
$x = \pm 15$

3. $x^2 = 121$
$x = \pm\sqrt{121}$
$x = \pm 11$

5. $x^2 = -36$
No real number

7. $15^2 = 225$

9. $\sqrt{49} = 7$

11. $\sqrt{1.96} = 1.4$

13. $\sqrt{0.01} = 0.1$

15. $\sqrt{75} = \sqrt{25 \cdot 3} = \sqrt{25} \cdot \sqrt{3} = 5\sqrt{3}$

17. $\sqrt{40} = \sqrt{4 \cdot 10} = \sqrt{4} \cdot \sqrt{10} = 2\sqrt{10}$

19. $\sqrt{200} = \sqrt{100 \cdot 2} = \sqrt{100} \cdot \sqrt{2} = 10\sqrt{2}$

21. $\sqrt{150} = \sqrt{25 \cdot 6} = \sqrt{25} \cdot \sqrt{6} = 5\sqrt{6}$

23. $1 + \sqrt{2} \approx 2.414$

25. $\dfrac{1-\sqrt{3}}{2} \approx -0.366$

27. $\dfrac{2+\sqrt{8}}{2} = \dfrac{2+\sqrt{4 \cdot 2}}{2} = \dfrac{2+2\sqrt{2}}{2} = \dfrac{2(1+\sqrt{2})}{2} = 1+\sqrt{2} \approx 2.414$

29. $\dfrac{3-\sqrt{27}}{6} = \dfrac{3-\sqrt{9 \cdot 3}}{3 \cdot 2} = \dfrac{3-3\sqrt{3}}{3 \cdot 2} = \dfrac{3(1-\sqrt{3})}{3(2)} = \dfrac{1-\sqrt{2}}{2} \approx -0.366$

31. $\dfrac{4+\sqrt{8}}{4} = \dfrac{4+2\sqrt{2}}{4} = \dfrac{2(2+\sqrt{2})}{2(2)} = \dfrac{2+\sqrt{2}}{2}$

33. $\dfrac{2+\sqrt{2}}{2}$, simplified

35. a. $\sqrt{0.81} = \sqrt{\dfrac{81}{100}} = \dfrac{\sqrt{81}}{\sqrt{100}} = \dfrac{9}{10} = 0.9$

b. $\sqrt{0.0025} = \sqrt{\dfrac{25}{10000}} = \dfrac{\sqrt{25}}{\sqrt{10000}} = \dfrac{5}{100} = 0.05$

c. $\sqrt{\dfrac{1}{36}} = \dfrac{\sqrt{1}}{\sqrt{36}} = \dfrac{1}{6}$

d. $\sqrt{\dfrac{121}{49}} = \dfrac{\sqrt{121}}{\sqrt{49}} = \dfrac{11}{7}$

e. $\sqrt{2.25} = \sqrt{\dfrac{9}{4}} = \dfrac{\sqrt{9}}{\sqrt{4}} = \dfrac{3}{2} = 1.5$

f. $\sqrt{640000} = \sqrt{64} \cdot \sqrt{10000} = 8(100) = 800$

g. $\dfrac{\sqrt{48}}{\sqrt{3}} = \dfrac{\sqrt{16} \cdot \sqrt{3}}{\sqrt{3}} = \sqrt{16} = 4$

h. $\dfrac{\sqrt{162}}{\sqrt{2}} = \dfrac{\sqrt{81} \cdot \sqrt{2}}{\sqrt{2}} = \sqrt{81} = 9$

In exercises 37 to 41 use the Pythagorean theorem, $a^2 + b^2 = c^2$.

37. a. $\left(\sqrt{8}\right)^2 + \left(\sqrt{6}\right)^2 = x^2$

$8 + 6 = x^2$

$x^2 = 14$

$x = \sqrt{14}$

b. $x^2 + 5^2 = 9^2$

$x^2 = 81 - 25$

$x^2 = 56$

$x = \sqrt{56} = 2\sqrt{14}$

39. a. $x^2 + 3^2 = 6^2$

$x^2 = 36 - 9$

$x^2 = 27$

$x = \sqrt{27} = 3\sqrt{3}$

b. $x^2 + 4^2 = 8^2$

$x^2 = 64 - 16$

$x^2 = 48$

$x = \sqrt{48} = 4\sqrt{3}$

41. a. $x^2 = 8^2 + 8^2$

$x^2 = 2(8^2)$

$x = 8\sqrt{2}$

b. $x^2 = 12^2 + 12^2$

$x^2 = 2(12^2)$

$x = 12\sqrt{2}$

c. $8^2 = 2(x^2)$

$32 = x^2$

$x = \sqrt{32} = 4\sqrt{2}$

43. $4.5^2 + 6^2 = 7.5^2$?

$20.25 + 36 = 56.25$?

$56.25 = 56.25$

Right triangle

45. $16^2 + 30^2 = 34^2$?

$256 + 900 = 1156$?

$1156 = 1156$

Right triangle

47. $\left(\sqrt{5}\right)^2 + \left(\sqrt{11}\right)^2 = 4^2$?

$5 + 11 = 16$?

$16 = 16$

Right triangle

49. $2(12^2) = d^2$

$12\sqrt{2} = d$

$d \approx 16.97$ ft

$d = 16$ ft 11.6 in.

51. $13^2 + 19^2 = d^2$

$\sqrt{530} = d$

$d \approx 23.02$ ft

$d = 23$ ft 0.3 in.

53. Distance from base of wall = 16 ÷ 4 = 4; ladder length = $\sqrt{16^2 + 4^2}$

Ladder length = 16.5 ft.

55. $25^2 + 16^2 = x^2$

$\sqrt{881} = x$

$x = 29.7$ ft

57. Distance from base of tower halfway up = 300 ÷ 2 = 150 ft; distance from tower on ground that cables are attached = 300 ÷ 4= 75 ft; each cable length = $\sqrt{75^2 + 150^2} \approx 168$ ft. So three cables together would be $3 \cdot 168 \text{ ft} = 504 \text{ ft}$.

59. $$\frac{\sqrt{125} + \sqrt{45} - \left(\sqrt{75} + \sqrt{27}\right)}{\sqrt{5} - \sqrt{3}} = \frac{\sqrt{25 \cdot 5} + \sqrt{9 \cdot 5} - \left(\sqrt{25 \cdot 3} + \sqrt{9 \cdot 3}\right)}{\sqrt{5} - \sqrt{3}} =$$

$$\frac{5\sqrt{5} + 3\sqrt{5} - \left(5\sqrt{3} + 3\sqrt{3}\right)}{\sqrt{5} - \sqrt{3}} = \frac{8\sqrt{5} - 8\sqrt{3}}{\sqrt{5} - \sqrt{3}} = \frac{8\left(\sqrt{5} - \sqrt{3}\right)}{\sqrt{5} - \sqrt{3}} = 8$$

Exercises 3.2: Solving Quadratic Equations with Square Roots

In exercise 1, solutions are found by tracing graphs on a graphing calculator. Answers are provided for reference only.

1. **a.** $x = -5, x = -1$ **b.** $x = -5, x = -1$

c. The graph is the same as that of $y = |x + 3|$. $\sqrt{w^2} = |w|$.

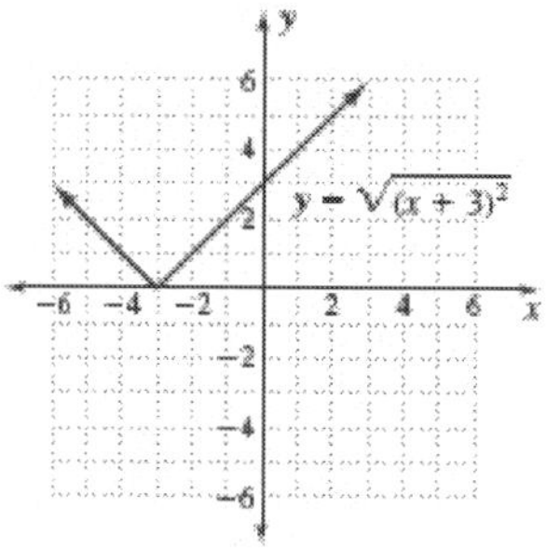

3. **a.**

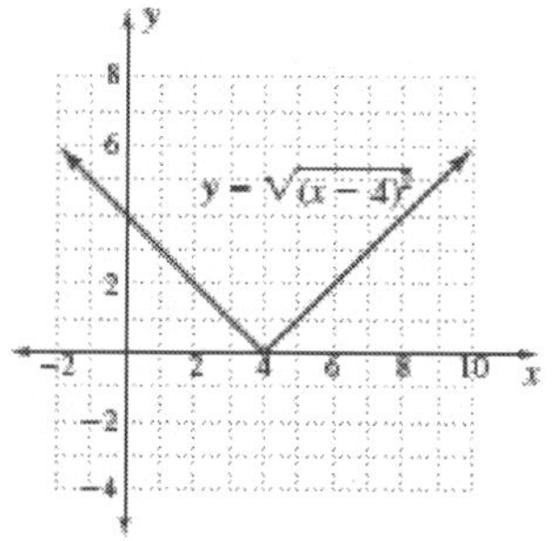

b.

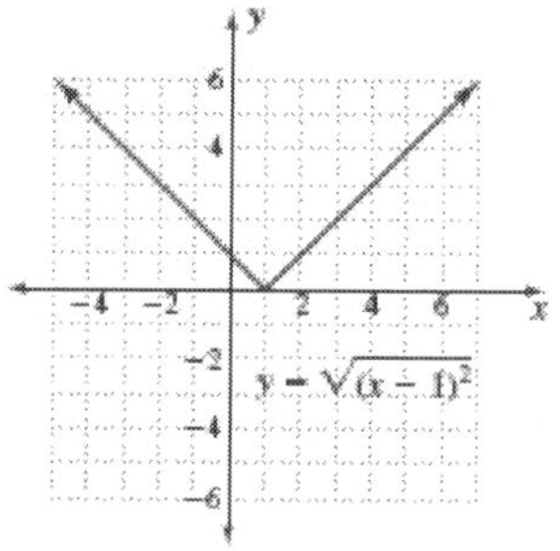

5.

$(x-4)^2 = 16$

$\sqrt{(x-4)^2} = \sqrt{16}$

$|x-4| = 4$

$x - 4 = \pm 4$

$x - 4 = 4$ or $x - 4 = -4$

$x = 8$ or $x = 0$

$\{0, 8\}$

7.

$(x+2)^2 = 4$

$\sqrt{(x+2)^2} = \sqrt{4}$

$|x+2| = 2$

$x + 2 = \pm 2$

$x + 2 = 2$ or $x + 2 = -2$

$x = 0$ or $x = -4$

$\{-4, 0\}$

9.

$x^2 - 10x + 25 = 36$

$(x-5)^2 = 36$

$\sqrt{(x-5)^2} = \sqrt{36}$

$|x-5| = 6$

$x - 5 = \pm 6$

$x - 5 = 6$ or $x - 5 = -6$

$x = 11$ or $x = -1$

$\{-1, 11\}$

11.

$x^2 - 4x + 4 = 16$

$(x-2)^2 = 16$

$\sqrt{(x+1)^2} = \sqrt{16}$

$|x-2| = 4$

$x - 2 = \pm 4$

$x - 2 = 4$ or $x - 2 = -4$

$x = 6$ or $x = -2$

$\{-2, 6\}$

13. $4x^2 - 1 = 0$
$4x^2 = 1$
$\sqrt{4x^2} = \sqrt{1}$
$|2x| = 1$
$2x = \pm 1$
$2x = 1$ or $2x = -1$
$x = \frac{1}{2}$ or $x = -\frac{1}{2}$
$\{\pm \frac{1}{2}\}$

15. $4x^2 + 4x + 1 = 5$
$(2x + 1)^2 = 0$
$2x + 1 = 0$
$x = -\frac{1}{2}$
$\{-\frac{1}{2}\}$

17. $4x^2 - 16 = 0$
$x^2 - 4 = 0$
$x^2 = 4$
$\sqrt{x^2} = \sqrt{4}$
$|x| = 2$
$x = \pm 2$
$\{\pm 2\}$

19. $x^2 = 225$
$|x| = 15$
$x = \pm 15$
$\{\pm 15\}$

21. $3x^2 - 48 = 0$
$3x^2 = 48$
$x^2 = 16$
$|x| = 4$
$x = \pm 4$
$\{\pm 4\}$

23. $16x^2 - 48 = 0$
$x^2 - 3 = 0$
$x^2 = 3$
$x = \pm 1.7$
$\{\pm 1.7\}$

25. $9x^2 - 45 = 0$
$x^2 - 5 = 0$
$x^2 = 5$
$x = \pm 2.2$
$\{\pm 2.2\}$

27. $2(20^2) = c^2$
$\sqrt{2(20^2)} = c$
$c = 20\sqrt{2}$ m

29. $2(\sqrt{2})^2 = c^2$
$2(2) = c^2$
$c^2 = 4$
$c = 2$ ft

31. $2(\sqrt{18})^2 = c^2$
$2(18) = c^2$
$c^2 = 36$
$c = 6$ in.

In exercises 33 to 56, use the formula for h from Example 7 in the text: $h = \dfrac{x\sqrt{3}}{2}$.

33. $x = 8;\ h = \dfrac{8\sqrt{3}}{2} = 4\sqrt{3}$ in

35. $x = 5;\ h = \dfrac{5\sqrt{3}}{2}$ in

37. $b = x;\ h = \dfrac{x\sqrt{3}}{2};\ A = \dfrac{1}{2}(x)\left(\dfrac{x\sqrt{3}}{2}\right);\ A = \dfrac{x^2\sqrt{3}}{4}$

39. a. $A = \dfrac{\pi d^2}{4};\ 4A = \pi d^2;\ d^2 = \dfrac{4A}{\pi};\ d = \sqrt{\dfrac{4A}{\pi}};\ d = 2\sqrt{\dfrac{A}{\pi}}$

b. $\dfrac{\pi}{4} = 0.7854$; numbers form a square in the upper left of a calculator keypad.

41. $d = 2\sqrt{\dfrac{5026}{\pi}};\ d \approx 80$ in.;$80 \text{ in.}\left(\dfrac{1\,ft}{12\,in.}\right) \approx 6.7$ ft

43. a. $h = 0,\ v_0 = 0:\ 0 = -\frac{1}{2}gt^2 + h_0;\ \frac{1}{2}gt^2 = h_0;\ t^2 = \dfrac{2h_0}{g};\ t = \sqrt{\dfrac{2h_0}{g}}$

b. $t = \sqrt{\dfrac{2(555)}{32.2}};\ t \approx 5.9$ sec

c. $t = \sqrt{\dfrac{2(49.05)}{9.81}};\ t \approx 3.2$ sec

d. $t = \sqrt{\dfrac{2(646.4)}{9.81}};\ t \approx 11.5$ sec

45. a. $d = \sqrt{\dfrac{3h}{2}};\ d^2 = \dfrac{3h}{2};\ 2d^2 = 3h;\ h = \dfrac{2d^2}{3}$

b. $d = \sqrt{\dfrac{3(500)}{2}};\ d \approx 27.4$ miles

c. $d = \sqrt{\dfrac{3(500)}{8}};\ d \approx 13.7$ miles

d. $h = \dfrac{2(9)^2}{3};\ h \approx 54$ ft

e. $h = \dfrac{8(9)^2}{3};\ h \approx 216$ ft

f. The curvature of the moon is greater, so we see ***half*** as far on the moon for a given height, and must be ***four*** times higher on the moon to see the same distance.

47. The ")" is placed at the end of the expression.

49. The square root function is increasing. It rises to the right for all values of x.

Exercises 3.3: Solving Quadratic Equations with the Quadratic Formula

1.

$x^2 + 22x = 23$

$22 \div 2 = 11$

$11^2 = 121$

$x^2 + 22x + 121 = 23 + 121$

$(x + 11)^2 = 144$

3.

$x^2 + 12x = 13$

$12 \div 2 = 6$

$6^2 = 36$

$x^2 + 12x + 36 = 13 + 36$

$(x + 6)^2 = 49$

5.

$x^2 - 11x = -18$

$-11 \div 2 = -\frac{11}{2}$

$\left(-\frac{11}{2}\right)^2 = \frac{121}{4}$

$x^2 - 11x + \frac{121}{4} = -18 + \frac{121}{4}$

$(x - \frac{11}{2})^2 = \frac{49}{4}$

7.

$x^2 + 15x = -14$

$15 \div 2 = \frac{15}{2}$

$\left(\frac{15}{2}\right)^2 = \frac{225}{4}$

$x^2 + 15x + \frac{225}{4} = -14 + \frac{225}{4}$

$(x + \frac{15}{2})^2 = \frac{169}{4}$

9.

$x^2 + 6x - 8 = 0$

$a = 1, b = 6, c = -8$

$$x = \frac{-6 \pm \sqrt{6^2 - 4(1)(-8)}}{2(1)}$$

$$x = \frac{-6 \pm \sqrt{68}}{2}$$

$$x = \frac{-6 \pm 2\sqrt{17}}{2}$$

$x = -3 \pm \sqrt{17}$; irrational

11.

$x^2 - 6x + 8 = 0$

$a = 1, b = -6, c = 8$

$$x = \frac{-(-6) \pm \sqrt{(-6)^2 - 4(1)(8)}}{2(1)}$$

$$x = \frac{6 \pm \sqrt{4}}{2}$$

$$x = \frac{6 \pm 2}{2} = 3 \pm 1$$

$x = 3 + 1$ or $x = 3 - 1$

$x = 4$ or $x = 2$

$\{2, 4\}$; rational

13.

$4x^2 - 4x + 1 = 0$

$a = 4, b = -4, c = 1$

$$x = \frac{-(-4) \pm \sqrt{(-4)^2 - 4(4)(1)}}{2(4)}$$

$$x = \frac{4 \pm \sqrt{0}}{8}$$

$x = \frac{1}{2}$; rational

15.

$2x^2 + 3x + 1 = 0$

$a = 2, b = 3, c = 1$

$$x = \frac{-3 \pm \sqrt{3^2 - 4(2)(1)}}{2(2)}$$

$$x = \frac{-3 \pm \sqrt{1}}{4}$$

$x = \frac{-3+1}{4} = -\frac{1}{2}$ or $x = \frac{-3-1}{4} = -1$

$\{-1, -\frac{1}{2}\}$; rational

17. $2x^2 + 3x + 2 = 0$

$a = 2, b = 3, c = 2$

$$x = \frac{-3 \pm \sqrt{3^2 - 4(2)(2)}}{2(2)}$$

$$x = \frac{-3 \pm \sqrt{-7}}{4}$$

Not in set of real numbers

19. $2x^2 - 5x + 3 = 0$

$a = 2, b = -5, c = 3$

$$x = \frac{-(-5) \pm \sqrt{(-5)^2 - 4(2)(3)}}{2(2)}$$

$$x = \frac{5 \pm \sqrt{1}}{4}$$

$x = \frac{5+1}{4} = \frac{3}{2}$ or $x = \frac{5-1}{4} = 1$

$\{1, \frac{3}{2}\}$; rational

21. $2x^2 - 20x + 50 = 0$

$a = 2, b = -20, c = 50$

$$x = \frac{-(-20) \pm \sqrt{(-20)^2 - 4(2)(50)}}{2(2)}$$

$$x = \frac{20 \pm \sqrt{0}}{4}$$

$x = 5$; rational

23. $x = \dfrac{-(-4) - \sqrt{(-4)^2 - 4(1)(-5)}}{2(1)} = \dfrac{4 - \sqrt{36}}{2} = \dfrac{4-6}{2} = -1$; $a = 1, b = -4, c = -5$

$x^2 - 4x - 5 = 0$

25. $x = \dfrac{-5 + \sqrt{5^2 - 4(2)(-12)}}{2(2)} = \dfrac{-5 + \sqrt{121}}{4} = \dfrac{-5+11}{4} = \dfrac{3}{2}$; $a = 2, b = 5, c = -12$

$2x^2 + 5x - 12 = 0$

27. $x = \dfrac{-1 + \sqrt{1^2 - 4(3)(-4)}}{2(3)} = \dfrac{-1 + \sqrt{49}}{6} = \dfrac{-1+7}{6} = 1$; $a = 3, b = 1, c = -4$

$3x^2 + x - 4 = 0$

29. $-0.12x^2 + 2.4x - 8 = 0$

$a = -0.12,\ b = 2.4,\ c = -8$

$$x = \frac{-2.4 \pm \sqrt{2.4^2 - 4(-0.12)(-8)}}{2(-0.12)}$$

$$x = \frac{-2.4 \pm \sqrt{1.92}}{-0.24}$$

$$x = \frac{-2.4 + \sqrt{1.92}}{-0.24} \text{ or } x = \frac{-2.4 - \sqrt{1.92}}{-0.24}$$

$x \approx 4.226$ or $x \approx 15.774$

31. $-0.12x^2 + 2.4x - 10 = 0$

$a = -0.12,\ b = 2.4,\ c = -10$

$$x = \frac{-2.4 \pm \sqrt{2.4^2 - 4(-0.12)(-10)}}{2(-0.12)}$$

$$x = \frac{-2.4 \pm \sqrt{0.96}}{-0.24}$$

$$x = \frac{-2.4 + \sqrt{0.96}}{-0.24} \text{ or } x = \frac{-2.4 - \sqrt{0.96}}{-0.24}$$

$x \approx 5.918$ or $x \approx 14.082$

33. $2000(r^2 + 2r + 1) = 3000$

$2r^2 + 4r + 2 = 3$

$2r^2 + 4r - 1 = 0$

$a = 2,\ b = 4,\ c = -1$

$$r = \frac{-4 \pm \sqrt{(4)^2 - 4(2)(-1)}}{2(2)}$$

$$r = \frac{-4 \pm \sqrt{24}}{4}$$

$$r = \frac{-4 + \sqrt{24}}{4} \text{ or } r = \frac{-4 - \sqrt{24}}{4}$$

$r \approx 0.2247$ or $r \approx -2.2247$

r is 22.47%, discard –2.2247 or –222.47%

Exercise 35 uses the graph from Example 6 in the text.

35. a. The y-intercept is the initial height of 32.8 ft.

b. The part of the parabola to the left of the y-axis has no meaning in the problem setting.

c. The diver is at 32.8 ft. at 0 sec. and ≈ 0.4 sec.

In exercise 37, assume the same initial height as in Example 6 in the text.

37. $h_o = 32.8$ ft. $v_o = 4$ ft/sec; $0 = -16.1t^2 + 4t + 32.8$; $a = -16.1,\ b = 4,\ c = 32.8$

$$t = \frac{-4 \pm \sqrt{4^2 - 4(-16.1)(32.8)}}{2(-16.1)};\ t = \frac{-4 + \sqrt{2128.32}}{-32.2} \text{ or } t = \frac{-4 - \sqrt{2128.32}}{-32.2}$$

$t \approx -1.3$ (discard negative) or $\mathbf{t \approx 1.55}$ **sec.**

39. $v_o = 0$, $h_o = 35$ m, $0 = -\frac{1}{2}(9.81)t^2 + 0t + 35$; $0 = -4.905t^2 + 35$; $a = -4.905$, $b = 0$, $c = 35$

$t = \dfrac{0 \pm \sqrt{0 - 4(-4.905)(35)}}{2(-4.905)}$; $t = \dfrac{\pm\sqrt{686.7}}{-9.81}$; $t \approx -2.7$ (discard negative) or $\mathbf{t \approx 2.7}$ **sec.**

41. $v_o = 1.5$ m/sec.; $0 = -4.905t^2 + 1.5t + 35$; $a = -4.905$, $b = 1.5$, $c = 35$

$t = \dfrac{-1.5 \pm \sqrt{1.5^2 - 4(-4.905)(35)}}{2(-4.905)}$; $t = \dfrac{-1.5 + \sqrt{688.95}}{-9.81}$ or $t = \dfrac{-1.5 - \sqrt{688.95}}{-9.81}$

$t \approx -2.5$ (discard negative) or $\mathbf{t \approx 2.8}$ **sec.**

43. a. $\sqrt{16x^2} = 4x$; $\sqrt{(w^2)} = |w|$, not w

b. $\sqrt{(16 + x^2)} = (4 + x)$; there is no square root of sums property.

c. $\sqrt{(4x^2 - 9y^2)} = (2x - 3y)$; there is no square root of differences property.

d. $\sqrt{(4x^2 - 4x + 1)} = |2x - 1|$; true, $(4x^2 - 4x + 1) = (2x - 1)^2$.

45. a. $2\square^2 + 3\square - 5 = 0$

$(2\square + 5)(\square - 1) = 0$

$2\square + 5 = 0$ or $\square - 1 = 0$

$\square = -2.5$ or $\square = 1$

$\{-2.5, 1\}$

b. $2\square^2 + 3\square - 2 = 0$

$(2\square - 1)(\square + 2) = 0$

$2\square - 1 = 0$ or $\square + 2 = 0$

$\square = 0.5$ or $\square = -2$

$\{-2, 0.5\}$

c. $A^2 - 2A + 1 = 0$

$(A - 1)^2 = 0$

$A = 1,$

$x^2 = 1$

$x = \pm 1$

d. $A^2 - 1 = 0$

$(A + 1)(A - 1) = 0$

$A = \pm 1$

$x^2 = +1$ (x^2 cannot $= -1$ in real numbers.)

$x = \pm 1$

e. $16x^4 - 8x^2 + 1 = 0$

$(4x^2 - 1)^2 = 0$

$x^2 = 0.25$

$x = \pm 0.5$

f. $x^4 + 5x^2 - 36 = 0$

$(x^2 + 9)(x^2 - 4) = 0$

$x^2 + 9 = 0$ or $x^2 - 4 = 0$

$x^2 = -9$ (not possible in real numbers) or $x^2 = 4$

$x = \pm 2$

g. If the equations contain only x^4, x^2, and constants, we can apply the quadratic formula.

Exercises 3.4: Solving Minimum and Maximum Problems

1. $0 = x(x - 2)$; $x = 0$ or $x = 2$
x-intercepts: (0, 0) and (2, 0)
$(0 + 2) \div 2 = 1$; $0 + 1 = 1$
Axis of symmetry: $x = 1$
$y = 1^2 - 2(1)$; $y = -1$
Vertex: (1, –1)

3. $0 = (-x - 7)(x - 3)$; $x = -7$ or $x = 3$
x-intercepts: (–7, 0) and (3, 0)
$(|-7| + 3) \div 2 = 5$; $-7 + 5 = -2$
Axis of symmetry: $x = -2$
$y = -(-2)^2 - 4(-2) + 21$; $y = 25$
Vertex: (–2, 25)

5. $x = \dfrac{-8}{2(1)} = -4$
$y = (-4)^2 + 8(-4) + 15$; $y = -1$
Vertex: (–4, –1)

7. $x = \dfrac{-(-4)}{2(1)} = 2$
$y = 2^2 - 4(2) + 5$; $y = 1$
Vertex: (2, 1)

9. $x = \dfrac{-(-1)}{2(2)} = \dfrac{1}{4}$
$y = 2(\frac{1}{4})^2 - (\frac{1}{4}) - 3$; $y = -\frac{25}{8}$
Vertex: $(\frac{1}{4}, -\frac{25}{8})$ or (0.25, –3.125)

11. $x = \dfrac{-6}{2(3)} = -1$
$y = 3(-1)^2 + 6(-1) - 2$; $y = -5$
Vertex: (–1, –5)

13. $x = \dfrac{-3}{2(-1)} = \dfrac{3}{2} = 1.5$
$y = -(1.5)^2 + 3(1.5) - 2$; $y = 0.25$
Vertex: (1.5, 0.25)

15. $x = \dfrac{-(-0.4)}{2(0.001)} = 200$
$y = 0.001(200)^2 - 0.4(200) + 50$; $y = 10$
Vertex: (200, 10)
The distance between the roadbed and the Vertex is the y-value of 10 feet.

17. $x = \dfrac{-\frac{4}{3}}{2(-\frac{2}{225})} = 75$
$y = -\frac{2}{225}(75)^2 + \frac{4}{3}(75) + 10$;
$y = 60$
Vertex: (75, 60)

a. Clearance is y value of vertex: 60 ft.

b. Clearance is $60 - 10 = 50$ ft.

Exercises 19 -23 use the equation from Example 2 in the text: $y = -\dfrac{32.2}{v^2}x^2 + x$.

19. Factor to find the x-intercepts and find the distance between them, use the quadratic formula to find the x-intercepts, and double the horizontal distance to the vertex. Using the x-intercept method: $0 = x(-0.005x + 1)$, $x = 0$ or $x = 200$. Distance between them is 200 feet.

21. $v = 90;\ y = -\frac{32.2}{90^2}x^2 + x;$

At the vertex, $x = -\frac{1}{2(\frac{32.2}{90^2})} = 125.8$.

$y = -\frac{32.2}{90^2}(125.8)^2 + 125.8$

$y \approx 62.9$ ft;

Ball travels $2(125.8) = 251.6$ ft.

23. $v = 200;\ y = -\frac{32.2}{200^2}x^2 + x$

At the vertex; $x = -\frac{1}{2(\frac{32.2}{200^2})} = 621.1$.

$y = -\frac{32.2}{200^2}(621.1)^2 + 621.1$

$y \approx 310.6$ ft., does not reach 500 ft.

Exercises 25 to 27 use the formula $h = -0.5gt^2 + v_ot + h_o$.

25. $v_o = 40$ ft/sec; $h_o = 5.5$ ft; $g = 32.2$ ft/sec^2; $h = -0.5(32.2)t^2 + 40t + 5.5$

Max. height is h-coordinate of vertex; t-coordinate $= \frac{-40}{2(-16.1)} \approx 1.242$;

$h = -16.1(1.242)^2 + 40(1.242) + 5.5$; $\mathbf{h \approx 30.3}$ **ft.**

27. $v_o = 115$ ft/sec; $h_o = 0$; $h = -16.1t^2 + 115t$;

$t = \frac{-115}{2(-16.1)} \approx 3.57$; $h = -16.1(3.57)^2 + 115(3.57)$; $h \approx 205.4$

Fireworks burst at (3.6 sec., 205.4 ft), the vertex of the equation.

29. $l + 2w = 60$; $l = 60 - 2w$

$A = lw$; substituting: $A = (60 - 2w)(w) = 60w - 2w^2$

Maximum area is at vertex: $w = \frac{-60}{2(-2)} = 15$; $60(15) - 2(15)^2$ $\mathbf{A = 450}$ **ft^2**

For maximum area, $\mathbf{w = 15}$ **ft**, $l = 60 - 2(15)$; $\mathbf{l = 30}$ **ft.**

31. $2l + 3w = 120$; $l = 60 - 1.5w$

$A = lw$; $A = (60 - 1.5w)(w) = 60w - 1.5w^2$

Maximum area is at the vertex: $w \doteq \frac{-60}{2(-1.5)} = 20$; $60(20) - 1.5(20)^2$ $\mathbf{A = 600}$ **ft^2**

For maximum area, $\mathbf{w = 20}$ **ft.**; $l = 60 - 1.5(20)$, $\mathbf{l = 30}$ **ft.**

33. Point L is where x = 2000: $y = \frac{7}{400{,}000}(2000)^2 - \frac{1}{25}(2000) + 700 \approx$ **690 ft.**

Storm drain is at the vertex: $x = \dfrac{-(-\frac{1}{25})}{2(\frac{7}{400{,}000})} \approx 1143;$

$y = \frac{7}{400{,}000}(1143)^2 - \frac{1}{25}(1143) + 700 \approx 677$

Storm drain is at **(1143, 677)**.

35. Point M is where y = 0, or at the y-intercept. From the equation elevation at M is **1188 ft**.

$x = \dfrac{-(-\frac{1}{20})}{2(\frac{11}{480{,}000})} \approx 1091;\ y = \frac{11}{480{,}000}(1091)^2 - \frac{1}{20}(1091) + 1188 \approx 1161$

Storm drain is at **(1091, 1161)**.

37. $\frac{1}{60{,}000}x^2 - \frac{3}{100}x + 1000 < 986.6$; multiply both sides by 60,000 to clear fractions.

$x^2 - 1800x + 60{,}000{,}000 < 59{,}196{,}000$

$x^2 - 1800x + 804{,}000 < 0$; use any method to solve for x-intercepts.

x < 977 and x > 822; length of puddle is distance between intercepts 977 – 822 = 155.

Puddle is approximately 155 ft long.

39. The other x-intercept will be the distance $|h - x|$ from the vertex on the opposite side.

41. $x = -\frac{b}{2a}$, the leading term of the quadratic formula when it is separated into two terms.

For exercise 43 the x-axis is time, the y-axis is height

43. a. x-coordinate of vertex is $\dfrac{-2\ \frac{m}{sec}}{-9.81\ \frac{m}{sec^2}} \approx 0.204$ sec.; $h = -4.905(0.204)^2 + 2(0.204) + 10;$

h = 10.2 m. ⇒ Vertex at (0.204 sec., 10.2 m).

b. x-coordinate of vertex is $\dfrac{-4\ \frac{m}{sec}}{-9.81\ \frac{m}{sec^2}} \approx 0.408$ sec; $h = -4.905(0.408)^2 + 2(0.408) + 10;$

h = 10.8 m. ⇒ Vertex at (0.408 sec., 10.8 m).

c. x-coordinate of vertex is $\dfrac{-8\ \frac{m}{sec}}{-9.81\ \frac{m}{sec^2}} \approx 0.815$ sec.; $h = -4.905(0.815)^2 + 2(0.815) + 10;$

h = 13.3 m. ⇒ Vertex at (0.815 sec., 13.3 m).

d. As initial velocity doubles, time at the vertex doubles.

e. An equation fit to the vertices shows that height changes quadratically:

$h \approx -0.054t^2 - 0.025t + 10$. Ten meters is the height of the platform.

Exercises 3.5: Complex Numbers and Solving Polynomial Equations

1. **a.** $\sqrt{-20} = i\sqrt{20} = i\sqrt{4 \cdot 5} = 2i\sqrt{5}$

b. $\sqrt{-40} = i\sqrt{40} = i\sqrt{4 \cdot 10} = 2i\sqrt{10}$

c. $\sqrt{-72} = i\sqrt{72} = i\sqrt{36 \cdot 2} = 6i\sqrt{2}$

3. **a.** $\sqrt{-3} = i\sqrt{3}$

b. $\sqrt{-54} = i\sqrt{54} = i\sqrt{9 \cdot 6} = 3i\sqrt{6}$

c. $\sqrt{-27} = i\sqrt{27} = i\sqrt{9 \cdot 3} = 3i\sqrt{3}$

5. **a.** $16 = 16 + 0i$

b. $\sqrt{-16} = i\sqrt{16} = 4i = 0 + 4i$

c. $\sqrt{36} + \sqrt{-6} = 6 + i\sqrt{6}$

7. **a.** $\dfrac{6+\sqrt{-12}}{2} = \dfrac{6+2i\sqrt{3}}{2} = \dfrac{2(3+i\sqrt{3})}{2} = 3 + i\sqrt{3}$

b. $\dfrac{2-\sqrt{-24}}{4} = \dfrac{2-2i\sqrt{6}}{4} = \frac{2}{4} - \frac{2}{4}i\sqrt{6} = \frac{1}{2} - \frac{1}{2}i\sqrt{6}$

9. $2 - 3i$

11. $i + 1 = 1 + i$, conjugate is $1 - i$.

13. $2i - 3 = -3 + 2i$, conjugate is $-3 - 2i$.

15. $4 + 2i$

17. $2i + 3i - 4i - 5i + 6i + 7i - 8i - 9i = -8i$

19. $6 + 4i - 7i + 5 - 12i =$
$6 + 5 + 4i - 7i - 12i = 11 - 15i$

21. $-a + bi - (a + bi) =$
$-a + bi - a - bi =$
$-a - a + bi - bi =$
$-2a$

23. $3 - 4(5 - i) =$
$3 - 20 + 4i =$
$-17 + 4i$

25. $(6 + 4i) - (3 - 2i) =$
$6 + 4i - 3 + 2i =$
$6 - 3 + 4i + 2i =$
$3 + 6i$

27. $(7 - 2i) + (6 - 4i) =$
$7 - 2i + 6 - 4i =$
$7 + 6 - 2i - 4i =$
$13 - 6i$

29. $4i - 3(5 - i) =$
$4i - 15 + 3i =$
$-15 + 4i + 3i =$
$-15 + 7i$

31. a. $(4 - 3i)(4 + 3i) =$
$16 - 9i^2 =$
$16 - 9(-1) =$
$16 + 9 = 25$

b. $(3 + 4i)(3 - 4i) =$
$9 - 16i^2 =$
$9 - 16(-1) =$
$9 + 16 = 25$

33. a. $(\sqrt{2} + i)(\sqrt{2} - i) =$
$2 - i^2 =$
$2 - (-1) =$
$2 + 1 = 3$

b. $(1 - i\sqrt{2})(1 + i\sqrt{2}) =$
$1 - 2i^2 =$
$1 - 2(-1) =$
$1 + 2 = 3$

35. $a = 4$, $b = 7$, $c = -2$
$7^2 - 4(4)(-2) =$
$49 + 32 = 81$
$4x^2 + 7x - 2 = 0$ has
2 real solutions.

37. $a = 5$, $b = 4$, $c = 1$
$4^2 - 4(5)(1) =$
$16 - 20 = -4$
$5x^2 + 4x + 1 = 0$ has
2 complex solutions.

39. $a = 4$, $b = 4$, $c = 1$
$4^2 - 4(4)(1) =$
$16 - 16 = 0$
$4x^2 + 4x + 1 = 0$ has
1 real solution.

41. $a = 2$, $b = 6$, $c = 5$
$6^2 - 4(2)(5) =$
$36 - 40 = -4$
$2x^2 + 6x + 5 = 0$ has
2 complex solutions.

43. $a = \frac{1}{2}$, $b = 0$, $c = -3$
$0 - 4(\frac{1}{2})(-3) = 6$
$\frac{1}{2}x^2 - 3 = 0$ has
2 real solutions.

45. $x^2 - 8 = 0$
$x^2 = 8$
$x = \pm\sqrt{8}$
$x = \pm 2\sqrt{2} \approx \{\pm 2.8\}$

47. $x^2 + 2x + 4 = 0$

$$x = \frac{-2 \pm \sqrt{2^2 - 4(1)(4)}}{2(1)}$$

$$x = \frac{-2 \pm \sqrt{-12}}{2}$$

$$x = \frac{-2 \pm 2i\sqrt{3}}{2} = -1 \pm i\sqrt{3}$$

$x \approx \{-1 \pm 1.7i\}$

49. $x^2 - 4x + 8 = 0$

$$x = \frac{-(-4) \pm \sqrt{(-4)^2 - 4(1)(8)}}{2(1)}$$

$$x = \frac{4 \pm \sqrt{-16}}{2} = \frac{4 \pm 4i}{2}$$

$x = 2 \pm 2i$

51. $(x + 1)^2 + 1 = 0$

$(x + 1)^2 = -1$

$x + 1 = \pm i$

$x = -1 \pm i$

$\{-1 \pm i\}$

53. $2x^2 + 10x + 13 = 0$

$$x = \frac{-10 \pm \sqrt{10^2 - 4(2)(13)}}{2(2)}$$

$$x = \frac{-10 \pm \sqrt{-4}}{4} = \frac{-10 \pm 2i}{4}$$

$$x = -\frac{5}{2} \pm \frac{i}{2}$$

$\{-2.5 \pm 0.5i\}$

55.

Factor	$\mathbf{x^2}$	**+2x**	**+4**
x	x^3	$\mathbf{+2x^2}$	**+4x**
−2	$\mathbf{-2x^2}$	**−4x**	−8

$x^3 - 8 = (x - 2)(x^2 + 2x + 4)$

57. $x^3 + 8 = 0$

$(x + 2)(x^2 - 2x + 4) = 0$

$x + 2 = 0;\ x^2 - 2x + 4 = 0$

$x = -2$ or

$$x = \frac{-(-2) \pm \sqrt{(-2)^2 - 4(1)(4)}}{2(1)}$$

$$x = \frac{2 \pm 2i\sqrt{3}}{2};\ x = 1 \pm i\sqrt{3}$$

$\{-2, 1 \pm i\sqrt{3}\} \approx \{-2, 1 \pm 1.7i\}$

59. $x^4 - 81 = 0$

$(x^2 - 9)(x^2 + 9) = 0$

$x^2 - 9 = 0$ or $x^2 + 9 = 0$

$(x + 3)(x - 3) = 0 \quad x = \pm 3$ or

$x^2 + 9 = 0,\ x^2 = -9,\ x = \pm 3i$

$\{\pm 3, \pm 3i\}$

61. $x^3 - 27 = 0$

$(x - 3)(x^2 + 3x + 9) = 0$

$x - 3 = 0$ or $x^2 + 3x + 9 = 0$

$x = 3$ or

$$x = \frac{-3 \pm \sqrt{3^2 - 4(1)(9)}}{2(1)}$$

$$x = \frac{-3 \pm 3i\sqrt{3}}{2} =$$

$\{3, -\frac{3}{2} \pm i\frac{3\sqrt{3}}{2}\} \approx \{3, -1.5 \pm 2.6i\}$

63. $x^2 - 8x^5 = 0$

$x^2(1 - 8x^3) = 0$

$x^2 = 0$ or $1 - 8x^3 = 0$

$x = 0$ (double root) or

$(1 - 2x)(1 + 2x + 4x^2) = 0$

$1 - 2x = 0$ or $1 + 2x + 4x^2 = 0$

$$x = \tfrac{1}{2}, \text{ or } x = \frac{-2 \pm \sqrt{2^2 - 4(4)(1)}}{2(4)}$$

$$x = \frac{-2 \pm 2i\sqrt{3}}{8} = \frac{-1 \pm i\sqrt{3}}{4}$$

$\{0, \frac{1}{2}, -\frac{1}{4} \pm i\frac{\sqrt{3}}{4}\} \approx \{0, 0.5, -0.25 \pm 0.43i\}$

65. a. $(1 + i\sqrt{3})^3 = (1 + 2i\sqrt{3} + 3i^2)(1 + i\sqrt{3}) = (1 + 2i\sqrt{3} + 3(-1))(1 + i\sqrt{3}) =$
$(1 - 3 + 2i\sqrt{3})(1 + i\sqrt{3}) = (-2 + 2i\sqrt{3})(1 + i\sqrt{3}) = -2 - 2i\sqrt{3} + 2i\sqrt{3} + 6i^2 =$
$-2 + 6(-1) = -8$

b. $(-1 - i\sqrt{3})^3 = (1 + 2i\sqrt{3} + 3i^2)(-1 - i\sqrt{3}) = (1 + 2i\sqrt{3} + 3(-1))(-1 - i\sqrt{3}) =$
$(1 - 3 + 2i\sqrt{3})(-1 - i\sqrt{3}) = (-2 + 2i\sqrt{3})(-1 - i\sqrt{3}) = 2 + 2i\sqrt{3} - 2i\sqrt{3} - 6i^2 =$
$2 - 6(-1) = 8$

c. $$\left(-\frac{3}{2} + \frac{3i\sqrt{3}}{2}\right)^3 = \left(\frac{9}{4} - \frac{9i\sqrt{3}}{2} + \frac{27i^2}{4}\right)\left(-\frac{3}{2} + \frac{3i\sqrt{3}}{2}\right) = \left(-\frac{9}{2} - \frac{9i\sqrt{3}}{2}\right)\left(-\frac{3}{2} + \frac{3i\sqrt{3}}{2}\right) =$$

$$\frac{27}{4} - \frac{81i}{4} + \frac{81i}{4} - \frac{81i^2}{4} = \frac{27}{4} + \frac{81}{4} = 27$$

d. $$\left(-\frac{1}{2} + \frac{i\sqrt{3}}{2}\right)^3 = \left(\frac{1}{4} - \frac{i\sqrt{3}}{2} + \frac{3i^2}{4}\right)\left(-\frac{1}{2} + \frac{i\sqrt{3}}{2}\right) = \left(-\frac{1}{2} - \frac{i\sqrt{3}}{2}\right)\left(-\frac{1}{2} + \frac{i\sqrt{3}}{2}\right) =$$

$$\frac{1}{4} - \frac{i\sqrt{3}}{4} + \frac{i\sqrt{3}}{4} - \frac{3i^2}{4} = \frac{1}{4} + \frac{3}{4} = 1$$

The cubes give integer answers, so the number being cubed is a complex cube root.

67. All powers equal one of these four: $i^1 = i$, $i^2 = -1$, $i^3 = -i$, $i^4 = +1$. Divide the exponent by 4. The remainder of $1 = i^1$, remainder of $2 = i^2$, remainder of $3 = i^3$, and remainder of $0 = i^4$.

$i^{16} = +1$, $i^{24} = +1$, $i^{35} = -i$, $i^{46} = -1$, $i^{97} = i$

Exercises 3.6: Shifts, Vertex Form, and Applications of Quadratic Functions

1. a. $y = (x + 2)^2$
H: shifted 2 units left

b. $y = (x - 3)^2$
H: shifted 3 units right

3. a. $y = x^2 + 2$
V: shifted up 2 units

b. $y = x^2 - 1$
V: shifted down 1 unit

5. a. $y = (x + 1)^2 - 2$
H: shifted left 1 unit
V: shifted down 2 units

b. $y = (x - 2)^2 + 1$
H: shifted right 2 units
V: shifted up 1 unit

7. Shift $y = x^2$ left 2 units.

9. Shift $y = -x^2$ right 4 units.

11. Shift $y = x^2$ right 3 units and up 4 units.

13. Shift $y = x^2$ left 4 units and up 3 units.

15. $y = x^2$ shifted right 1 unit; new vertex is at (1, 0).

17. $y = -x^2$ shifted left 3 and up 4; new vertex is at (–3, 4).

19. $y = x^2$ shifted right 2 and down 3; new vertex is at (2, –3).

21.
$y = a(x - 3)^2 + 1$
$3 = a(-1 - 3)^2 + 1$
$3 = 16a + 1$
$2 = 16a,\ a = \frac{1}{8}$
$y = \frac{1}{8}(x - 3)^2 + 1$

23.
$y = a(x + 1)^2 - 1$
$7 = a(1 + 1)^2 - 1$
$7 = 4a - 1$
$8 = 4a,\ a = 2$
$y = 2(x + 1)^2 - 1$

25.
$y = a(x - 3)^2 - 2$
$-10 = a(5 - 3)^2 - 2$
$-10 = 4a - 2$
$-8 = 4a,\ \ a = -2$
$y = -2(x - 3)^2 - 2$

27. a. $y = -0.0022x^2$

b. $y = -0.0022x^2 + 359$

29. The ordered pair for point C is (10, 0). Using $y = a(x - x_1)(x - x_2)$ and $a = -0.24$, $y = -0.24(x - 0)(x - 10)$ then leads to $y = -0.24x^2 + 2.4x$.

31. a. A (0, –25), C (20, –15)

b. $y = a(x - 20)^2 - 15;$
$0 = a(0 - 20)^2 - 15$
$15 = 400a;\ a = 0.0375$
$y = 0.0375(x - 20)^2 - 15$ or $y = 0.0375x^2 - 1.5x$

c. y-intercepts indicate a vertical shift of 25 feet.

33. Vertex is (30, 100); $y = a(x - 30)^2 + 100$
$0 = a(0 - 30)^2 + 100$
$-100 = 900a;\ a = -\frac{1}{9}$
$y = -\frac{1}{9}(x - 30)^2 + 100$ or $y = -\frac{1}{9}x^2 + \frac{20}{3}x$

35. $y = x^2 + 10x + 25$
Perfect square trinomial
$y = (x + 5)^2$
H: shift left 5 units

37. $y = x^2 - 6x + 9$
Perfect square trinomial
$y = (x - 3)^2$
H: shift right 3 units

39. $y = x^2 + 10x + 30$
$y - 30 = x^2 + 10x$
$y - 30 + 25 = x^2 + 10x + 25$
$y - 5 = (x + 5)^2$
$y = (x + 5)^2 + 5$
H: shift left 5 units
V: shift up 5 units

41. $y = x^2 - 6x + 8$
$y - 8 = x^2 - 6x$
$y - 8 + 9 = x^2 - 6x + 9$
$y + 1 = (x - 3)^2$
$y = (x - 3)^2 - 1$
H: shift right 3 units
V: shift down 1 unit

43. a. Vertex is on the x-axis at x = r.

b. $y = (x + 1)^2$ is the same as $y = (x - (-1))^2$.
Vertex is at (–1, 0).

c. Vertex is at (4, 0).

45. a. Shift is right $\left|\frac{t}{s}\right|$ when t > 0, left $\left|\frac{t}{s}\right|$ when t < 0.

b. t = 1, s = 2; shift is right $\frac{1}{2}$ unit.

c. $(2x + 3)^2 = (2x - (-3))^2$; s = 2, t = –3; shift is left $\frac{3}{2}$ units

47. For origin at A, the domain is $0 \le x \le 40$ and the range is $10 \le y \le 25$.

49. There is a third point on the parabola on the other side of the axis of symmetry from (p, q). The horizontal distance between (p, q) and the axis of symmetry is |h – p|. With absolute value we describe two points, (h |h – p|, q). One is (p, q) and the other is the unknown.

Exercises 3.7: Solving Quadratic Inequalites

	Inequality	Interval	Words	Line Graph
1.	**$x < -3$ or $x > 2$**	**$(-\infty, -3)$ or $(2, +\infty)$**	Set of numbers less than −3 or greater than 2	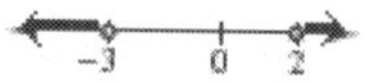
3.	**$x \geq 4$**	**$[4, +\infty)$**	**Set of numbers greater than or equal to 4**	see text
5.	$x > -3$	**$(-3, +\infty)$**	**Set of number greater than −3**	
7.	**$x \leq 4$**	$(-\infty, 4]$	**Set of numbers less than or equal to 4**	

9. $x^2 - 5x - 14 \leq 0$

$(x - 7)(x + 2) \leq 0$

x-intercepts are $(7, 0)$ and $(-2, 0)$.

Graph opens upward.

$f(x) \leq 0$ for $-2 \leq x \leq 7$.

11. $x^2 - 5x - 6 \geq 0$

$(x - 6)(x + 1) \geq 0$

x-intercepts are $(6, 0)$ and $(-1, 0)$.

Graph opens upward.

$f(x) \geq 0$ for $x \leq -1$ or $x \geq 6$.

13. $x^2 + 6x + 9 > 0$

$(x + 3)^2 > 0$

x-intercept is $(-3, 0)$.

Graph opens upward.

$f(x) > 0$ for all $\Re$, $x \neq -3$.

15. $x^2 + 10x - 24 \geq 0$

$(x + 12)(x - 2) \geq 0$

x-intercepts are $(-12, 0)$ and $(2, 0)$.

Graph opens upward.

$f(x) \geq 0$ for $x \leq -12$ or $x \geq 2$.

17. $x^2 - 3x - 18 \leq 0$

$(x - 6)(x + 3) \leq 0$

x-intercepts are $(6, 0)$ and $(-3,0)$.

Graph opens upward.

$f(x) \leq 0$ for $-3 \leq x \leq 6$.

19. $x^2 - 10x + 25 < 0$

$(x - 5)^2 < 0$

x-intercept is $(5, 0)$.

Graph opens upward.

$f(x) < 0$, never, always ≥ 0.

21. $x^2 - x > x + 3$

$x^2 - 2x - 3 > 0$

$(x - 3)(x + 1) > 0$

Intercepts are $x = 3$ and $x = -1$.

$x^2 - x > x + 3$, for $x < -1$ or $x > 3$.

$(-\infty, -1)$ or $(3, +\infty)$

23. $x^2 + 2x - 9 \leq 1 - x$

$x^2 + 3x - 10 \leq 0$

$(x + 5)(x - 2) \leq 0$

Intercepts are $x = -5$ and $x = 2$.

$x^2 + 2x - 9 \leq 1 - x$, for $-5 \leq x \leq 2$.

$[-5, 2]$

25. $x^2 + x < x - 1$

$x^2 < -1$

Not possible in real numbers.

Graph both sides; no intersection.

$x^2 + x$ is always above $x - 1$.

{ }or ∅

27. $x^2 - x \geq x - 1$

$x - 2x + 1 \geq 0$

$(x - 1)^2 \geq 0$

Intercept is $x = 1$.

$x^2 - x \geq x - 1$, for all $\Re$.

$(-\infty, +\infty)$

29. $2x^2 - 2 < x^2 - x$

$x^2 + x - 2 < 0$

$(x + 2)(x - 1) < 0$

Intercepts are $x = -2$ and $x = 1$.

$2x^2 - 2 < x^2 - x$, for $-2 < x < 1$

$(-2, 1)$

31. $3x^2 - 3 \geq 4x - x^2$

$4x^2 - 4x - 3 \geq 0$

$(2x - 3)(2x + 1) \geq 0$

Intercepts are $x = 1.5$ and $x = -0.5$.

$3x^2 - 3 \geq 4x - x^2$, for $x \leq -0.5$ or $x \geq 1.5$.

$(-\infty, -0.5]$ or $[1.5, +\infty)$

Answers for Exercise 33 are taken from the graph. Answers are recorded for reference.

33. a. $\approx 12 < x < 100$

b. $\approx 0 < x < 12$ or $100 < x < 112$

c. ≈ 28 ft.

d. No. The graph shows the ball's path; both axes are in feet.

e. 50 ft.

f. $\approx 25 < x < 85$

35.

n	$v_x = \sqrt{2(32.2)(n)}$
0	0
0.5	$\sqrt{2(32.2)(0.5)} \approx 5.67$
1.0	$\sqrt{2(32.2)(1.0)} \approx 8.02$
1.5	$\sqrt{2(32.2)(1.5)} \approx 9.83$
2.0	$\sqrt{2(32.2)(2.0)} \approx 11.35$
2.5	$\sqrt{2(32.2)(2.5)} \approx 12.69$
3.0	$\sqrt{2(32.2)(3.0)} \approx 13.90$
3.5	$\sqrt{2(32.2)(3.5)} \approx 15.01$
4.0	$\sqrt{2(32.2)(4.0)} \approx 16.05$
4.5	$\sqrt{2(32.2)(4.5)} \approx 17.02$
5.0	$\sqrt{2(32.2)(5.0)} \approx 17.94$
5.5	$\sqrt{2(32.2)(5.5)} \approx 18.82$
6.0	$\sqrt{2(32.2)(6.0)} \approx 19.66$

With the water 6 ft above the hole, the horizontal water speed is 19.7 ft/sec as it exits the hole.

$h - h_o$ is the distance from the hole to the top of the rustler's head, $h_o = 9$, $h = 0$; $h - h_o = 0 - 9 = -9$.

v_y is the initial velocity of the water. The water is dropping from the hole, so $v_y = 0$. Substituting into the vertical motion equation: $0 = -16.1t^2 + 9$, $t \approx 0.75$ sec. This time remains constant, as the distance from the hole to the top of the rustler's head does not change.

In 0.75 sec. the water can travel 19.7 ft/sec · 0.75 sec ≈ 14.8 ft. The rustler can be 14.8 feet from the tank and still get wet.

When the water is 4 ft above the hole, the horizontal water speed is 16 ft/sec as it exits the hole.

In 0.75 sec. the water can travel 16 ft/sec · 0.75 sec ≈ 12 ft

The region where the rustler will get wet is between 14.8 and 12 ft from the tank, or 12 ft ≤ wet region ≤ 14.8 ft.

Chapter 3 Review

1. **a.** $\sqrt{75}=\sqrt{25\cdot 3}=5\sqrt{3}$

b. $\sqrt{8}=\sqrt{4\cdot 2}=2\sqrt{2}$

c. $\sqrt{32}=\sqrt{16\cdot 2}=4\sqrt{2}$

d. $\sqrt{48}=\sqrt{16\cdot 3}=4\sqrt{3}$

2. **a.** $\dfrac{3-3\sqrt{6}}{3}=\dfrac{3(1-\sqrt{6})}{3}=1-\sqrt{6}$

b. $\dfrac{3+3\sqrt{6}}{3}=\dfrac{3(1+\sqrt{6})}{3}=1+\sqrt{6}$

c. $\dfrac{3+\sqrt{18}}{3}=\dfrac{3+3\sqrt{2}}{3}=\dfrac{3(1+\sqrt{2})}{3}=1+\sqrt{2}$

d. $\dfrac{3+\sqrt{12}}{2}=\dfrac{3+2\sqrt{3}}{2}$

3. **a.** $\sqrt{14400}=\sqrt{144\cdot 100}=12\cdot 10=120$

b. $\sqrt{1.44}=\sqrt{\dfrac{144}{100}}=\dfrac{12}{10}=1.2$

c. $\sqrt{0.0121}=\sqrt{\dfrac{121}{10{,}000}}=\dfrac{11}{100}=0.11$

d. $\sqrt{0.000\,000\,01}=\sqrt{\dfrac{1}{100{,}000{,}000}}=\dfrac{1}{10{,}000}=0.0001$

4. There are two square roots, but only the nonnegative one is the principal square root.

5. **a.** $7.5^2+10^2=12.5^2$

$56.25+100=156.25$

b. $18^2+24^2=30^2$

$324+576=900$

c. $(\sqrt{5})^2+(\sqrt{8})^2=(\sqrt{13})^2$

$5+8=13$

d. $4^2+(\sqrt{20})^2=6^2$

$16+20=36$

6. **a.** $x^2+5^2=8^2$

$x^2=64-25$

$x^2=39;\ x=\sqrt{39}$

b. $x^2+4^2=17$

$x^2=17-16$

$x=1$

c. $2(\sqrt{2})^2=x^2$

$x^2=2(2);\ x=2$

d. $x^2+3^2=6^2$

$x^2=36-9$

$x^2=27$

$x=3\sqrt{3}$

e. $(3\sqrt{2})^2 + x^2 = 6^2$

$x^2 = 36 - 18$

$x^2 = 18$

$x = 3\sqrt{2}$

7. $2x^2 = z^2$; $z = x\sqrt{2}$; $\frac{z}{x} = \sqrt{2}$

8. a. $x^2 = 225$

$x = \pm\sqrt{225}$

$x = \pm 15$

b. $x = \sqrt{225}$

$x = 15$

c. $x^2 = 0.81$

$x = \pm\sqrt{0.81}$

$x = \pm 0.9$

d. $x = \sqrt{0.81}$

$x = 0.9$

e. $|x + 3| = 2$

$x + 3 = \pm 2$

$x + 3 = 2$ or $x + 3 = -2$

$x = -1$ or $x = -5$

$\{-5, -1\}$

f. $|x - 5| = 3$

$x - 5 = \pm 3$

$x - 5 = 3$ or $x - 5 = -3$

$x = 8$ or $x = 2$

$\{2, 8\}$

9. False. The graph of $y = \sqrt{x}$ does not exist in the 3rd or 4th quadrants.

10. $T = 2\pi\sqrt{\frac{L}{g}}$; $\frac{T}{2\pi} = \sqrt{\frac{L}{g}}$; $\frac{T^2}{4\pi^2} = \frac{L}{g}$; $L = \frac{gT^2}{4\pi^2}$

11. $C = 2\pi\sqrt{\frac{a^2+b^2}{2}}$; $\frac{C}{2\pi} = \sqrt{\frac{a^2+b^2}{2}}$; $\frac{C^2}{4\pi^2} = \frac{a^2+b^2}{2}$; $\frac{C^2}{2\pi^2} = a^2 + b^2$;

$\frac{C^2}{2\pi^2} - b^2 = a^2$; $a = \sqrt{\frac{C^2}{2\pi^2} - b^2}$; *or* $a = \sqrt{\frac{C^2 - 2\pi^2 b^2}{2\pi^2}}$

12. a. $x^2 - 8x + (-8 \div 2)^2 = 3 + (-8 \div 2)^2$

$x^2 - 8x + \underline{\mathbf{16}} = 3 + \underline{\mathbf{16}}$

b. $x^2 - 10x + (-10 \div 2)^2 = 2 + (-10 \div 2)^2$

$x^2 - 10x + \underline{\mathbf{25}} = 2 + \underline{\mathbf{25}}$

c. $x^2 - 3x + (-3 \div 2)^2 = 4 + (-3 \div 2)^2$

$x^2 - 3x + \underline{\tfrac{9}{4}} = 4 + \underline{\tfrac{9}{4}}$

d. $x^2 + 7x + (7 \div 2)^2 = 5 + (7 \div 2)^2$

$x^2 + 7x + \underline{\tfrac{49}{4}} = 5 + \underline{\tfrac{49}{4}}$

13. a. $\left(x+\dfrac{b}{2a}\right)^2 = x^2 + 2x\dfrac{b}{2a} + \left(\dfrac{b}{2a}\right)^2 = x^2 + \dfrac{bx}{a} + \dfrac{b^2}{4a^2}$

b. $\dfrac{b^2}{4a^2} - \dfrac{c}{a} = \dfrac{b^2}{4a^2} - \dfrac{c(4a)}{a(4a)} = \dfrac{b^2 - 4ac}{4a^2}$

c. $\sqrt{\left(x+\dfrac{b}{2a}\right)^2} = \left|x + \dfrac{b}{2a}\right|$

14.

$x^2 - x - 1 = 0$	
$x^2 - x = 1$	***Add 1 to both sides.***
$x^2 - 1x + \frac{1}{4} = 1 + \frac{1}{4}$	***Add $\frac{1}{4}$ to both sides.***
$(x - \frac{1}{2})^2 = 1 + \frac{1}{4}$	***Factor left side.***
$(x - \frac{1}{2})^2 = \frac{5}{4}$	***Add terms on right side.***
$\sqrt{(x-\frac{1}{2})^2} = \sqrt{\frac{5}{4}}$	***Take square root.***
$\lvert x - \frac{1}{2} \rvert = \frac{\sqrt{5}}{2}$	***Use definition of principal square root.***
$x - \frac{1}{2} = \pm\frac{\sqrt{5}}{2}$	***Use definition of absolute value.***
$x = \frac{1}{2} \pm \frac{\sqrt{5}}{4}$	***Add $\frac{1}{2}$ to both sides, evaluate, and round,***
$x \approx 1.618$ or $x \approx -0.618$	

15. For $ax^2 + bx + c = 0$, $x = \dfrac{\left(-b \pm \sqrt{(b^2 - 4ac)}\right)}{(2a)}$

16. $25x^2 - 5x = 0$

$5x(5x - 1) = 0$

$5x = 0$ or $5x - 1 = 0$

$x = 0$ or $x = \frac{1}{5}$

$\{\frac{1}{5}, 0\}$

17. $-4.905x^2 + 6x + 10 = 11$

$-4.905x^2 + 6x - 1 = 0$

$x = \dfrac{-6 \pm \sqrt{6^2 - 4(-4.905)(-1)}}{2(-4.905)}$

$x = \dfrac{-6 \pm \sqrt{16.38}}{-9.81}$

$x \approx 0.199$ or $x \approx 1.024$

$\{0.199, 1.024\}$

18. $-4.905x^2 + 6x + 10 = 0$

$x = \dfrac{-6 \pm \sqrt{6^2 - 4(-4.905)(10)}}{2(-4.905)}$

$x = \dfrac{-6 \pm \sqrt{232.2}}{-9.81}$

$x \approx -0.942$ or $x \approx 2.165$

$\{-0.942,\ 2.165\}$

19. $(x - 5)^2 = 1$

$|x - 5| = 1$

$x - 5 = \pm 1$

$x - 5 = 1$ or $x - 5 = -1$

$x = 6$ or $x = 4$

$\{4, 6\}$

20. $9x^2 = 4$

$9x^2 - 4 = 0$

$(3x + 2)(3x - 2) = 0$

$3x + 2 = 0$ or $3x - 2 = 0$

$x = -\frac{2}{3}$ or $x = \frac{2}{3}$

$\{-\frac{2}{3}, \frac{2}{3}\}$

21. $9x^2 - 18x - 25 = 47$

$9x^2 - 18x - 72 = 0$

$9(x^2 - 2x - 8) = 0$

$(x - 4)(x + 2) = 0$

$x - 4 = 0$ or $x + 2 = 0$

$x = 4$ or $x = -2$

$\{-2, 4\}$

22. $(x + 6)^2 = 9$

$|x + 6| = 3$

$x + 6 = \pm 3$

$x + 6 = 3$ or $x + 6 = -3$

$x = -3$ or $x = -9$

$\{-9, -3\}$

23. $4x^2 + 4x + 1 = 25$

$(2x + 1)^2 = 25$

$|2x + 1| = 5$

$2x + 1 = \pm 5$

$2x + 1 = 5$ or $2x + 1 = -5$

$x = 2$ or $x = -3$

$\{-3, 2\}$

24. $9x^2 + 6x + 1 = 9$

$(3x + 1)^2 = 9$

$|3x + 1| = 3$

$3x + 1 = \pm 3$

$3x + 1 = 3$ or $3x + 1 = -3$

$x = \frac{2}{3}$ or $x = -\frac{4}{3}$

$\{-\frac{4}{3}, \frac{2}{3}\}$

25. $3x^2 - 6x + 3 = 0$

$3(x^2 - 2x + 1) = 0$

$(x - 1)^2 = 0$

$x - 1 = 0$

$x = 1$

$\{1\}$

26. $5x^2 + 7x + 4 = 0$; $x = \dfrac{-7 \pm \sqrt{7^2 - 4(5)(4)}}{2(5)}$; $x = \dfrac{-7 \pm \sqrt{-31}}{10}$

No real number solutions or x-intercepts; negative number under the radical sign.

27. a. x-coordinate is $\dfrac{-3}{2(1)} = -\dfrac{3}{2}$; y-coordinate is $y = (-\frac{3}{2})^2 + 3(-\frac{3}{2}) + 4 = \frac{7}{4}$;

vertex is $(-\frac{3}{2}, \frac{7}{4}) \Rightarrow y = \left(x + \frac{3}{2}\right)^2 + \frac{7}{4}$.

b. x-coordinate is $\dfrac{-(-2)}{2(0.5)} = 2$; y-coordinate is $y = 0.5(2^2) - 2(2) + 1 = -1$;

vertex is $(2, -1) \Rightarrow y = 0.5(x-2)^2 - 1$.

c. x-coordinate is $\dfrac{-(-4)}{2(3)} = \dfrac{2}{3}$; y-coordinate is $y = 3(\frac{2}{3})^2 - 4(\frac{2}{3}) = -\frac{4}{3}$;

vertex is $(\frac{2}{3}, -\frac{4}{3}) \Rightarrow y = 3\left(x - \frac{2}{3}\right)^2 - \frac{4}{3}$.

28. No; the axis of symmetry is $x = \frac{-b}{2a}$ and does not contain c.

29.

$x^2 - x - 2 = 0$

$(x - 2)(x + 1) = 0$

$x - 2 = 0;\ x = 2$

$x + 1 = 0;\ x = -1$

$\{-1, 2\}$

Vertex is half-way between the intercepts at x = 0.5.

$y = (0.5)^2 - 0.5 - 2$

$y = -2.25$ vertex: $(0.5, -2.25)$

30.

$x^2 - x - 6 = 0$

$(x - 3)(x + 2) = 0$

$x - 3 = 0;\ x = 3$

$x + 2 = 0;\ x = -2$

$\{-2, 3\}$

Vertex is half-way between the intercepts at x = 0.5.

$y = (0.5)^2 - 0.5 - 6$

$y = -6.25$ vertex: $(0.5, -6.25)$

31.

$2x^2 - 4x + 5 = 0$

$$x = \frac{-(-4) \pm \sqrt{(-4)^2 - 4(2)(5)}}{2(2)}$$

$$x = \frac{4 \pm \sqrt{-24}}{4}$$

No x-intercepts.

Vertex: x = 1

$y = 2(1)^2 - 4(1) + 5$

$y = 3$

Vertex: (1, 3)

32.

$3x^2 - 6x + 5 = 0$

$$x = \frac{-(-6) \pm \sqrt{(-6)^2 - 4(3)(5)}}{2(3)}$$

$$x = \frac{6 \pm \sqrt{-24}}{6}$$

No x-intercepts.

Vertex: x = 1

$y = 3(1)^2 - 6(1) + 5$

$y = 2$

Vertex: (1, 2)

33. a. v = 132 ft/sec. $y = -\frac{32.2}{132^2}x^2 + x$; $x = \frac{-1}{2\left(\frac{32.2}{132^2}\right)} \approx 271$, $y = \frac{32.2}{132^2}(271)^2 + 271 \approx 135$

Vertex ≈ (271, 135)

b. v = 220 ft/sec. $y = -\frac{32.2}{220^2}x^2 + x$; $x = \frac{-1}{2\left(\frac{32.2}{220^2}\right)} \approx 752$, $y = \frac{32.2}{220^2}(752)^2 + 752 \approx 376$

Vertex ≈ (752, 376)

c. v = 400 ft/sec. $y = -\frac{32.2}{400^2}x^2 + x$; $x = \frac{-1}{2\left(\frac{32.2}{400^2}\right)} \approx 2484$, $y = \frac{32.2}{400^2}(2484)^2 + 2484 \approx 1242$

Vertex ≈ (2484, 1242)

d. v = 500 ft/sec. $y = -\frac{32.2}{500^2}x^2 + x$; $x = \frac{-1}{2\left(\frac{32.2}{500^2}\right)} \approx 3882$, $y = \frac{32.2}{500^2}(3882)^2 + 3882 \approx 1941$

Vertex ≈ (3882, 1941)

34. a. $y = a(x - 75)^2 + 40$; $0 = a(0 - 75)^2 + 40$; $-40 = 5{,}625a$; $a \approx -0.0071$
$y = -0.0071(x - 75)^2 + 40$ or
$y = -0.0071(x^2 - 150x + 5{,}625) + 40$; $y = -0.0071x^2 + 1.067x - 40 + 40$;
$y = -0.0071x^2 + 1.067x$

b. Vertex can be found from the problem, y = 40, x = 150 ÷ 2 = 75; (75, 40).

c. 40 + 12 = 52 ft clearance

35. a. $x = \frac{-(-\frac{1}{50})}{2(\frac{1}{48{,}000})} \approx 480$; $y = \frac{1}{48{,}000}(480)^2 - \frac{1}{50}(480) + 700 \approx 695.2$

Drain location: (480, 695.2)

b. Elevation of puddle is 695.2 + 0.3 = 695.5;
$695.5 = \frac{1}{48{,}000}x^2 - \frac{1}{50}x + 700$;
$0 = \frac{1}{48{,}000}x^2 - \frac{1}{50}x + 4.5$ clear fractions
$0 = x^2 - 960x + 216{,}000$
$0 = (x - 360)(x - 600)$; x = 360 or x = 600
600 – 360 = 240; puddle is 240 ft. long.

36. $2l + 2w = 157$; $l = 78.5 - w$
$A = lw$; $A = (78.5 - w)w = 78.5w - w^2$
vertex: $w = \frac{-78.5}{2(-1)} = 39.25$; $A = 78.5(39.25) - (39.25)^2 = 1541$; (39.25, 1541)
The rectangular shape with the maximum area is a square: $l = w$; $4w = 157$, $w = 39.25$.
$A = 39.25^2 \approx 1541$

37. $2l + 2w = 80;\ l = 40 - w$

$A = lw;\ A = (40 - w)w = 40w - w^2$

vertex: $w = \dfrac{-40}{2(-1)} = 20$; $A = 40(20) - (20)^2 = 400;\ (20, 400)$

The rectangular shape with the maximum area is a square: $l = w$; $4w = 80$, $w = 20$.

$A = 20^2 = 400$

38. a. $\sqrt{-16} = i\sqrt{16} = 4i$

b. $\sqrt{-50} = i\sqrt{2 \cdot 25} = 5i\sqrt{2}$

c. $\sqrt{-64} = i\sqrt{64} = 8i$

d. $\sqrt{-18} = i\sqrt{2 \cdot 9} = 3i\sqrt{2}$

39. a. $\dfrac{8+6i}{2} = \dfrac{2(4+3i)}{2} = 4+3i$

b. $\dfrac{6+3i}{3} = \dfrac{3(2+i)}{3} = 2+i$

c. $\dfrac{4-\sqrt{-48}}{2} = \dfrac{4-i\sqrt{3\cdot 16}}{2} = \dfrac{4-4i\sqrt{3}}{2} = \dfrac{2(2-2i\sqrt{3})}{2} = 2-2i\sqrt{3}$

d. $\dfrac{9+\sqrt{-18}}{6} = \dfrac{9+i\sqrt{2\cdot 9}}{6} = \dfrac{9+3i\sqrt{2}}{6} = \dfrac{3(3+i\sqrt{2})}{3(2)} = \dfrac{3+i\sqrt{2}}{2}$

40. a. $4 + 3i$ **b.** $-6 - 2i$ **c.** $5i$

41. a. $(3 + 2i) - (6 + 2i) =$

$3 + 2i - 6 - 2i = -3$

b. $(4 - 2i) - 2(3 - 4i) =$

$4 - 2i - 6 + 8i = -2 + 6i$

42. a. $(4 + 3i)(4 + 3i) =$

$16 + 24i + 9i^2 =$

$16 + 24i + 9(-1) =$

$16 - 9 + 24i = 7 + 24i$

b. $(4 - 3i)(4 + 3i) =$

$16 - 9i^2 =$

$16 - 9(-1) =$

$16 + 9 = 25$

c. $(3 - 4i)(3 + 4i) =$

$9 - 16i^2 =$

$9 - 16(-1) =$

$9 + 16 = 25$

d. $(1 - 3i)(1 + 3i) =$

$1 - 9i^2 =$

$1 - 9(-1) =$

$1 + 9 = 10$

e. $(3 - i)(3 + i) =$

$9 - i^2 =$

$9 - (-1) =$

$9 + 1 = 10$

f. $(1 + 3i)(1 + 3i) =$

$1 + 6i + 9i^2 =$

$1 + 6i + 9(-1) =$

$1 - 9 + 6i = -8 + 6i$

43. $(3)^2 - 4(2)(4) = 9 - 32 = -23$; 2 complex number solutions, no x-intercepts.

$f(x) = 2x^2 + 3x + 4$

$2x^2 + 3x + 4 = 0$

$$x = \frac{-3 \pm \sqrt{-23}}{4} = \frac{-3 \pm i\sqrt{23}}{4}$$

$$\{\frac{-3}{4} - \frac{i\sqrt{23}}{4}, \frac{-3}{4} + \frac{i\sqrt{23}}{4}\}$$

44. $(2)^2 - 4(3)(1) = 4 - 12 = -8$; 2 complex number solutions, no x-intercepts.

$f(x) = 3x^2 + 2x + 1$

$3x^2 + 2x + 1 = 0$

$$x = \frac{-2 \pm \sqrt{-8}}{6} = \frac{-2 \pm 2i\sqrt{2}}{6}$$

$$\{\frac{-1}{3} - \frac{i\sqrt{2}}{3}, \frac{-1}{3} + \frac{i\sqrt{2}}{3}\}$$

45. $(1)^2 - 4(-2)(3) = 1 + 24 = 25$; 2 real number solutions, 2 x-intercepts.

$f(x) = -2x^2 + x + 3$

$-2x^2 + x + 3 = 0$

$3 + x - 2x^2 = 0$

$(1 + x)(3 - 2x) = 0$

$x + 1 = 0$ or $3 - 2x = 0$

$x = -1$ or $x = 1.5$

$\{-1, 1.5\}$

46. $(2)^2 - 4(-1)(-1) = 4 - 4 = 0$; 1 real number solution (a double root), 1 x-intercept.

$f(x) = -1x^2 + 2x - 1$

$-1x^2 + 2x - 1 = 0$

$-1(x^2 - 2x + 1) = 0$

$(x - 1)^2 = 0$

$x - 1 = 0, x = 1; \{1\}$

47. $x^3 - 27 = 0$

$(x - 3)(x^2 + 3x + 9) = 0$

$x - 3 = 0;\ x = 3$

$x^2 + 3x + 9 = 0$

$$x = \frac{-3 \pm \sqrt{3^2 - 4(1)(9)}}{2(1)}$$

$$x = \frac{-3 \pm \sqrt{-27}}{2}$$

$x = -\frac{3}{2} \pm \frac{3i\sqrt{3}}{2}$

$\{3,\ -\frac{3}{2} \pm \frac{3i\sqrt{3}}{2}\}$

48. $x^3 + 8 = 0$

$(x + 2)(x^2 - 2x + 4) = 0$

$x + 2 = 0;\ x = -2$

$x^2 - 2x + 4 = 0$

$$x = \frac{-(-2) \pm \sqrt{(-2)^2 - 4(1)(4)}}{2(1)}$$

$$x = \frac{2 \pm \sqrt{-12}}{2}$$

$$x = \frac{2 \pm 2i\sqrt{3}}{2} = 1 \pm i\sqrt{3}$$

$\{-2,\ 1 \pm i\sqrt{3}\}$

49. $x^3 + x^2 + x = 0$

$x(x^2 + x + 1) = 0$

$x = 0$

$x^2 + x + 1 = 0$

$$x = \frac{-1 \pm \sqrt{1^2 - 4(1)(1)}}{2(1)} = \frac{-1 \pm \sqrt{-3}}{2}$$

$x = -\frac{1}{2} \pm i\frac{\sqrt{3}}{2}$

$\{0,\ -\frac{1}{2} \pm i\frac{\sqrt{3}}{2}\}$

50. $x^3 + 3x^2 + 4x = 0$

$x(x^2 + 3x + 4) = 0$

$x = 0$

$x^2 + 3x + 4 = 0$

$$x = \frac{-3 \pm \sqrt{3^2 - 4(1)(4)}}{2(1)} = \frac{-3 \pm \sqrt{-7}}{2}$$

$x = -\frac{3}{2} \pm i\frac{\sqrt{7}}{2}$

$\{0,\ -\frac{3}{2} \pm i\frac{\sqrt{7}}{2}\}$

51. $2x^3 - 2x^2 = 0$

$2x^2(x - 1) = 0$

$2x^2 = 0;\ x = 0$

$x - 1 = 0;\ x = 1$

{0 (double root), 1}

52. $x^3 + 9x = 0$

$x(x^2 + 9) = 0$

$x = 0$

$x^2 + 9 = 0$

$x^2 = -9;\ x = \pm 3i$

$\{0, \pm 3i)$

53. $x^4 - 16 = 0$

$(x^2 + 4)(x^2 - 4) = 0$

$x^2 + 4 = 0;\ x^2 = -4,\ x = \pm 2i$

$x^2 - 4 = 0;\ x^2 = 4,\ x = \pm 2$

$\{\pm 2, \pm 2i\}$

54. $x^2 - 9x^4$

$x^2(1 - 9x^2) = 0$

$x^2 = 0,\ x = 0$

$1 - 9x^2 = 0$

$(1 + 3x)(1 - 3x) = 0$

$1 + 3x = 0;\ x = -\frac{1}{3}$

$1 - 3x = 0;\ x = \frac{1}{3}$

{0 (double root), $\pm \frac{1}{3}$}

55. H: shift right 3 units
V: shift up 2 units

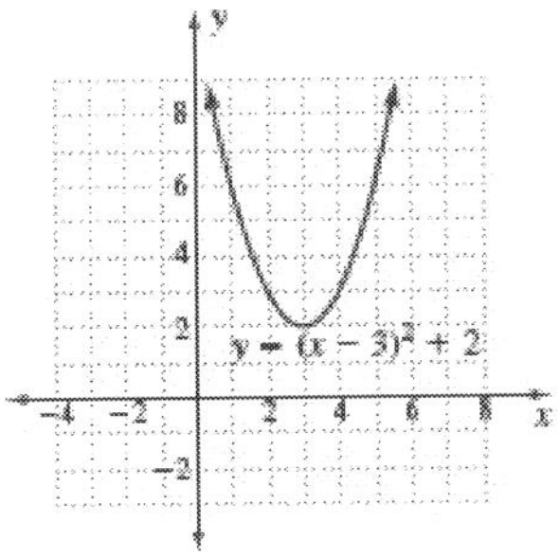

56. H: shift left 3 units
V: shift up 1 unit

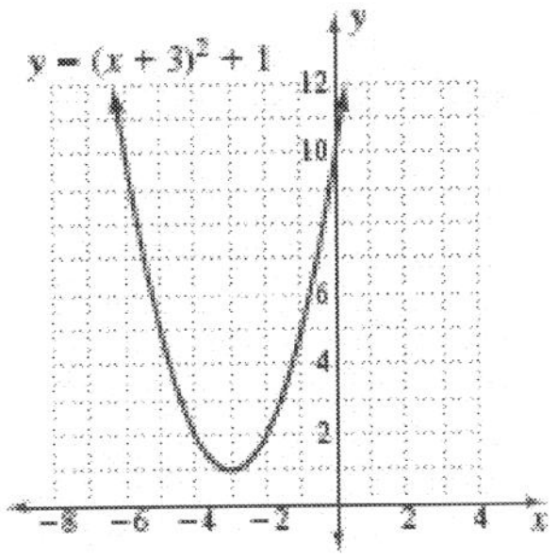

57. H: shift left 2 units
V: shift down 1 unit

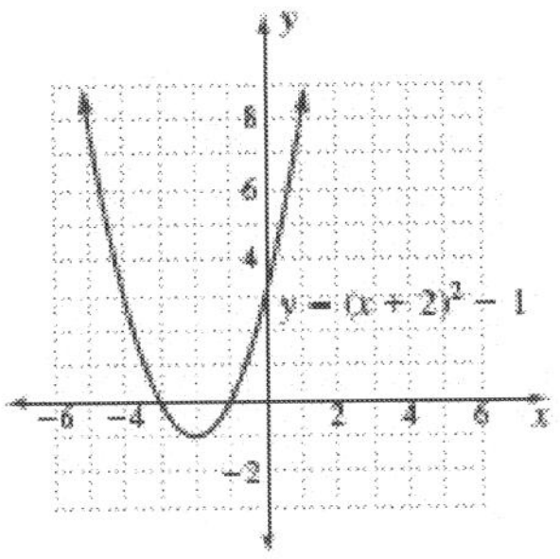

58. H: shift right 4 units
V: shift down 2 units

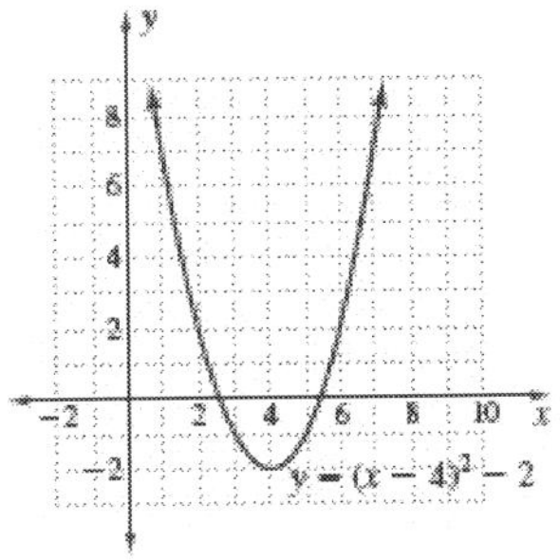

59.
$y = a(x - 8)^2 + 2$
$0 = a(0 - 8)^2 + 2$
$-2 = 64a,\ a = -\frac{1}{32}$
$y = -\frac{1}{32}(x - 8)^2 + 2$

60.
$y = a(x - 4)^2 - 3$
$0 = a(0 - 4)^2 - 3$
$3 = 16a,\ a = \frac{3}{16}$
$y = \frac{3}{16}(x - 4)^2 - 3$

61.
$y = a(x - 0)^2 + 0$
$2 = a(8)^2$
$a = \frac{1}{32}$
$y = \frac{1}{32}x^2$

62.
$y = a(x - 0)^2 + 0$
$-3 = a(4)^2$
$a = -\frac{3}{16}$
$y = -\frac{3}{16}x^2$

63.
$y = a(x + 2)^2 + 3$
$-3 = a(2 + 2)^2 + 3$
$-3 = 16a + 3$
$a = -\frac{3}{8}$
$y = -\frac{3}{8}(x + 2)^2 + 3$

64.
$y = a(x - 2)^2 - 3$
$3 = a(-2 - 2)^2 - 3$
$3 = 16a - 3$
$a = \frac{3}{8}$
$y = \frac{3}{8}(x - 2)^2 - 3$

65. Vertex is (100, 250).

$y = a(x - 100)^2 + 250$

$0 = a(0 - 100)^2 + 250$

$-250 = 10{,}000a;\ \ a = -0.025$

$y = -0.025(x - 100)^2 + 250$

66. Vertex is (250, −100).

$y = a(x - 250)^2 - 100$

$0 = a(0 - 250)^2 - 100$

$100 = 62{,}500a;\ \ a = 0.0016$

$y = 0.0016(x - 250)^2 - 100$

67. $y = x^2 + 4x + 9$

$y - 9 + 4 = x^2 + 4x + 4$

$y - 5 = (x + 2)^2$

$y = (x + 2)^2 + 5$

H: shift left 2 units

V: shift up 5 units

68. $y = x^2 - 6x + 6$

$y - 6 + 9 = x^2 - 6x + 9$

$y + 3 = (x - 3)^2$

$y = (x - 3)^2 - 3$

H: shift right 3 units

V: shift down 3 units

69. a. $x = \dfrac{-\frac{20}{3}}{2(-\frac{1}{9})} = 30,\ y = -\frac{1}{9}(30)^2 + \frac{20}{3}(30) = \mathbf{100\ ft.}$

b. $-\frac{1}{9}x^2 + \frac{20}{3}x < 50 \qquad -\frac{1}{9}x^2 + \frac{20}{3}x - 50 < 0 \quad x = \dfrac{-\frac{20}{3} \pm \sqrt{(\frac{20}{3})^2 - 4(-\frac{1}{9})(-50)}}{2(-\frac{1}{9})}$

$x = \dfrac{-\frac{20}{3} \pm \sqrt{\frac{200}{9}}}{-\frac{2}{9}}$, $x \approx 8.8$ or $x \approx 51.2$ ft. Graph turns down $x < 8.8$ ft or $x > 51.2$ ft.

70. a. $x = \frac{-6}{2(-0.12)} = 25,\ y = -0.12(25)^2 + 6(25) = \mathbf{75\ ft.}$

b. $-0.12x^2 + 6x > 60 \qquad -0.12x^2 + 6x - 60 > 0 \qquad x = \dfrac{-6 \pm \sqrt{6^2 - 4(-0.12)(-60)}}{2(-0.12)}$

$x = \dfrac{-6 \pm \sqrt{7.2}}{-0.24}$, $x \approx 13.8$ or $x \approx 36.2$ Graph turns down $13.8\text{ ft} < x < 36.2$ ft.

71. $x^2 + 6x - 8 \le 0$

Use quadratic formula to find intercepts.

$x = \dfrac{-6 \pm \sqrt{36 - 4(1)(8)}}{2(1)}$

$x = 1.123$ or $x = -7.123$

Graph turns upward.

Function is ≤ 0 for $-7.123 \le x \le 1.123$.

72. $2x^2 - 5x - 3 > 0$

$(2x + 1)(x - 3) > 0$

$x = -\frac{1}{2}$ or $x = 3$

Graph turns upward.

Function is > 0 for $x < -\frac{1}{2}$ or $x > 3$.

73. $3x^2 + 5x - 3 > 19$
$3x^3 + 5x - 22 > 0$
$(3x + 11)(x - 2) > 0$
$x = -3.667$ or $x = 2$
Graph turns upward.
Function is > 19 for $x < -3.667$ or $x > 2$.

74. $-4x^2 + 5x + 2 < -7$
$-4x^2 + 5x + 9 < 0$
$(-4x + 9)(x + 1) < 0$
$x = 2.25$ or $x = -1$
Graph turns downward.
Function is < -7 for $x < -1$ or $x > 2.25$.

75. $x^2 + 5x + 2 > x - 3$
$x^2 + 4x + 5 > 0$
Use quadratic formula to find intercepts.
$$x = \frac{-4 \pm \sqrt{16 - 4(1)(5)}}{2(1)}$$
$$x = \frac{-4 \pm \sqrt{-4}}{2}$$
No intercepts. For both sides of the graph,
$x^2 + 5x + 2 > x - 3$ for all x, $\Re$.

76. $x^2 - 3 > 5 - x^2$
$2x^2 - 8 > 0$
$2(x^2 - 4) > 0$
$(x + 2)(x - 2) > 0$

$x = \pm 2$

$x^2 - 3 > 5 - x^2$ for $x < -2$ or $x > 2$

77. $x^2 + 3 < 2 - x^2$
$2x^2 + 1 < 0$
$2x^2 < -1$
No real number solution.

78. $x^2 + 3x + 2 > 2 - x^2$
$2x^2 + 3x > 0$
$x(2x + 3) > 0$
$x = 0$ or $x = -1.5$
$x^2 + 3x + 2 > 2 - x^2$ for $x < -1.5$ or $x > 0$

79. $x^2 + 5x + 2 \leq 2 - x$
$x^2 + 6x \leq 0$
$x(x + 6) \leq 0$
$x = 0$ or $x = -6$
$x^2 + 5x + 2 \leq 2 - x$ for
$-6 \leq x \leq 0$

80. $4x^2 + 6x - 3 \leq 1 + x - 2x^2$
$6x^2 + 5x - 4 \leq 0$
$(3x + 4)(2x - 1) \leq 0$
$x = -1.333$ or $x = 0.50$
$4x^2 + 6x - 3 \leq 1 + x - 2x^2$ for
$-1.333 \leq x \leq 0.50$

81. a. By the Pythagorean Theorem

$$d^2 + r^2 = \left(r + \frac{h}{5280}\right)^2$$

$$d^2 + r^2 = r^2 + 2\left(\frac{rh}{5280}\right) + \left(\frac{h}{5280}\right)^2$$

$$d^2 = \left(\frac{rh}{2640}\right) + \left(\frac{h}{5280}\right)^2$$

Now for $h < 1000$, $\frac{h}{5280} < 0.19$

and $\left(\frac{h}{5280}\right)^2 < 0.0361 \approx 0$

Thus: $d^2 \approx \frac{rh}{2640}$

$$d \approx \sqrt{\frac{r}{2640} \cdot h}$$

For the earth, $d \approx \sqrt{\frac{3963}{2640} \cdot h} \approx \sqrt{1.5h}$

b. For the moon, $d \approx \sqrt{\frac{1090}{2640} \cdot h} \approx \sqrt{0.41h}$ which is accurate to one decimal place under the radical.

82. a. Vertex is at (15, 15); x-intercepts are (0, 0) and (0, 30).

$y = a(x - 15)^2 + 15$; $0 = a(-15)^2 + 15$; $a = -\frac{1}{15}$;

$y = -\frac{1}{15}(x - 15)^2 + 15$; $y = -\frac{1}{15}(x^2 - 30x + 225) + 15$

$y = -\frac{1}{15}x^2 + 2x - 15 + 15$; $y = -\frac{1}{15}x^2 + 2x$

b. When y = 14.4 ft, x = 12 ft and 18 ft; the distance between them is 18 – 12 = 6 ft.

A = 14.4(6); A = 86.4 ft^2

c. When y = 12.6 ft, x = 9 ft and 21 ft; distance between is 21 – 9 = 12 ft

A = 12.6(12) = 151.2 ft^2

d. Continue to guess and check. Maximum area is 174.8 ft^2 when y = 10.1 ft and horizontal distance is 17.3 ft ($x \approx 6.3$ and $x \approx 23.6$).

Chapter 3 Test

1. $x = \dfrac{-b \pm \sqrt{b^2 - 4ac}}{2a}$

2. **a.** $\sqrt{45} = \sqrt{9 \cdot 5} = 3\sqrt{5}$

b. $\sqrt{98} = \sqrt{49 \cdot 2} = 7\sqrt{2}$

3. $\dfrac{6+\sqrt{18}}{12} = \dfrac{6+3\sqrt{2}}{12} = \dfrac{3(2+\sqrt{2})}{3(4)} = \dfrac{2+\sqrt{2}}{4}$

4.
$x^2 = 196$
$\sqrt{x^2} = \sqrt{196}$
$|x| = 14$
$x = \pm 14$
$\{\pm 14\}$

5.
$(x-4)^2 = 25$
$\sqrt{(x-4)^2} = \sqrt{25}$
$|x-4| = 5$
$x - 4 = 5$ or $x - 4 = -5$
$x = 9$ or $x = -1$
$\{-1, 9\}$

6.
$x^2 + 10x + 5^2 = 5^2 + 24$
$(x+5)^2 = 49$
$\sqrt{(x+5)^2} = \sqrt{49}$
$|x+5| = 7$
$x + 5 = 7$ or $x + 5 = -7$
$x = 2$ or $x = -12$
$\{-12, 2\}$

7.
$0.04x^2 - 169 = 0$
$0.04x^2 = 169$
$\sqrt{0.04x^2} = \sqrt{169}$
$|0.2x| = 13$
$0.2x = \pm 13$
$x = \pm 65$
$\{\pm 65\}$

8.
$x^2 + 2x + 1 = 16$
$(x+1)^2 = 16$
$\sqrt{(x+1)^2} = \sqrt{16}$
$|x+1| = 4$
$x + 1 = \pm 4$
$x + 1 = 4$ or $x + 1 = -4$
$x = 3$ or $x = -5$
$\{-5, 3\}$

9.
$|x-2| = 1$
$x - 2 = \pm 1$
$x - 2 = 1$ or $x - 2 = -1$
$x = 3$ or $x = 1$
$\{1, 3\}$

10. $a = 2, b = -7, c = 6$

$$x = \frac{-(-7) \pm \sqrt{(-7)^2 - 4(2)(6)}}{2(2)}$$

$$x = \frac{7 \pm \sqrt{49 - 48}}{4}$$

$$x = \frac{7 \pm \sqrt{1}}{4}$$

$$x = \frac{7+1}{4} \text{ or } x = \frac{7-1}{4}$$

$x = 2$ or $x = 1.5$

11. a. $9^2 + x^2 = (2x)^2$

$81 + x^2 = 4x^2$

$81 = 3x^2$

$x^2 = 27$

$x = \sqrt{27} = 3\sqrt{3}$

b. $2x^2 = 7^2$

$$x^2 = \frac{49}{2}$$

$$x = \frac{7}{\sqrt{2}} = \frac{7\sqrt{2}}{\sqrt{2}\sqrt{2}} = \frac{7\sqrt{2}}{2}$$

12. $x^2 - x - 2 = 0$

$(x - 2)(x + 1) = 0$

$x = 2$ or $x = -1$

Intercepts $\{-1, 2\} \Rightarrow (-1,0)(2,0)$

Axis of symmetry is $x = \dfrac{-(-1)}{2(1)} = \dfrac{1}{2}$.

$y = (0.5)^2 - 0.5 - 2$

$y = -2.25$

Vertex is $(0.5, -2.25)$.

13. a. $\sqrt{-36} = i\sqrt{36} = 6i$

b. $\sqrt{-75} = i\sqrt{3 \cdot 25} = 5i\sqrt{3} \approx 8.7i$

14. a. $\dfrac{5+10i}{5} = \dfrac{5(1+2i)}{5} = 1+2i$

b. $\dfrac{8+4i}{2} = \dfrac{2(4+2i)}{2} = 4+2i$

15. a. $(1 + 5i)(1 - 5i) =$
$1 - 25i^2 = 1 - 25(-1) =$
$1 + 25 = 26$

b. $(5 - i)(5 + i) =$
$25 - i^2 = 25 - (-1) =$
$25 + 1 = 26$

c. $(1 - 5i)(1 - 5i) =$
$1 - 10i + 25i^2 =$
$1 - 10i + 25(-1) =$
$1 - 25 - 10i = -24 - 10i$

16. $5(2 - 3i) - 3(6 - 7i) =$
$10 - 15i - 18 + 21i =$
$-8 + 6i$

17. a. $2^2 - 4(-1)(3) = 4 + 12 = 16$

b. The discriminant is a positive real number; there will be two real number solutions and two x-intercepts.

c. $-x^2 + 2x + 3 = 0$
$(-x - 1)(x - 3) = 0$
$-x - 1 = 0;\ x = -1$ or $x - 3 = 0;\ x = 3$
$\{-1, 3\}$

18. $(x - 1)(x^2 + x + 1) = 0$
$x - 1 = 0$ or $x^2 + x + 1 = 0$

$$x = 1 \text{ or } x = \frac{-1 \pm \sqrt{1^2 - 4(1)(1)}}{2(1)};\ x = \frac{-1 \pm \sqrt{-3}}{2};\ x = \frac{-1 \pm i\sqrt{3}}{2}$$

$\{1, -\frac{1}{2} \pm i\frac{\sqrt{3}}{2}\}$

19. The graph of $y = (x + 2)^2$ has the same shape as $y = x^2$, but is shifted left 2 units. The vertex and the x-intercept are at (0, 0) for $y = x^2$ and at (–2, 0) for $y = (x + 2)^2$. The y-intercept of $y = x^2$ is also at (0, 0), while the y-intercept of $y = (x + 2)^2$ is at (0, 4).

20. **a.** The graphs of $f(x)$ and $g(x)$ have the same shape. $g(x)$ is m units below $f(x)$.

b. The graphs of $f(x)$ and $h(x)$ have the same shape. $h(x)$ is n units right of $f(x)$.

21. The graph of $y = (x - 3)^2 - 1$ is $y = x^2$ shifted right 3 units and down 1 unit.

22. $y = a(x + 1)^2 - 8;$

$19 = a(2 + 1)^2 - 8$

$27 = 9a;\ a = 3$

$y = 3(x + 1)^2 - 8$ or $y = 3(x^2 + 2x + 1) - 8 = 3x^2 + 6x + 3 - 8$

$y = 3x^2 + 6x - 5$

23. $y = x^2 + 4x - 1$

$y + 1 = x^2 + 4x$

$y + 1 + 4 = x^2 + 4x + 4$

$y + 5 = (x + 2)^2$

$y = (x + 2)^2 - 5$

24. $x^2 + 5x - 6 \leq 0$

$(x + 6)(x - 1) \leq 0$

$x = -6$ or $x = 1$

Graph opens upward.

$x^2 + 5x - 6 \leq 0$ for $-6 \leq x \leq 1$

25. $x^2 + 5x - 6 > 0$

$(x + 6)(x - 1) > 0$

$x = -6$ or $x = 1$

Graph opens upward.

$x^2 + 5x - 6 > 0$ for $x < -6$ or $x > 1$

26. $x^2 + x < x + 1$

$x^2 - 1 < 0$

$(x + 1)(x - 1) < 0$

$x = \pm 1$

$x^2 + x < x + 1$ for $-1 < x < 1$

27. $4 - x^2 < x^2 + 6$

$-2 < 2x^2$

$-1 < x^2$

True for all x,

$4 - x^2 < x^2 + 6$ for

All real numbers.

28. $5x^2 + 15x - 16 \geq 4 - 4x - x^2$

$6x^2 + 19x - 20 \geq 0$

$(6x - 5)(x + 4) \geq 0$

$x = \frac{5}{6}$ or $x = -4$

$5x^2 + 15x - 16 \geq 4 - 4x - x^2$ for

$x \leq -4$ or $x \geq \frac{5}{6}$

29. $v = \sqrt{2gs}$

$v^2 = 2gs$

$s = \dfrac{v^2}{2g}$

30. Let x = steps for one side of the block.

$2x^2 = 212^2$

$x^2 = \dfrac{44{,}944}{2} = 22{,}472$

$x = \sqrt{22{,}472}$; $x \approx 150$ steps for one side

≈ 300 steps to go around 2 sides

31. $x = \sqrt{4}$; $x = 2$ (principal square root)

$\sqrt{x^2} = 2$; $|x| = 2$; $x = \pm 2$; $\sqrt{w^2} = |w|$; if $|w| = z$ then $w = \pm z$

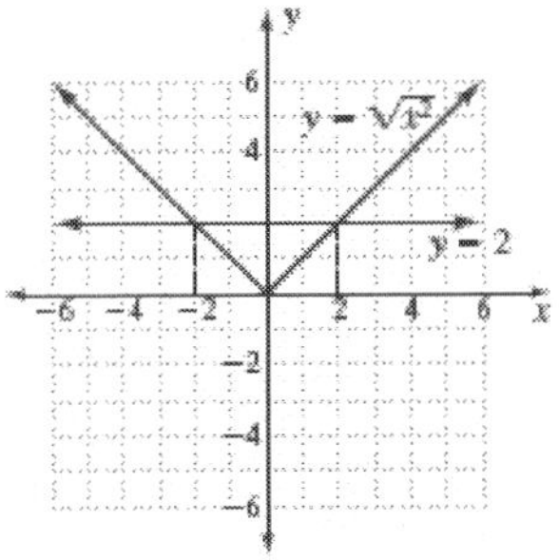

32. a.

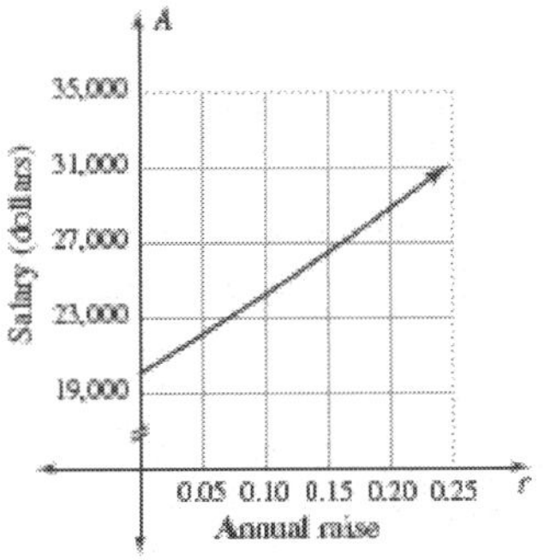

r	A = 20,000(1 + r)^2
0	$20{,}000(1+0)^2 = 20{,}000$
0.02	$20{,}000(1+0.02)^2 = 20{,}808$
0.04	$20{,}000(1+0.04)^2 = 21{,}632$
0.06	$20{,}000(1+0.06)^2 = 22{,}472$
0.07	$20{,}000(1+0.07)^2 = 22{,}898$
0.08	$20{,}000(1+0.08)^2 = 23{,}328$
0.10	$20{,}000(1+0.10)^2 = 24{,}200$

b. 7%

c. $\approx 22.5\%$

33. a. $x = \dfrac{-6}{2(-4.905)} \approx 0.612$; $y = -4.905(0.612)^2 + 6(0.612) + 3 \approx 4.83$

(0.61 sec, 4.83 m)

b. $0 = -4.905t^2 + 6t + 3$;

$$t = \frac{-6 \pm \sqrt{6^2 - 4(-4.905)(3)}}{2(-4.905)} = \frac{-6 \pm \sqrt{94.86}}{-9.81} \approx \frac{-6 \pm 9.74}{-9.81}$$

$t = -0.38$ (discard negative) or $t = 1.6$

1.6 sec. to hit water.

34. a. $y = lw$

b. $2w + l = 15$; $l = 15 - 2w$; $y = (15 - 2w)w$; $y = 15w - 2w^2$

c. $w = \dfrac{-15}{2(-2)} = \dfrac{15}{4}$, $y = 15(\frac{15}{4}) - 2(\frac{15}{4})^2$; $y = \frac{225}{4} - \frac{225}{8} = \frac{450}{8} - \frac{225}{8} = \frac{225}{8}$

Vertex: $\left(\frac{15}{4}, \frac{225}{8}\right)$; area, y, is largest at the peak of the graph.

Cumulative Review: Chapters 0 to 3

1. $8(3\frac{1}{4}) = 8(3) + 8(\frac{1}{4}) = 24 + 2 = 26$

2. $(25\%)(6)(8)(4) = (25\%)(4)(6)(8) = (1)(6)(8) = 48$

3. $5 - 3(x - 2) = 3(x - 1)$

$5 - 3x + 6 = 3x - 3$

$11 - 3x = 3x - 3$

$11 + 3 = 3x + 3x$

$14 = 6x$

$x = \frac{14}{6} = \frac{7}{3}$

4. $3x - 2y = 6$

$3x - 6 = 2y$

$y = 1.5x - 3$; slope = m = 1.5

Lines are parallel; slopes equal and given (0, 2); y-intercept = (0, 2).

Equation of line that passes through (0, 2) and parallel to given equation is $y = 1.5x + 2$.

5. a. $\sqrt{(-2)^2} = \sqrt{4} = 2$

b. $\sqrt{-2^2} = \sqrt{-4} = i\sqrt{4} = 2i$

c. $\sqrt{-2} = i\sqrt{2}$

d. $(\sqrt{-2})^2 = -2$

6. a. $\sqrt{108} = \sqrt{3 \cdot 36} = 6\sqrt{3}$

b. $\sqrt{24} = \sqrt{4 \cdot 6} = 2\sqrt{6}$

7. $(3 - i)(3 + i) = 9 - i^2 = 9 - (-1) = 9 + 1 = 10$

8. a. $|-1 - 4| = |-5| = 5$

b. $|(a + b) - 4| = |a + b - 4|$

9. a. $(x + 1)^2 = x^2 + 2x + 1$

b. $(a + b)^2 = a^2 + 2ab + b^2$

10. Does $4.5^2 + 20^2 = 20.5^2$? $20.25 + 400 = 420.25$; since they are equal these sides would form a right triangle.

11. a.

y	1st diff	2nd diff
3		
	−3	
0		4
	1	
1		4
	5	
6		4
	9	
15		

Next number: 9 + 4 + 15 = 28.
Quadratic
2a = 4; a = 2; 3(2) + b = −3
b = −9; c − 3 − 4 = 3; c = 10
$y = 2x^2 - 9x + 10$

b.

y	1st diff	2nd diff
8		
	3	
11		0
	3	
14		0
	3	
17		0
	3	
20		

Next number: 20 + 3 = 23.
Linear
m = 3; b = 8 − 3 = 5
y = 3x + 5

c.

y	1st diff	2nd diff
3		
	1	
4		2
	3	
7		1
	4	
11		3
	7	
18		

Each number is the sum of the two previous numbers. Next number is 11 + 18 = 29.
Sequence is neither quadratic nor linear.

12. $\frac{x-6}{2} < x+1$

$x - 6 < 2x + 2$

$-8 < x$

$x > -8$

13. $x^2 - 2x - 3 \geq -3$

$x^2 - 2x \geq 0$

$x(x - 2) \geq 0$

$x \leq 0$ or $x \geq 2$

14. $x^2 - 1 > 1 - x^2$

$2x^2 - 2 > 0$

$2(x^2 - 1) > 0$

$(x - 1)(x + 1) > 0$

$x < -1$ or $x > 1$

15. $x^2 - 4x + 4 \geq 8 - x$

$x^2 - 3x - 4 \geq 0$

$(x - 4)(x + 1) \geq 0$

$x \leq -1$ or $x \geq 4$

16. $|x - 5| = 8$

$x - 5 = 8$ or $-(x - 5) = 8$

$x = 13$ or $-x + 5 = 8$

$x = -3$

17. $|x + 6| < 5$

$x + 6 < 5$ and $-(x + 6) < 5$

$x < -1$ and $-x - 6 < 5$

$-6 - 5 < x$

$-11 < x$

$-11 < x < -1$

18. $x + 2y = 6;\ x = 6 - 2y$

$3x - 4y = -32$

$3(6 - 2y) - 4y = -32$

$18 - 6y - 4y = -32$

$18 - 10y = -32$

$50 = 10y;$ $\mathbf{y = 5}$

$x = 6 - 2(5)$

$\mathbf{x = -4}$

19. $2a - 3b = -8;\ -2(2a - 3b) = -2(-8)$

$-4a + 6b = 16$

$4a + 5b = 17$

$0 + 11b = 33$

$\mathbf{b = 3}$

$2a - 3(3) = -8$

$2a = 1;$

$\mathbf{a = \frac{1}{2}}$

20. $A + B = 180°;\ A - B = 25°;$ (add equations) $2A = 205°;$ $\mathbf{A = 102.5°}$

$102.5 + B = 180;$ $\mathbf{B = 77.5°}$

21.

\$	Rate	QV
450	0.22	450(0.22) = 99
1200	x	1200x
1650	0.10	1650(0.10) = 165

$1200x + 99 = 165$; $1200x = 66$; $x = 0.055$

Interest rate = 5.5%

22. $y = a(x-2)^2 + 4$

$2 = a(-3-2)^2 + 4$

$-2 = 25a$; $a = -\frac{2}{25}$

$y = -\frac{2}{25}(x-2)^2 + 4$ or

$y = -\frac{2}{25}(x^2 - 4x + 4) + 4$

$y = -\frac{2}{25}x^2 + \frac{8}{25}x - \frac{8}{25} + 4$

$y = -\frac{2}{25}x^2 + \frac{8}{25}x + \frac{92}{25}$

23. $x^2 + 3x + \underline{\left(\frac{9}{4}\right)} = \left(x + \frac{3}{2}\right)^2$

24. $h = -\frac{1}{2}gt^2$

$\frac{-2h}{g} = t^2$

$t = \sqrt{\frac{-2h}{g}}$

g must be a negative number because you cannot take a square root of a negative number.

25. $d = \sqrt{\frac{4A}{\pi}}$

$d^2 = \frac{4A}{\pi}$

$A = \frac{\pi d^2}{4}$

CHAPTER 4

Ratios, Variation, and Rational Functions

Exercises 4.1: Ratios, Units of Measure, and Proportions

1. $\dfrac{186\,\text{cm}^2}{6\,\text{cm}} = \dfrac{6\,\text{cm}\cdot 31\,\text{cm}}{6\,\text{cm}} = 31\,\text{cm}$

3. $\dfrac{1500\text{ft}\cdot\text{lbs}}{25\,\text{ft}} = \dfrac{25\,\text{ft}\cdot 60\,\text{lbs}}{25\,\text{ft}} = 60\,\text{lbs}$

5. $\dfrac{1200\,\text{kW hr}}{24\,\text{hr}} = \dfrac{24\,\text{hr}\cdot 50\,\text{kW}}{24\,\text{hr}} = 50\,\text{kW}$

7. a. $\dfrac{\frac{1}{2}\,\text{ft}}{2\,\text{in}}\cdot\dfrac{12\,\text{in}}{1\,\text{ft}} = \dfrac{6}{2} = \dfrac{3}{1}$

b. $\dfrac{3000\,\text{g}}{6\,\text{kg}}\cdot\dfrac{1\,\text{kg}}{1000\,\text{g}} = \dfrac{3000}{6000} = \dfrac{1}{2}$

c. $\dfrac{32\,\text{oz}}{6\,\text{lb}}\cdot\dfrac{1\,\text{lb}}{16\,\text{oz}} = \dfrac{32}{96} = \dfrac{1}{3}$

9. $300\,\text{mL}\cdot\dfrac{1\,\text{L}}{1000\,\text{mL}} = \dfrac{300\,\text{L}}{1000} = 0.3\,\text{L}$

11. $150\,\text{ft}^3\cdot\left(\dfrac{1\,\text{yd}}{3\,\text{ft}}\right)^3 = \dfrac{150\,\text{ft}^3\cdot 1\,\text{yd}^3}{9\,\text{ft}^3} \approx 5.6\,\text{yd}^3$

13. $100\,\text{in}^3\cdot\left(\dfrac{1\,\text{ft}}{12\,\text{in}}\right)^3 = \dfrac{100\,\text{in}^3\cdot 1\,\text{ft}^3}{1728\,\text{in}^3} \approx 0.058\,\text{ft}^3$

15. $\dfrac{\$9.95}{0.18\text{ fl oz}}\cdot\dfrac{8\text{ fl oz}}{1\text{c}}\cdot\dfrac{4\text{c}}{1\text{qt}}\cdot\dfrac{4\text{qt}}{1\text{gal}} = \dfrac{\$1{,}273.6}{0.18\,\text{gal}} \approx \$7{,}075.56\text{ per gallon}$

17. $\dfrac{1\,\text{gal}}{5\,\text{mi}}\cdot\dfrac{55\,\text{mi}}{1\,\text{hr}}\cdot\dfrac{\$3.35}{1\,\text{gal}}\cdot\dfrac{10\,\text{hr}}{1\,\text{day}} = \dfrac{55\cdot\$3.35\cdot 10}{5\,\text{day}} = \$368.50/\text{day}$

19. $1\,\text{yr}\cdot\dfrac{12\,\text{mo}}{1\,\text{yr}}\cdot\dfrac{\text{adult}}{150\,\text{mo}}\cdot\dfrac{500\,\text{mg}}{\text{adult}} = \dfrac{12\cdot 500\,\text{mg}}{150} = 40\,\text{mg}$

21. Minimum: 11 to 1.67, or approximately 6.6 to 1
Maximum: 15 to 1.67, or approximately 9 to 1

23. a. $14x \cdot \frac{3}{x} = \frac{5}{14} \cdot 14x$

$14 \cdot 3 = 5x$

$x = 8.4$

b. $25 \cdot 15 \cdot \frac{4}{25} = \frac{x}{15} \cdot 25 \cdot 15$

$15 \cdot 4 = 25x$

$x = 2.4$

c. $24x \cdot \frac{7}{24} = \frac{16}{x} \cdot 24x$

$7x = 16 \cdot 24$

$x \approx 54.9$

25. $\frac{1}{12} = \frac{15}{x}$

$12x \cdot \frac{1}{12} = \frac{15}{x} \cdot 12x$

$x = 15 \cdot 12 \text{ in}$

$x = 180 \text{ in} \cdot \frac{1 \text{ ft}}{12 \text{ in}} = 15 \text{ ft}$

27. $\frac{a}{12} = \frac{5}{9}$

$9a = 60$

$a = \frac{60}{9} = 6\frac{2}{3}$

$\frac{b}{9} = \frac{8}{12}$

$12b = 72$

$b = 6$

29. a. $\frac{d}{6} = \frac{4\sqrt{3}}{4}$

$d = 6\sqrt{3}$

$f = \sqrt{(6\sqrt{3})^2 + 6^2}$

$f = 12$

b. $\frac{e}{10} = \frac{4}{4\sqrt{3}}$

$e = \frac{10\sqrt{3}}{3}$

$f = \sqrt{10^2 + \left(\frac{10\sqrt{3}}{3}\right)^2}$

$f = \sqrt{\frac{400}{3}} = \frac{20}{\sqrt{3}} = \frac{20\sqrt{3}}{3}$

c. $\frac{d}{\sqrt{5}} = \frac{4\sqrt{3}}{4}$

$d = \sqrt{15}$

$f = \sqrt{(\sqrt{15})^2 + (\sqrt{5})^2}$

$f = \sqrt{20} = 2\sqrt{5}$

31. $n^2 = 2^2 + 2^2$

$n = \sqrt{8} = 2\sqrt{2}$

$\frac{x}{2} = \frac{y}{2} = \frac{8}{n}$

$x = y = \frac{16}{2\sqrt{2}} = \frac{8\sqrt{2}}{\sqrt{2}\cdot\sqrt{2}} = \frac{8\sqrt{2}}{2} = 4\sqrt{2}$

33. Total length of $b_1 = 18 + b_2$

$\frac{25}{18+b_2} = \frac{5}{b_2}$

$25b_2 = 5(18 + b_2)$

$25b_2 = 90 + 5b_2$

$20b_2 = 90$

$b_2 = 4.5$ ft

35. Angle is $180 - 90^\circ - 45^\circ = 45^\circ$

Since the angles are the same, this

is an isosceles triangle. So the two sides are the same length.

$\sin 45^\circ = \frac{x}{30}$

$\frac{\sqrt{2}}{2} = \frac{x}{30}$

$\frac{\sqrt{2}}{2}\cdot 30 = \frac{x}{30}\cdot 30$

$15\sqrt{2} = x$

$x \approx 21.21$ in

37. Angle is $180 - 90^\circ - 40^\circ = 50^\circ$

$\cos 40^\circ = \frac{\sqrt{3}}{c}$

$c \cdot \cos 40^\circ = c \cdot \frac{\sqrt{3}}{c}$

$\frac{c \cdot \cos 40^\circ}{\cos 40^\circ} = \frac{\sqrt{3}}{\cos 40^\circ}$

$c = \frac{\sqrt{3}}{\cos 40^\circ}$

$c \approx 2.26$ ft

$\tan 40^\circ = \frac{a}{\sqrt{3}}$

$\sqrt{3}\cdot \tan 40^\circ = \sqrt{3}\cdot\frac{a}{\sqrt{3}}$

$\sqrt{3}\cdot \tan 40^\circ = a$

$a \approx 1.45$ ft

Exercises 4.2: Direct Variation

1. a. x = weight in lb.; y = distance stretched in inches; $m = \dfrac{6-4}{15-10} = \dfrac{2}{5}$;

$y = \frac{2}{5}x + b$; $4 = \frac{2}{5}(10) + b$; $4 = 4 + b$; $b = 0$; $y = \frac{2}{5}x$

b. Relationship is a direct variation. **c.** Intercept is zero.

3. a. x = number of credit hours; y = cost in \$; $m = \dfrac{345-185}{4-2} = 80$;

$y = 80x + b$; $185 = 80(2) + b$; $185 = 160 + b$; $b = 25$; $y = 80x + 25$

b. Relationship is not proportional. **c.** b = \$25 student fees

5. a. x = purchase price in \$; y = tax in \$; $m = \dfrac{2.25-1.20}{30-16} = 0.075$;

$y = 0.075x + b$; $1.20 = 0.075(16) + b$; $1.20 = 1.20 + b$; $b = 0$; $y = 0.075x$

b. Relationship is a direct variation. **c.** Intercept is zero.

7. a. $A = \frac{1}{2}bh$; $k = \frac{1}{2}$

b. $A = lw$; $k = 1$

c. $V = \frac{1}{3}\pi r^2 h$; $k = \frac{1}{3}\pi$

9. $C = kd$; $k = \pi$

11. $A = kr^2$; $k = \pi$

13. $A = klw$; $k = 1$

15. $V = khr^2$; $k = \pi$

17. $V = \frac{1}{3}bh$; $b \approx 53{,}095\ m^2$; $h \approx 146.6$ m; $V = \frac{1}{3}(53{,}095)(146.6)\ m^3$; $V \approx 2{,}594{,}576\ m^3$

19. $L = ks^2$; $\dfrac{k40^2}{k25^2} = \dfrac{1600}{625} = 2.56$. The length of the 40-mph skid mark will be 2.56 times the length of the 25 mph skid mark.

21. $E = kh^2$; $\dfrac{k10^2}{k2^2} = \dfrac{100}{4} = 25$. The 10-foot wave has 25 times the energy of the 2-foot wave.

23. a. Comparing ratios of $\frac{w}{p}$ and $\frac{w}{p^2}$ shows that $\frac{12}{4}=3$ is not the same as $\frac{75}{10}=7.5$ whereas $\frac{12}{4^2}=\frac{75}{10^2}=\frac{3}{4}$ so the variation is quadratic.

b. $k=\frac{3}{4}$, so $w=\frac{3}{4}p^2$

c.

p	W
Ans. 2	3
4	12
7	**Ans.** $\frac{147}{4}$
10	75
14	**Ans. 147**
Ans. 18	243

25. Let t be the time to get a sun burn and S be the SPF rating of the sun screen lotion.

$$t = kS \qquad t = 3.75S$$

$$30 = k \cdot 8 \qquad 120 = 3.75S$$

$$\frac{30}{8} = \frac{k \cdot 8}{8} \qquad \frac{120}{3.75} = \frac{3.75S}{3.75}$$

$$3.75 = k \qquad 32 = S$$

So the SPF32 would be needed.

27. a. $b=0$

b. $c=0$

c. $y=ax^2$

d. $\frac{y}{x^2}=a$

e. a

Exercises 4.3: Inverse Variation and Related Graphs

1. x = no. of problems, y = no. of points per problem; k = 100 points; $y = \frac{100}{x}$

3. x = copper usage in metric tons/yr; y = time in yrs; k = 940,000,000 metric tons;
$y = \frac{940,000,000}{x}$

5. $160 \cdot 35 = 5600$; $200 \cdot 28 = 5600$; inverse variation; $k = 5600 \frac{\text{tons}}{\text{day}} \cdot \text{years}$

7. $15 \cdot 6 = 90$; $25 \cdot 3.6 = 90$; inverse variation; $k = 90$ mi

9. $2 \cdot \$2200 = \$4400$; $3 \cdot \$3300 = \$9900$; not inverse variation
$\$2200 \div 2 = \$1100$; $\$3300 \div 3 = \$1100$; direct variation; $k = \$1100$ per semester

11. $35 \cdot 6 = 4x$
x = 52.5 lb
k = 35 lb · 6 ft = 210 lb·ft

13. $80 \cdot 0.75 = x \cdot 1$
x = 60 people
k = 60 people · 1 cup/person = 60 cups

15. $4 \cdot 900 = 3x$
x = 1200 ft/gal
k = 4 in · 900 ft/gal
k = 3600 in·ft/gal · 1ft/12 in
k = 300 ft^2/gal

17.

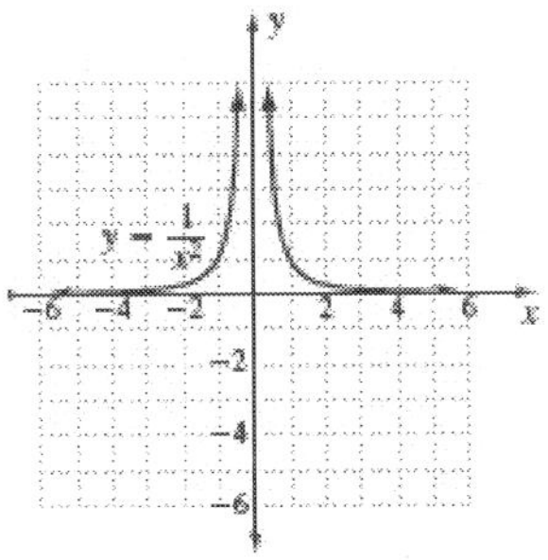

a. *f*(0) is undefined; division by zero.
b. x = 0
c. The graph approaches the x-axis.

19. $y_1 = \frac{k}{x_1}$; $y_2 = \frac{k}{x_2}$; solve 2nd equation for k and substitute into the 1st;

$k = x_2y_2$; $y_1 = \frac{x_2y_2}{x_1}$; multiply both sides by x_1; $x_1y_1 = x_2y_2$.

21. Multiply both sides by x_1x_2; $x_1x_2\frac{y_1}{x_1} = \frac{y_2}{x_2}x_1x_2$; $x_2y_1 = x_1y_2$; the subscripts are not alike on each side of the equation.

23. No, if there were steps, the equations would be the same variation. The first is inverse variation; the second is direct variation.

25. a. (10 lbs, 3 inches), (x lbs, 1 inch) **b.** $10 \cdot 3 = 1x$; $x = 30$ lbs

c. k = 10 lb. · 3 in. = 30 lb·in

27. **b.** $H = \frac{M}{r^3}$; distance cubed term is in the denominator.

29. $\frac{\frac{k}{2^2}}{\frac{k}{6^2}} = \frac{36}{4} = 9$ times greater

31. a. $T_1T_2\frac{V_1}{T_1} = \frac{V_2}{T_2}T_1T_2$; $T_2V_1 = T_1V_2$; direct; equal ratios

b. $C_1C_2\frac{V_1}{C_2} = \frac{V_2}{C_1}C_1C_2$; $C_1V_1 = C_2V_2$; inverse; equal products

c. $V_1V_2\frac{P_1}{V_2} = \frac{P_2}{V_1}V_1V_2$; $P_1V_1 = P_2V_2$; inverse; equal products

d. $T_1T_2\frac{P_1V_1}{T_1} = \frac{P_2V_2}{T_2}T_1T_2$; $P_1T_2V_1 = P_2T_1V_2$; P and V inverse because have equal products;

P and T direct because they have equal ratios.

e. $\frac{C_1V_1}{V_1V_2} = \frac{C_2V_2}{V_1V_2}$; $\frac{C_1}{V_2} = \frac{C_2}{V_1}$; inverse

f. $\frac{V_1T_2}{V_1V_2} = \frac{V_2T_1}{V_1V_2}$; $\frac{T_2}{V_2} = \frac{T_1}{V_1}$; direct

g. $\frac{P_1V_1}{P_1P_2} = \frac{P_2V_2}{P_1P_2}$; $\frac{V_1}{P_2} = \frac{V_2}{P_1}$; inverse

Exercises 4.4: Simplification, Multiplication, and Division of Rational Expressions

1. Undefined when $x + 3 = 0$ or $x = -3$

3. Undefined when $4 - a = 0$ or $a = 4$

5. Undefined when $x^2 - 3x + 28 = 0$

$(b^2 - 4ac) = (-3)^2 - 4(1)(28) = -103$

No real roots; defined for all $\Re$.

7. **a.** $x - 3, x \neq 3;\ \dfrac{x-3}{x-3} = 1$

b. $-2 - x, x \neq -2;\ \dfrac{x+2}{-2-x} = -1$

c. $b - a, b \neq a;\ \dfrac{b-a}{b-a} = 1$

d. $x - 3, x \neq 3;\ \dfrac{3-x}{x-3} = -1$

9. **a.** $\dfrac{16a^2}{10a} = \dfrac{2\cdot 8\cdot a\cdot a}{2\cdot 5\cdot a} = \dfrac{8a}{5},\ a \neq 0$

b. $\dfrac{2x^2y}{10xy^2} = \dfrac{2\cdot x\cdot x\cdot y}{2\cdot 5\cdot x\cdot y\cdot y} = \dfrac{x}{5y},\ x \neq 0, y \neq 0$

c. $\dfrac{-49x^3y^5}{63x^2y^2} = \dfrac{-7\cdot 7\cdot x^2\cdot x\cdot y^2\cdot y^3}{7\cdot 9\cdot x^2\cdot y^2} = -\dfrac{7xy^3}{9},\ x \neq 0, y \neq 0$

d. $\dfrac{64a^3b^2c}{-48a^5b^3} = \dfrac{2\cdot 4\cdot 8\cdot a^3\cdot b^2\cdot c}{-3\cdot 2\cdot 8\cdot a^3\cdot a^2\cdot b^2\cdot b} = -\dfrac{4c}{3a^2b};\ a \neq 0, b \neq 0$

11. **a.** $\dfrac{xy}{2x+y}$; simplified; $y \neq -2x$

b. $\dfrac{xy}{2x+xy} = \dfrac{xy}{x(2+y)} = \dfrac{y}{x+y};\ x \neq 0, y \neq -2$

c. $\dfrac{xy - x^2y - x^3}{xy} = \dfrac{x\left(y - xy - x^2\right)}{xy} = \dfrac{y - xy - x^2}{y},\ x \neq 0, y \neq 0$

d. $\dfrac{x^3y - x^2y^2 + xy^3}{xy} = \dfrac{xy\left(x^2 - xy + y^2\right)}{xy} = x^2 - xy + y^2,\ x \neq 0, y \neq 0$

13. **a.** $\dfrac{3-x}{(x+3)(x-3)} = \dfrac{-(x-3)}{(x+3)(x-3)} = -\dfrac{1}{x+3};\ x \neq \pm 3$

b. $\dfrac{x^2-9}{x^2+5x+6} = \dfrac{(x+3)(x-3)}{(x+3)(x+2)} = \dfrac{x-3}{x+2};\ x \neq -3, -2$

15. a. $\dfrac{2ac+4bc}{4ad+8bd}=\dfrac{2c(a+2b)}{4d(a+2b)}=\dfrac{c}{2d}$; $a\neq -2b,\ d\neq 0$;

b. $\dfrac{6x^2+3x}{12x^2-6x}=\dfrac{3x(2x+1)}{6x(2x-1)}=\dfrac{2x+1}{2(2x-1)}$; $x\neq 0,\ \frac{1}{2}$

17. a. $\dfrac{x^2+x-6}{2-x}=\dfrac{(x+3)(x-2)}{-(x-2)}=-(x+3)$; $x\neq 2$

b. $\dfrac{x-3}{6-2x}=\dfrac{x-3}{-2(x-3)}=-\dfrac{1}{2}$; $x\neq 3$

19. a. $\dfrac{1}{x}\cdot\dfrac{x^2}{1}=\dfrac{x\cdot x}{x}=x$

b. $\dfrac{1}{a}\div\dfrac{a^2b^2}{1}=\dfrac{1}{a}\cdot\dfrac{1}{a^2b^2}=\dfrac{1}{a^3b^2}$

c. $\dfrac{a}{b}\div\dfrac{a^2b}{b^2}=\dfrac{a}{b}\cdot\dfrac{b^2}{a^2b}=\dfrac{ab^2}{a^2b^2}=\dfrac{1}{a}$

d. $\dfrac{18}{cd}\div\dfrac{4d^2}{9c}=\dfrac{18}{cd}\cdot\dfrac{9c}{4d^2}=\dfrac{9\cdot 9}{2d^3}=\dfrac{81}{2d^3}$

21. a. $\dfrac{a^2+7a+12}{a^2-4}\div\dfrac{a^2+4a}{a-2}=\dfrac{(a+4)(a+3)}{(a+2)(a-2)}\cdot\dfrac{(a-2)}{a(a+4)}=\dfrac{a+3}{a(a+2)}$

b. $\dfrac{x^2-2x}{x}\cdot\dfrac{x^2-1}{x^2-3x+2}=\dfrac{x(x-2)}{x}\cdot\dfrac{(x+1)(x-1)}{(x-2)(x-1)}=x+1$

23. a. $\dfrac{x^2+3x}{x}\cdot\dfrac{x^2-x-6}{x^2-9}=\dfrac{x(x+3)}{x}\cdot\dfrac{(x-3)(x+2)}{(x+3)(x-3)}=x+2$

b. $\dfrac{4a^2+4a+1}{4-9a^2}\div\dfrac{4a^2+2a}{3a-2}=\dfrac{(2a+1)(2a+1)}{(2-3a)(2+3a)}\cdot\dfrac{(3a-2)}{2a(2a+1)}=\dfrac{-(2a+1)}{2a(2+3a)}=\dfrac{-2a-1}{2a(2+3a)}$

25. a. $\dfrac{x+2}{x^2-4x+4}\cdot\dfrac{x^2-2x}{x+2}=\dfrac{(x+2)}{(x-2)(x-2)}\cdot\dfrac{x(x-2)}{(x+2)}=\dfrac{x}{x-2}$

b. $\dfrac{a^2-5a}{a^2+5a}\div\dfrac{a^2-10a+25}{a}=\dfrac{a(a-5)}{a(a+5)}\cdot\dfrac{a}{(a-5)(a-5)}=\dfrac{a}{(a+5)(a-5)}$

27. a. $\dfrac{a^2-b^2}{a^2+2ab+b^2}\cdot\dfrac{a+b}{a^3-b^3}=\dfrac{(a+b)(a-b)}{(a+b)(a+b)}\cdot\dfrac{(a+b)}{(a-b)(a^2+ab+b^2)}=\dfrac{1}{a^2+ab+b^2}$

b. $\dfrac{x^2-2xy+y^2}{x^2-y^2}\cdot\dfrac{x-y}{y-x}=\dfrac{(x-y)(x-y)}{(x+y)(x-y)}\cdot\dfrac{(x-y)}{-(x-y)}=\dfrac{-(x-y)}{x+y}=\dfrac{y-x}{x+y}$

29. a. $\frac{a^3-b^3}{a+b}\cdot\frac{a^2+2ab+b^2}{a^2-b^2}=\frac{(a-b)(a^2+ab+b^2)}{(a+b)}\cdot\frac{(a+b)(a+b)}{(a+b)(a-b)}=a^2+ab+b^2$

b. $\frac{x^2+2xy+y^2}{x^3+y^3}\cdot\frac{x^2-y^2}{x+y}=\frac{(x+y)(x+y)}{(x+y)(x^2-xy+y^2)}\cdot\frac{(x+y)(x-y)}{(x+y)}=\frac{(x+y)(x-y)}{x^2-xy+y^2}$

31. a. $\frac{6+x-2x^2}{9-4x^2}\cdot\frac{15-7x-2x^2}{x^2-2x}=\frac{(2-x)(3+2x)}{(3-2x)(3+2x)}\cdot\frac{(5+x)(3-2x)}{x(x-2)}=-\frac{(5+x)}{x}$

b. $\frac{25-9x^2}{6x^2+7x-5}\div\frac{5+7x-6x^2}{2+2x-4x^2}=\frac{(5+3x)(5-3x)}{(3x+5)(2x-1)}\cdot\frac{-2(1+2x)(-1+x)}{(5-3x)(1+2x)}=\frac{-2(x-1)}{2x-1}$

33. a. $\frac{\frac{1}{a}}{\frac{1}{b}}=\frac{1}{a}\cdot\frac{b}{1}=\frac{b}{a}$

b. $\frac{\frac{1}{b}}{a}=\frac{1}{b}\cdot\frac{1}{a}=\frac{1}{ab}$

35. $\frac{\frac{x^2-3x+2}{x}}{\frac{x^2-1}{x^2}}=\frac{(x-2)(x-1)}{x}\cdot\frac{x^2}{(x+1)(x-1)}=\frac{x(x-2)}{x+1}$

37. $\frac{\frac{x-2}{x^2+3x-4}}{\frac{2-x}{x-1}}=\frac{(x-2)}{(x+4)(x-1)}\cdot\frac{(x-1)}{-(x-2)}=\frac{-1}{x+4}$

39. $\frac{5280\text{ ft}}{88\ \frac{\text{ft}}{\text{sec}}}=60\frac{\text{ft}}{1}\cdot\frac{\text{sec}}{\text{ft}}=60\ sec$

41. $\frac{65\ \frac{\text{mi}}{\text{hr}}}{15\ \frac{\text{mi}}{\text{gal}}}=\frac{13}{3}\cdot\frac{\text{mi}}{\text{hr}}\cdot\frac{\text{gal}}{\text{mi}}=4\tfrac{1}{3}\ \text{gal}/\text{hr}$

43. $\frac{24\ \frac{\text{cans}}{\text{case}}}{3.98\ \frac{\$}{\text{case}}}\approx 6\cdot\frac{\text{cans}}{\text{case}}\cdot\frac{\text{case}}{\$}\approx 6\ \frac{\text{cans}}{\$}$

45. $\frac{\frac{8\text{ stitches}}{1\text{ inch}}}{\frac{1\text{ foot}}{12\text{ inches}}}=\frac{8\text{ stitches}}{1\text{ inch}}\cdot\frac{12\text{ inches}}{1\text{ foot}}=\frac{96\text{ stitches}}{\text{foot}}$

47. $\frac{95\ \frac{\text{words}}{\text{min}}}{300\ \frac{\text{words}}{\text{page}}}\approx 0.3\ \frac{\text{words}}{\text{min}}\cdot\frac{\text{page}}{\text{words}}\approx 0.3\ \text{page}/\text{min}$

49. Use division to remove common unit of gallons: $\frac{3.50\ \frac{\$}{\text{gal}}}{7\ \frac{\text{mi}}{\text{gal}}}=0.50\frac{\$}{\text{gal}}\cdot\frac{\text{gal}}{\text{mi}}=\$0.50/\text{mi}$

or $\frac{7\ \frac{\text{mi}}{\text{gal}}}{1.40\ \frac{\$}{\text{gal}}}=5\frac{\text{mi}}{\text{gal}}\cdot\frac{\text{gal}}{\$}=5\text{ mi}/\$$

51. Use multiplication to remove the common unit of days: $\frac{4\text{ days}}{1\text{ week}} \cdot \frac{2\text{ hrs}}{1\text{ day}} = 8\text{ hrs}/\text{week}$

53. exercise 13a.

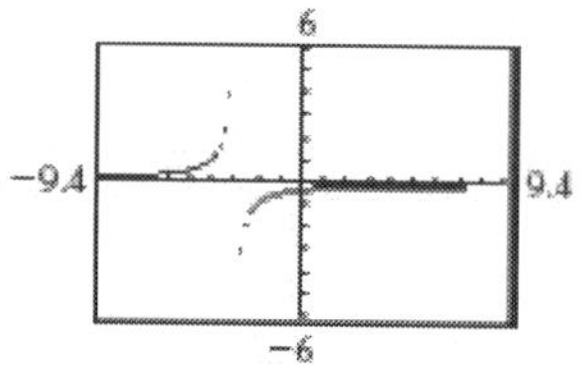

a, d. hole at x = 3 before simplifying

b. not linear

c. nearly vertical around x = −3

exercise 13b

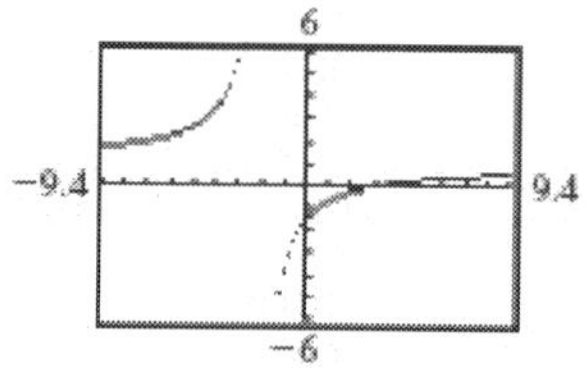

a, d. hole at x = −3 before simplifying

b. not linear

c. nearly vertical around x = −2

exercise 17a

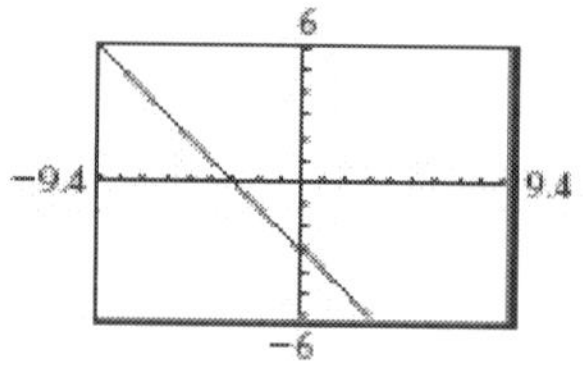

a., d. hole at x = 2 before simplifying

b. linear; m = −1

c. no

Exercises 4.5: Division of Polynomials and Related Graphs

1. Comparing the degree of the numerator and the degree of the denominator, we would predict the outcome to be of degree 2, or quadratic.

$$\begin{array}{r} x^2+2x+4 \\ (x-2)\overline{\big)\,x^3+0x^2+0x-8} \\ \underline{-(x^3-2x^2)} \\ 2x^2+0x \\ \underline{-(2x^2-4x)} \\ 4x-8 \\ \underline{-(4x-8)} \\ 0 \end{array}$$

3. Comparing the degree of the numerator and the degree of the denominator, we would predict the outcome to be of degree 1, or linear.

$$\begin{array}{r} 2x-5 \\ (6x+5)\overline{\big)\,12x^2-20x-25} \\ \underline{-(12x^2+10x)} \\ -30x-25 \\ \underline{-(-30x-25)} \\ 0 \end{array}$$

5. Comparing degrees we predict the outcome will be degree 3, or cubic. Dividing the answer would give an outcome of degree 2, or quadratic.

a.

$$\begin{array}{r} x^3+6x^2-x-6 \\ (x-6)\overline{\big)\,x^4+0x^3-37x^2+0x+36} \\ \underline{-(x^4-6x^3)} \\ +6x^3-37x^2 \\ \underline{-(6x^3-36x^2)} \\ -x^2+0x \\ \underline{-(-x^2+6x)} \\ -6x+36 \\ \underline{-(-6x+36)} \\ 0 \end{array}$$

b.

$$\begin{array}{r} x^2-1 \\ (x+6)\overline{\big)\,x^3+6x^2-x-6} \\ \underline{-(x^3+6x^2)} \\ 0-x-6 \\ \underline{-(-x-6)} \\ 0 \end{array}$$

c. $x^2-1=(x-1)(x+1)$

7. Defined for all $\Re$ except when $x-4=0$ or $x=4$.

$$\frac{x^2-3x-4}{x-4}=\frac{(x-4)(x+1)}{(x-4)}=x+1$$

9. Defined for all $\Re$ except when $x-3=0$ or $x=3$.

$$\begin{array}{r} x \\ (x-3)\overline{\big)\,x^2-3x-4} \\ \underline{-(x^2-3x)} \\ 0-4 \end{array} \quad ; \quad \frac{x^2-3x-4}{x-3}=x-\frac{4}{x-3}$$

11. Defined for all $\Re$ except when

$x-2=0$ or $x=2$.

$$\begin{array}{r} x-1 \\ (x-2)\overline{\big)\, x^2-3x-4} \\ \underline{-(x^2-2x)} \\ -x-4 \\ \underline{-(-x+2)} \\ -6 \end{array}$$

$$\frac{x^2-3x-4}{x-2}=x-1-\frac{6}{x-2}$$

13. a. $x^2-4x-12=(x-6)(x+2)$

b. zero

c. The graph will be a straight line with a hole in the graph at the location where the factor is equal to zero.

15. a. The graph will be curves approaching an asymptote at $x=1$ from both sides.

b.

$$\begin{array}{r} x-3 \\ (x-1)\overline{\big)\, x^2-4x-12} \\ \underline{-(x^2-x)} \\ -3x-12 \\ \underline{-(-3x+3)} \\ -15 \end{array}$$

$$\Rightarrow x-3-\frac{15}{x-1}$$

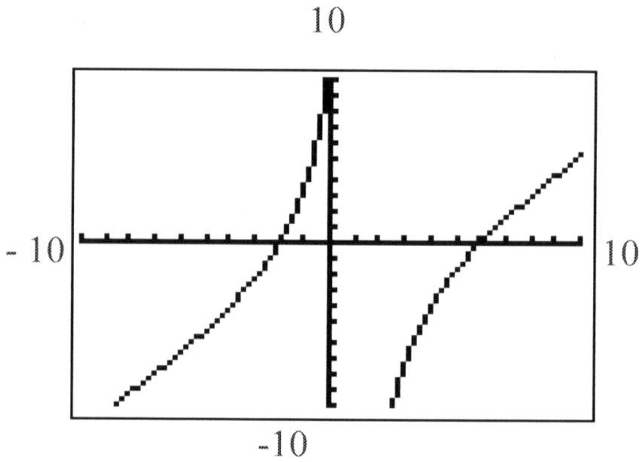

The graph will approach a line, $y=x-3$, at the ends of the graph but will be approaching an asymptote at $x=1$ from both sides.

17. $x^2-3x-4=(x-4)(x+1)$; $x^2+3x+2=(x+2)(x+1)$

a. Exercise 9; asymptote, $x=3$

X	Y1
2.7	16.033
2.8	22.8
2.9	42.9
3	ERR:
3.1	-36.9
3.2	-16.8
3.3	-10.03

X=3.3

b. Exercise 7; hole, (4, 5)

X	Y1
3.7	4.7
3.8	4.8
3.9	4.9
4	ERR:
4.1	5.1
4.2	5.2
4.3	5.3

X=4.3

19. a.

$$\begin{array}{r} x^2+2x+4 \\ (x-2)\overline{)x^3-0x^2-0x-1} \\ \underline{-(x^3-2x^2)} \\ 2x^2-0x \\ \underline{-(2x^2-4x)} \\ 4x-1 \\ \underline{-(4x-8)} \\ 7 \end{array}$$

$$\frac{x^3-1}{x-2}=x^2+2x+4+\frac{7}{x-2}$$

b. Asymptote is $x=2$.

21. a. Denominator is a factor of the numerator.

$$\begin{array}{r} x^2+x+1 \\ (x-1)\overline{)x^3-0x^2-0x-1} \\ \underline{-(x^3-x^2)} \\ x^2-0x \\ \underline{-(x^2-x)} \\ x-1 \\ \underline{-(x-1)} \\ 0 \end{array}$$

b. Hole at $(1, 3)$.

23. a. Denominator is a factor of the numerator.

$$\begin{array}{r} x^2-x+1 \\ (x+1)\overline{)x^3+0x^2+0x+1} \\ \underline{-(x^3+x^2)} \\ -x^2+0x \\ \underline{-(-x^2-x)} \\ x+1 \\ \underline{-(x+1)} \\ 0 \end{array}$$

b. Hole at $(-1, 3)$.

25. a.

$$\begin{array}{r} x^3-x^2+x-1 \\ (x+1)\overline{)x^4+0x^3+0x^2+0x+1} \\ \underline{-(x^4+x^3)} \\ -x^3+0x^2 \\ \underline{-(-x^3-x^2)} \\ x^2+0x \\ \underline{-(x^2+x)} \\ -x+1 \\ \underline{-(-x-1)} \\ 2 \end{array}$$

$$\frac{x^4+1}{x+1}=x^3-x^2+x-1+\frac{2}{x+1}$$

b. Asymptote is $x=-1$.

27. a.

$$\begin{array}{r} 3x^2-2x+2 \\ (x+1)\overline{)\,3x^3+x^2+0x-1} \\ -(3x^3+3x^2) \\ \hline -2x^2+0x \\ -(-2x^2-2x) \\ \hline 2x-1 \\ -(2x+2) \\ \hline -3 \end{array}$$

$$\frac{x^2+3x^3-1}{x+1}=3x^2-2x+2-\frac{3}{x+1}$$

b. Asymptote is $x=-1$.

29.

$$\begin{array}{r} x^2-xy+y^2 \\ (x+y)\overline{)\,x^3+0x^2y+0xy^2+y^3} \\ -(x^3+x^2y) \\ \hline -x^2y+0xy^2 \\ -(-x^2y-xy^2) \\ \hline xy^2+y^3 \\ -(xy^2+y^3) \\ \hline 0 \end{array}$$

31. True. If the divisor $(x-a)$ is a factor of the dividend, there is a hole in the graph at $x=a$.

33. a. Asymptote is $x=0$. **b.** Hole at $(-1,\,0)$.

Exercises 4.6: Addition and Subtraction of Rational Expressions

1. a. $\frac{3\cdot 7}{5\cdot 7}=\frac{21}{35}$ **b.** $\frac{5\cdot 6}{8\cdot 6}=\frac{30}{48}$ **c.** $\frac{8\cdot 16}{3\cdot 16}=\frac{128}{48}$

3. a. $\frac{3\cdot 5}{8\cdot 5}=\frac{15}{40}$ **b.** $\frac{5\cdot 9}{6\cdot 9}=\frac{45}{54}$ **c.** $\frac{5\cdot 8}{4\cdot 8}=\frac{40}{32}$

5. a. $\frac{4\cdot 2x}{xy\cdot 2x}=\frac{8x}{2x^2y}$ **b.** $\frac{3\cdot 2b}{a^2b\cdot 2b}=\frac{6b}{2a^2b^2}$

7. a. $\frac{x\cdot(x+1)}{(x-1)(x+1)}=\frac{x^2+x}{x^2-1}$ **b.** $\frac{a(a-b)}{(a+b)(a-b)}=\frac{a^2-ab}{a^2-b^2}$

9. a. $\frac{3}{4}-\frac{x}{4}=\frac{3-x}{4}$ **b.** $\frac{2}{5x}-\frac{7}{5x}=\frac{2-7}{5x}=\frac{-5}{5x}=-\frac{1}{x};\ x\neq 0$

c. $\frac{4}{x+2}-\frac{x^2}{x+2}=\frac{4-x^2}{x+2}=\frac{(2+x)(2-x)}{(x+2)}=2-x;\ x\neq -2$

d. $\frac{2x}{x^2-1}-\frac{x-1}{x^2-1}=\frac{2x-(x-1)}{x^2-1}=\frac{2x-x+1}{(x-1)(x+1)}=\frac{x+1}{(x-1)(x+1)}=\frac{1}{x-1};\ x\neq -1,\,1$

11. a. Common denominator is 12: $\frac{1\cdot 3}{4\cdot 3}+\frac{7}{12}=\frac{3+7}{12}=\frac{10}{12}=\frac{5\cdot 2}{6\cdot 2}=\frac{5}{6}$

b. Common denominator is 2a: $\frac{2\cdot 2}{a\cdot 2}-\frac{5}{2a}=\frac{4-5}{2a}=-\frac{1}{2a}$

c. Common denominator is a^2b^2: $\frac{b\cdot b^2}{a^2b^2}-\frac{c\cdot a}{ab^2\cdot a}=\frac{b^3-ac}{a^2b^2}$

d. $x^2-2x=x(x-2)$; $x^2-4=(x+2)(x-2)$; common denominator is $x(x+2)(x-2)$;

$$\frac{5(x+2)}{x(x-2)(x+2)}+\frac{3x}{x(x-2)(x+2)}=\frac{5x+10+3x}{x(x-2)(x+2)}=\frac{8x+10}{x(x-2)(x+2)}$$

e. $x^2-6x+9=(x-3)^2$; $x^2-9=(x-3)(x+3)$; common denominator is $(x+3)(x-3)^2$;

$$\frac{5x(x+3)}{(x-3)^2(x+3)}-\frac{3x(x-3)}{(x+3)(x-3)(x-3)}=\frac{5x^2+15x-3x^2+9x}{(x+3)(x-3)^2}=\frac{2x^2+24x}{(x+3)(x-3)^2}$$

13. $$\frac{5\cdot 3b}{4a\cdot 3b}+\frac{3\cdot 2a}{6b\cdot 2a}=\frac{15b+6a}{12ab}=\frac{3(2a+5b)}{3(4ab)}=\frac{2a+5b}{4ab}$$

15. $$\frac{x(x)}{(x-3)x}+\frac{1(x-3)}{x(x-3)}=\frac{x^2+x-3}{x(x-3)}$$

17. $$\frac{5x(x)}{(x-1)x}-\frac{(8+x)(x-1)}{x(x-1)}=\frac{5x^2-(x^2+7x-8)}{x(x-1)}=\frac{4x^2-7x+8}{x(x-1)}$$

19. $$\frac{x(x-3)}{(x-3)(x-3)}+\frac{3}{(x-3)^2}=\frac{x^2-3x+3}{(x-3)^2}$$

21. $$\frac{2b}{(b+1)(b-1)}-\frac{3(b+1)}{-(b-1)(b+1)}=\frac{2b+3b+3}{(b+1)(b-1)}=\frac{5b+3}{(b+1)(b-1)}$$

23. $$\frac{2b}{a(2+b)b}-\frac{3a}{b(2+b)a}=\frac{2b-3a}{ab(2+b)}$$

25. $$\frac{x(x)}{(x-3)^2(x)}+\frac{3(x-3)}{x(x-3)(x-3)}=\frac{x^2+3x-9}{x(x-3)^2}$$

27. $$\frac{(x+2)(x)}{(2x+3)(3x+2)(x)}-\frac{1(3x+2)}{x(2x+3)(3x+2)}=\frac{x^2+2x-3x-2}{x(2x+3)(3x+2)}=\frac{x^2-x-2}{x(2x+3)(3x+2)}=$$
$$\frac{(x-2)(x+1)}{x(2x+3)(3x+2)}$$

29. $\dfrac{RT(v^2)}{v^2(v-b)} - \dfrac{a(v-b)}{v^2(v-b)} = \dfrac{RTv^2 - av + ab}{v^2(v-b)}$

31. $\dfrac{t_2}{t_1t_2} + \dfrac{t_1}{t_1t_2} = \dfrac{t_1+t_2}{t_1t_2}$

33. $\dfrac{7+1}{\frac{1}{7}+1} = \dfrac{8}{\frac{1}{7}+\frac{7}{7}} = \dfrac{8}{\frac{8}{7}} = \dfrac{8}{1}\cdot\dfrac{7}{8} = 7$

35. $a = \dfrac{V}{\frac{4}{3}\pi b^2} = \dfrac{V}{\frac{4\pi b^2}{3}} = \dfrac{V}{1}\cdot\dfrac{3}{4\pi b^2} = \dfrac{3V}{4\pi b^2}$

37. $\dfrac{x-\frac{x}{2}}{2+\frac{x}{3}} = \dfrac{\frac{2x}{2}-\frac{x}{2}}{\frac{6}{3}+\frac{x}{3}} = \dfrac{\frac{x}{2}}{\frac{x+6}{3}} = \dfrac{x}{2}\cdot\dfrac{3}{(x+6)} = \dfrac{3x}{2(x+6)}$

39. $I = \dfrac{E}{R+\frac{r}{2}} = \dfrac{E}{\frac{2R}{2}+\frac{r}{2}} = \dfrac{E}{\frac{2R+r}{2}} = \dfrac{E}{1}\cdot\dfrac{2}{2R+r} = \dfrac{2E}{2R+r}$

41. $v = \dfrac{2x}{\frac{x}{v_1}+\frac{x}{v_2}} = \dfrac{2x}{\frac{v_2x}{v_1v_2}+\frac{v_1x}{v_1v_2}} = \dfrac{2x}{\frac{v_1x+v_2x}{v_1v_2}} = \dfrac{2x}{1}\cdot\dfrac{v_1v_2}{(v_1+v_2)x} = \dfrac{2v_1v_2}{v_1+v_2}$

43. $284 \div 55 \approx 5.16$ hr; $284 \div 65 \approx 4.37$ hr; $5.16 + 4.37 \approx 9.5$ hrs total

Using the formula from example 12 to find average speed, $\dfrac{2(55)(65)}{55+65} \approx 59.6$ mph.

45. Answer is wrong; did not convert to common denominator before adding.

47. Answer is right; common denominator is not "least common denominator (LCD)," so additional simplification may be necessary.

49. a. At 15 mph, Marsha will reach the bike path in a total elapsed time of

$$\frac{12\text{ mi}}{15\text{ mph}} = 0.8\text{ hr} = 0.8\text{ hr}\cdot\frac{60\text{ min}}{1\text{ hr}} = 48\text{ min}$$

At 35 mph, Jon will reach the bike path in

$$\frac{12\text{ mi}}{35\text{ mph}} = 0.34\text{ hr} + 0.5\text{ hr (time since Marsha left home)} = 0.84\text{ hr.}$$

$$0.84\text{ hr} = 0.84\text{ hr}\cdot\frac{60\text{ min}}{1\text{ hr}} = 50.4\text{ min}$$

Jon will not catch her.

b. With t = total time and x = speed in mph; $t = 0.5\,hr + \dfrac{12\,mi}{x}$

c. Let t equal Marsha's time of 0.8 hr.

$$0.8 = 0.5 + \frac{12}{x};\ 0.3 = \frac{12}{x};\ 0.3x = 12;\ x = \frac{12}{0.3} = 40\text{ mph}$$

Exercises 4.7: Solving Rational Equations

1. $\frac{3}{4}x + 5 = 23$

$\frac{3}{4}x = 23 - 5 = 18$

$x = 18 \cdot \frac{4}{3} = 24$

3. $L = \dfrac{\pi r\theta}{180}$

$180L = \pi r\theta$

$r = \dfrac{180L}{\pi\theta},\ \theta \neq 0$

5. $x(\frac{1}{10} + \frac{1}{12}) = 1$

$x(\frac{6}{60} + \frac{5}{60}) = 1$

$x\frac{11}{60} = 1$

$x = \frac{60}{11} = 5\frac{5}{11}$

7. $\dfrac{2n+4}{7} = \dfrac{3n-7}{4}$

$4(2n + 4) = 7(3n - 7)$

$8n + 16 = 21n - 49$

$16 + 49 = 21n - 8n$

$65 = 13n,\quad n = 5$

9. From the equation, $x \neq -3, 0$; from the graph, $x = -6, -2$.

11. $\dfrac{x+1}{2} = \dfrac{x-3}{1}$

$x + 1 = 2(x - 3)$

$x + 1 = 2x - 6$

$1 + 6 = 2x - x$

$7 = x$

13. $\dfrac{x-8}{5} = \dfrac{-6}{x+5}$

$(x + 5)(x - 8) = (-6)(5)$

$x^2 - 3x - 40 = -30$

$x^2 - 3x - 10 = 0$

$(x - 5)(x + 2) = 0$

$\{-2, 5\},\ x \neq -5$

15. $\dfrac{x-3}{6} = \dfrac{3}{2x-1}$

$(2x - 1)(x - 3) = (3)(6)$

$2x^2 - 7x + 3 = 18$

$2x^2 - 7x - 15 = 0$

$(x - 5)(2x + 3) = 0$

$\{-1.5, 5\},\ x \neq 0.5$

17. $\dfrac{x}{32} + \dfrac{x}{8} = 10$

$32 \cdot \left(\dfrac{x}{32} + \dfrac{x}{8}\right) = 10 \cdot 32$

$x + 4x = 320$

$5x = 320;\ x = 64$

19. $6x^2\left(\dfrac{1}{2x^2} - \dfrac{1}{3x}\right) =$

$\dfrac{6x^2}{2x^2} - \dfrac{6x^2}{3x} = 3 - 2x$

21. $2x^2\left(\dfrac{1}{2x^2} + \dfrac{3}{x}\right) =$

$\dfrac{2x^2}{2x^2} + \dfrac{6x^2}{x} =$

$1 + 6x$

23. $x(x+1)\left(\frac{1}{x+1}-\frac{2}{x}\right)=$

$\frac{x(x+1)}{x+1}-\frac{2x(x+1)}{x}=$

$x - 2(x + 1) = x - 2x - 2 = -x - 2$

25. $x \neq 0;\ 15x\left(\frac{3}{5}+\frac{2}{3}\right)=\left(\frac{1}{x}\right)15x$

$\frac{15x\cdot 3}{5}+\frac{15x\cdot 2}{3}=\frac{15x}{x}$

$3x \cdot 3 + 5x \cdot 2 = 15$

$9x + 10x = 15$

$19x = 15;\ x = \frac{15}{19}$

27. $x \neq 0;\ 6x\left(\frac{1}{3}+\frac{1}{x}\right)=\left(\frac{1}{2}\right)6x$

$\frac{6x}{3}+\frac{6x}{x}=\frac{6x}{2}$

$2x + 6 = 3x$

$6 = 3x - 2x;\ x = 6$

29. $x \neq 0;\ x\left(\frac{5}{x}+\frac{2}{x}\right)=\left(\frac{4}{x}\right)x$

$5 + 2 = 4$; contradiction;

{ } or $\varnothing$

31. $x \neq \frac{1}{2}, 1;\ (2x-1)(x-1)\frac{4}{2x-1}=\frac{1}{x-1}(2x-1)(x-1)$

$4(x - 1) = 2x - 1$

$4x - 4 = 2x - 1$

$4x - 2x = -1 + 4$

$2x = 3;\ x = \frac{3}{2}$

33. $x \neq -2;\ 2(x+2)\frac{(x+1)}{2}=\frac{15}{(x+2)}(x+2)2$

$(x + 2)(x + 1) = 30$

$x^2 + 3x + 2 = 30$

$x^2 + 3x - 28 = 0$

$(x + 7)(x - 4) = 0 \qquad \{-7, 4\}$

35. $x \neq 1;\ 10(x-1)\frac{4}{(x-1)}=\frac{(x+2)}{10}(x-1)10$

$40 = (x + 2)(x - 1)$

$40 = x^2 + x - 2$

$x^2 + x - 42 = 0$

$(x + 7)(x - 6) = 0 \qquad \{-7, 6\}$

37. $x \neq 0;\ x^2\left(\frac{18}{x^2}+\frac{9}{x}-2\right)=x^2\cdot 0$

$18 + 9x - 2x^2 = 0$

$(6 - x)(3 + 2x) = 0 \qquad \{-1.5, 6\}$

39. $x \neq 2;\ (x-2)\left(\frac{1}{(x-2)} - 4\right) = \frac{(3-x)}{(x-2)}(x-2)$

$1 - 4(x - 2) = 3 - x$

$1 - 4x + 8 = 3 - x$

$9 - 3 = -x + 4x$

$6 = 3x$

$x = 2$; violates restriction on variable; { } or $\varnothing$

41. $x \neq 3, 6;\ (6-x)(x-3)\frac{x}{(6-x)} = \frac{2}{(x-3)}(6-x)(x-3)$

$x(x - 3) = 2(6 - x)$

$x^2 - 3x = 12 - 2x$

$x^2 - x - 12 = 0$

$(x + 3)(x - 4) = 0 \qquad \{-3, 4\}$

43. $x \neq 0, 2;\ 2x(x-2)\left(\frac{1}{(x-2)} + \frac{(x-1)}{x}\right) = \frac{2x+1}{2x}(x-2)2x$

$2x + 2(x - 2)(x - 1) = (2x + 1)(x - 2)$

$2x + 2(x^2 - 3x + 2) = 2x^2 - 3x - 2$

$2x + 2x^2 - 6x + 4 = 2x^2 - 3x - 2$

$2x^2 - 4x + 4 = 2x^2 - 3x - 2$

$4 + 2 = 2x^2 - 2x^2 - 3x + 4x \qquad \Rightarrow 6 = x \Rightarrow \{6\}$

45. $\frac{x^2}{1.000 - x} = 4;\ x^2 = 4(1.000 - x);\ x^2 = 4 - 4x;\ x^2 + 4x - 4 = 0;\ a = 1, b = 4, c = -4$

$x = \frac{-4 \pm \sqrt{4^2 - 4(1)(-4)}}{2(1)};\ x = \frac{-4 \pm \sqrt{32}}{2} = \frac{-4 \pm 4\sqrt{2}}{2}$

$x = -2 + 2\sqrt{2}$ or $x = -2 - 2\sqrt{2}$;

$x \approx 0.828$ or $x \approx -4.828$ (discard - violates restriction)

47. $\frac{(3.000 + 2x)^2}{(1.000 - x)(2.000 - x)} = 42;\ 9 + 12x + 4x^2 = 42(2 - 3x + x^2)$

$4x^2 + 12x + 9 = 84 - 126x + 42x^2;\ 42x^2 - 4x^2 - 126x - 12x + 84 - 9 = 0$

$38x^2 - 138x + 75 = 0;\ x = \frac{-(-138) \pm \sqrt{(-138)^2 - 4(38)(75)}}{2(38)}$

$x = \frac{138 \pm \sqrt{7644}}{76} = \frac{138 \pm 14\sqrt{39}}{76};\ x = \frac{69 + 7\sqrt{39}}{38}$ or $x = \frac{69 - 7\sqrt{39}}{38}$

$x \approx 2.966$ (discard - violates restriction) or **$x \approx 0.665$**

49. $abc\left(\frac{1}{a}+\frac{1}{b}\right)=\frac{1}{c}abc$

bc + ac = ab; bc = ab – ac

bc = a(b – c); $a=\frac{bc}{b-c}$

51. $I=\frac{E}{R+r}$; I(R + r) = E;

IR + Ir = E; IR = E – Ir;

$R=\frac{E-Ir}{I}=\frac{E}{I}-r$

Formulas for exercises 53 to 59 can be found in the chapter text.

53. $\frac{1}{6}+\frac{1}{x}=\frac{1}{2}$

$6x\left(\frac{1}{6}+\frac{1}{x}\right)=\frac{1}{2}(6x)$

x + 6 = 3x

6 = 2x; x = 3 min

55. $\frac{1}{20}+\frac{1}{x}=\frac{1}{12}$

$60x\left(\frac{1}{20}+\frac{1}{x}\right)=\frac{1}{12}(60x)$

3x + 60 = 5x

60 = 2x; x = 30 min

57. $\frac{1}{10{,}000}+\frac{1}{4000}=\frac{1}{R}$

$40{,}000R\left(\frac{1}{10{,}000}+\frac{1}{4000}\right)=\frac{1}{R}40{,}000R$

4R + 10R = 40,000

14R = 40,000; R ≈ 2857 ohms

59. $\frac{1}{5}+\frac{1}{3}+\frac{1}{3}=\frac{1}{x}$

$15x\left(\frac{1}{5}+\frac{1}{3}+\frac{1}{3}\right)=\frac{1}{x}15x$

3x + 5x + 5x = 15

13x = 15; x ≈ 1.15 min.

61. Did not distribute 15x(x + 2) over $\frac{2}{x}$; should be 15x + 2 · 15(x + 2) = 11x(x + 2).

63. Forgot the 2 in the numerator of the 2nd term; should be
$\frac{15x(x+2)}{x+2}+\frac{2\cdot 15x(x+2)}{x}=11x(x+2)$

65. a. 2 – 1 = **1**; 3 – 2 = **1**; 5 – 3 = **2**; 8 – 5 = **3**; 13 – 8 = **5**; 21 – 13 = **8**; 34 – 21 = **13**
21 + 34 = 55; 34 + 55 = 89; 55 + 89 = 144; 89 + 144 = 233; 144 + 233 = 377;
233 + 377 = 610; 377 + 610 = 987; 610 + 987 = 1597; 987 + 1597 = 2584;
1597 + 2584 = 4181

b. Each term is the sum of the two prior terms. The sequence starts with 1, 1.

c. (1 ÷ 1) = 1; (2 ÷ 1) = 2; (3 ÷ 2) = 1.5; (5 ÷ 3) ≈ 1.67; (8 ÷ 5) = 1.6; (13 ÷ 8) = 1.625;
(21 ÷ 13) ≈ 1.615; (34 ÷ 21) ≈ 1.619; (55 ÷ 34) ≈ 1.6176; (89 ÷ 55) ≈ 1.6182;
(144 ÷ 89) ≈ 1.61798; (233 ÷ 144) ≈ 1.61806
Quotients get closer to each other, approaching 1.618.

Chapter 4 Review Exercises

1. $\frac{x}{4}=\frac{6}{5};\ \frac{y}{4}=\frac{7}{5}$

$x=\frac{6\cdot 4}{5}=5.6;\ y=\frac{7\cdot 4}{5}=4.8$

2. $\frac{x}{6}=\frac{20}{15};\ \frac{y}{15}=\frac{8}{6}$

$x=\frac{6\cdot 20}{15}=8;\ y=\frac{15\cdot 8}{6}=20$

3. $\frac{r}{7.5}=\frac{8}{6};\quad s=\sqrt{6^2+8^2};\ \ t=\sqrt{7.5^2+r^2}$

$r=\frac{8\cdot 7.5}{6}=10;\ s=\sqrt{100}=10;$

$t=\sqrt{56.25+100}=\sqrt{156.25}=12.5$

4. $\frac{x}{10}=\frac{12}{18};\ y=\sqrt{12^2+18^2};\ w=\sqrt{10^2+x^2}$

$x=\frac{10\cdot 12}{18}=\frac{20}{3}\approx 6.667;\ \ y=\sqrt{144+324}=\sqrt{468}=6\sqrt{13}\approx 21.633$

$w=\sqrt{100+\frac{400}{9}}=\sqrt{\frac{1300}{9}}=\frac{10}{3}\sqrt{13}\approx 12.019$

5. **a.** $\tan A=\frac{10}{x};\ \sin A=\frac{10}{w};\ \cos A=\frac{x}{w}$

b. $\tan A=\frac{18}{12}=\frac{3}{2}=1.5$

6. $\sin A=\frac{24}{n};\ \cos A=\frac{7}{n};\tan A=\frac{24}{7}$

7. $\frac{78\text{ ft}}{156\text{ mph}};\ 78\text{ ft}\cdot\frac{1\text{ hr}}{156\text{ mi}}\cdot\frac{1\text{ mi}}{5280\text{ ft}}\cdot\frac{3600\text{ sec}}{1\text{ hr}}\approx 0.341\text{ sec}$

8. $60\text{ ft}\cdot\frac{1\text{mile}}{5280\text{ft}}\cdot\frac{1\text{ hr}}{90\text{ miles}}\cdot\frac{3600\text{sec}}{1\text{ hr}}=\frac{5}{11}\text{sec}\approx 0.4545\text{ sec}$

9. $18\text{ yrs}\cdot\frac{21\text{ in}}{9\text{ mo}}\cdot\frac{12\text{ mo}}{1\text{ yr}}=504\text{ in (growth)};\ 21\text{ in.}+504\text{ in}=525\text{ in (total height)}$

$504\text{ in}\cdot\frac{1\text{ ft}}{12\text{ in}}=42\text{ ft};\ \ 525\text{ in}\cdot\frac{1\text{ ft}}{12\text{ in}}=43.75\text{ ft}$

10. 69 in – 21 in = 48 inches in 18 yrs or $\frac{8}{3}$ in per yr. $21\text{in.}\cdot\dfrac{1\text{yr}}{\frac{8}{3}\text{in.}}\approx 7.9\text{yr}$

11. Let x = gallons and y = cost; $m=\dfrac{312-192}{65-40}=\dfrac{120}{25}=4.8$;

192 = 4.8(40) + b; b = 192 – 192 = 0

y = 4.8x; direct variation; y-intercept is possibly basic cost of using no gallons.

12. Let x = GB and y = cost; $m=\dfrac{19.99-12.99}{8-4}=\dfrac{7}{4}=1.75$;

12.99 = 1.75(4) + b; b = 12.99 – 7 = 5.99

y = 1.75x + 5.99; not direct since the line does not pass through origin; y-intercept is possibly basic cost of memory card.

13. Let x = TB and y = cost; $m=\dfrac{129.99-79.99}{2-1}=\dfrac{50}{1}=50$; $\dfrac{179.99-129.99}{3-2}=\dfrac{50}{1}=50$;

The data does appear linear since the slope is constant.

79.99 = 50(1) + b; b = 79.99 – 50 = 29.99

y = 50x + 29.99; not direct since the line does not pass through origin; y-intercept is possibly basic cost of hard drive.

14. Let x = # of years, y = cost; $m=\dfrac{98-54.50}{2-1}=43.50$;

54.50 = 43.50(1) + b; b = 54.50 – 43.50 = 11

y = 43.50x + 11; not direct; not meaningful.

15. Let D = distance; h = height; $D=k\sqrt{h}$.

16. Let E = earnings, w = wage, h = hours; k = 1; E = kwh.

17. Let V = volume, l = length, w = width, h = height, k = 1; V = $klwh$.

18. Let C = cost, w = weight; k = 1; C = kw.

19. $d=ks^2$

20. $s=k\sqrt{d}$

21. $s = k\sqrt{df}$

22. $t = k\dfrac{\sqrt{l}}{\sqrt{f}} = k\sqrt{\dfrac{l}{f}}$

23. Energy = k(height)^2; $\dfrac{k12^2}{k3^2} = \dfrac{144}{9} = 16$; the 12-ft wave has 16 times the energy of a 3-ft wave.

24. Length = k(speed)^2; $\dfrac{k30^2}{k20^2} = \dfrac{900}{400} = \dfrac{9}{4} = 2.25$; the skid mark at 30 mph is 2.25 times the skid mark at 20 mph.

25. Divide total credits by number of credits per quarter: $y = \dfrac{90}{x}$; k = 90 credits.

26. Divide total earnings by number of hours worked: $y = \dfrac{10{,}000}{x}$; k = \$10,000.

27. $\dfrac{1}{5.5} + \dfrac{1}{7} = \dfrac{1}{x}$; $38.5x\left(\dfrac{1}{5.5} + \dfrac{1}{7}\right) = \dfrac{1}{x}38.5x$: $7x + 5.5x = 38.5$; $12.5x = 38.5$; $x = 3.08$

Working together it takes 3.08 hours, longer than required by code.

28. $\dfrac{1}{7} + \dfrac{1}{x} = \dfrac{1}{3}$; $(21x)\left(\dfrac{1}{7} + \dfrac{1}{x}\right) = \dfrac{1}{3}(21x)$; $3x + 21 = 7x$; $21 = 4x$; $x = 5.25$ min.

29. a. $\dfrac{27xy^2}{15x^2y} = \dfrac{3 \cdot 9xyy}{3 \cdot 5xxy} = \dfrac{9y}{5x}$

b. $\dfrac{2-x}{x-2} = \dfrac{-(x-2)}{x-2} = -1$

c. $\dfrac{a - ac}{a} = \dfrac{a(1-c)}{a} = 1 - c$

30. a. $\dfrac{12x^3yz}{39xy^2z^2} = \dfrac{3 \cdot 4xxxyz}{3 \cdot 13xyyzz} = \dfrac{4x^2}{13yz}$

b. $\dfrac{x-5}{5-x} = \dfrac{(x-5)}{-(x-5)} = -1$

c. $\dfrac{ab + b^2}{ab} = \dfrac{b(a+b)}{ab} = \dfrac{a+b}{a}$

31. a. $\dfrac{x^2-5x-6}{x^2-4x-5}=\dfrac{(x+1)(x-6)}{(x+1)(x-5)}=\dfrac{x-6}{x-5}$ **b.** $\dfrac{4x^2-1}{2x^2+5x+2}=\dfrac{(2x+1)(2x-1)}{(2x+1)(x+2)}=\dfrac{2x-1}{x+2}$

c. $\dfrac{x^2+3x-4}{x^2-16}=\dfrac{(x+4)(x-1)}{(x+4)(x-4)}=\dfrac{x-1}{x-4}$

32. a. $\dfrac{21ab+7b^2}{6a+2b}=\dfrac{7b(3a+b)}{2(3a+b)}=\dfrac{7b}{2}$ **b.** $\dfrac{(x-2)(x+3)}{(x+3)(x+2)}=\dfrac{x-2}{x+2}$

c. $x(2-x)\left(\dfrac{1}{x-2}+\dfrac{3}{-x}\right)=\dfrac{x(2-x)}{(x-2)}-\dfrac{3x(2-x)}{x}=-x-6+3x=2x-6$

33. 4 gal · 4 qts/1 gal = 16 qts; 3 to 16 **34.** 5 yd · 3 ft/1 yd = 15 yd; 5 to 15 = 1 to 3

35. Undefined when x – 1 = 0 or x + 3 = 0, or when x = 1 or x = –3.

36. Undefined when x + 1 = 0, or when x = –1.

37. $\dfrac{x^2+5x+4}{x^2-16}\cdot\dfrac{2x-8}{1-x^2}=\dfrac{(x+1)(x+4)}{(x-4)(x+4)}\cdot\dfrac{2(x-4)}{(1+x)(1-x)}=\dfrac{2}{1-x}$

38. $\dfrac{2x^2+x-3}{x-3}\div\dfrac{x^2-2x+1}{3-x}=\dfrac{(2x+3)(x-1)}{(x-3)}\cdot\dfrac{-(x-3)}{(x-1)^2}=\dfrac{-(2x+3)}{x-1}$

39. $\dfrac{x}{x-1}-\dfrac{1}{1-x}=\dfrac{x}{x-1}-\dfrac{1}{-(x-1)}=\dfrac{x}{x-1}+\dfrac{1}{x-1}=\dfrac{x+1}{x-1}$

40. $\dfrac{x+2}{2-x}-\dfrac{x+2}{x-2}=\dfrac{x+2}{(2-x)}-\dfrac{x+2}{-(2-x)}=\dfrac{x+2}{2-x}+\dfrac{x+2}{2-x}=\dfrac{2x+4}{2-x}$

41. $\dfrac{x-2}{x^2-1}-\dfrac{2}{x-1}=\dfrac{x-2}{(x-1)(x+1)}-\dfrac{2(x+1)}{(x-1)(x+1)}=\dfrac{x-2-2(x+1)}{x^2-1}=\dfrac{x-2-2x-2}{x^2-1}=\dfrac{-x-4}{x^2-1}$

42. $\dfrac{1-x}{x^2-4}+\dfrac{2}{x-2}=\dfrac{1-x}{(x+2)(x-2)}+\dfrac{2(x+2)}{(x+2)(x-2)}=\dfrac{1-x+2x+4}{x^2-4}=\dfrac{x+5}{x^2-4}$

43. $\dfrac{1}{x^2-2x}-\dfrac{1}{3x^2-10x+8}=\dfrac{1}{x(x-2)}-\dfrac{1}{(x-2)(3x-4)}=\dfrac{3x-4}{x(x-2)(3x-4)}-\dfrac{x}{x(x-2)(3x-4)}$

$=\dfrac{3x-4-x}{x(x-2)(3x-4)}=\dfrac{2x-4}{x(x-2)(3x-4)}=\dfrac{2(x-2)}{x(x-2)(3x-4)}=\dfrac{2}{x(3x-4)}$

44. $\dfrac{1}{3x-x^2}-\dfrac{1}{-9+9x-2x^2}=\dfrac{1}{x(3-x)}-\dfrac{1}{(3-x)(-3+2x)}$

$\dfrac{2x-3}{x(3-x)(2x-3)}-\dfrac{x}{x(3-x)(2x-3)}=\dfrac{2x-3-x}{x(3-x)(2x-3)}=\dfrac{x-3}{-x(x-3)(2x-3)}=\dfrac{-1}{x(2x-3)}$

45. $\dfrac{A}{\frac{1}{2}h}=\dfrac{A}{1}\cdot\dfrac{2}{h}=\dfrac{2A}{h}$

46. $\dfrac{V}{\frac{1}{3}\pi r^2}=\dfrac{V}{1}\cdot\dfrac{3}{\pi r^2}=\dfrac{3V}{\pi r^2}$

47. $\dfrac{2500\text{ miles}}{12.5\text{ miles per gal}}=\dfrac{2500\text{ miles}}{1}\cdot\dfrac{1\text{ gal}}{12.5\text{ miles}}=200\text{ gal}$

48. $40\text{ months}\cdot\dfrac{\frac{1}{150}\text{grain}}{150\text{ months}}=\dfrac{40}{150}\cdot\dfrac{1}{150}\text{grain}=\dfrac{40}{150^2}\text{grain}\approx 0.00178\text{grain}$

49. $\dfrac{x+\frac{x}{3}}{3+\frac{2}{x}}=\dfrac{\frac{3x}{3}+\frac{x}{3}}{\frac{3x}{x}+\frac{2}{x}}=\dfrac{\frac{4x}{3}}{\frac{3x+2}{x}}=\dfrac{4x}{3}\cdot\dfrac{x}{3x+2}=\dfrac{4x^2}{9x+6}$

50. $\dfrac{x-\frac{4}{x}}{1-\frac{2}{x}}=\dfrac{\frac{x^2}{x}-\frac{4}{x}}{\frac{x}{x}-\frac{2}{x}}=\dfrac{\frac{x^2-4}{x}}{\frac{x-2}{x}}=\dfrac{x^2-4}{x}\cdot\dfrac{x}{x-2}=\dfrac{(x+2)(x-2)}{(x-2)}=x+2$

51. $\dfrac{\frac{x^2-x}{2}}{\frac{x^2-1}{x}}=\dfrac{x^2-x}{2}\cdot\dfrac{x}{x^2-1}=\dfrac{x(x-1)x}{2(x-1)(x+1)}=\dfrac{x^2}{2(x+1)}$

52. $\dfrac{\frac{x-1}{2x^2-3x-5}}{\frac{x^2-1}{4x^2-20x+25}}=\dfrac{x-1}{2x^2-3x-5}\cdot\dfrac{4x^2-20x+25}{x^2-1}=\dfrac{(x-1)(2x-5)(2x-5)}{(2x-5)(x+1)(x+1)(x-1)}=\dfrac{2x-5}{(x+1)^2}$

53. x ≠ −3; asymptote is x = −3

$$\begin{array}{r} x^2-3x+9-\frac{26}{x+3} \\ x+3\overline{)\,x^3+0x^2+0x+1} \\ -(x^3+3x^2) \\ \hline -3x^2+0x \\ -(-3x^2-9x) \\ \hline 9x+1 \\ -(9x+27) \\ \hline -26 \end{array}$$

54. x ≠ 2; asymptote is x = 2

$$\begin{array}{r} x^2-x+1+\frac{1}{x-2} \\ x-2\overline{)\,x^3-3x^2+3x-1} \\ -(x^3-2x^2) \\ \hline -x^2+3x \\ -(-x^2+2x) \\ \hline x-1 \\ -(x-2) \\ \hline +1 \end{array}$$

55. x ≠ −1; hole at (−1, −4)

$$\begin{array}{r} x^3-x^2+x-1 \\ x+1\overline{)\,x^4+0x^3+0x^2+0x-1} \\ -(x^4+x^3) \\ \hline -x^3+0x^2 \\ -(-x^3-x^2) \\ \hline x^2+0x \\ -(x^2+x) \\ \hline -x-1 \\ -(-x-1) \\ \hline 0 \end{array}$$

56. x ≠ 1; hole at (1, 4)

$$\begin{array}{r} x^2+x+2 \\ x-1\overline{)\,x^3+0x^2+x-2} \\ -(x^3-x^2) \\ \hline x^2+x \\ -(x^2-x) \\ \hline 2x-2 \\ -(2x-2) \\ \hline 0 \end{array}$$

57. $\frac{x}{2}=\frac{7}{x+5}$

$2(x+5)\frac{x}{2}=\frac{7}{x+5}2(x+5)$

$x^2+5x=14$;

$x^2+5x-14=0$

$(x+7)(x-2)=0$; $\{-7, 2\}$; $x \neq -5$

58. $\frac{x-6}{8}=\frac{5}{x}$

$(8x)\frac{x-6}{8}=\frac{5}{x}(8x)$

$x^2-6x=40$

$x^2-6x-40=0$

$(x-10)(x+4)=0$; $\{-4, 10\}$, $x \neq 0$

59. $\dfrac{x+3}{1}=\dfrac{-5}{2x}$

$(2x)\dfrac{x+3}{1}=\dfrac{-5}{2x}(2x)$

$2x^2+6x=-5$

$2x^2+6x+5=0$

$x=\dfrac{-6\pm\sqrt{6^2-4(2)(5)}}{2(2)}$

$x=\dfrac{-6\pm\sqrt{-4}}{4}$; $x=\dfrac{-6\pm 2i}{4}$

$x=\dfrac{-3}{2}\pm\dfrac{1}{2}i$; $\{-1.5\pm 0.5i\}$; $x\neq 0$

60. $\dfrac{x^2+1}{x}=\dfrac{6}{5}$

$(5x)\dfrac{x^2+1}{x}=\dfrac{6}{5}(5x)$

$5x^2+5=6x$

$5x^2-6x+5=0$

$x=\dfrac{-(-6)\pm\sqrt{(-6)^2-4(5)(5)}}{2(5)}$

$x=\dfrac{6\pm\sqrt{-64}}{10}$; $x=\dfrac{6\pm 8i}{10}$

$x=\dfrac{3}{5}\pm\dfrac{4}{5}i$; $\{0.6\pm 0.8i\}$; $x\neq 0$

61. $\dfrac{1}{x}+\dfrac{5}{x+2}=\dfrac{13}{3x}$

$3x(x+2)\left(\dfrac{1}{x}+\dfrac{5}{x+2}\right)=\dfrac{13}{3x}(x+2)3x$

$3(x+2)+15x=13(x+2)$

$3x+6+15x=13x+26$

$18x-13x=26-6$

$5x=20$; $x=4$; $x\neq -2, 0$

62. $\dfrac{2}{x-4}-\dfrac{1}{x}=\dfrac{11}{3x}$

$3x(x-4)\left(\dfrac{2}{x-4}-\dfrac{1}{x}\right)=\dfrac{11}{3x}(x-4)3x$

$6x-3(x-4)=11(x-4)$

$6x-3x+12=11x-44$

$12+44=11x-3x$

$56=8x$; $x=7$; $x\neq 0, 4$

63. $\dfrac{1}{x-1}+\dfrac{2}{x}=\dfrac{7}{6}$

$6x(x-1)\left(\dfrac{1}{x-1}+\dfrac{2}{x}\right)=\dfrac{7}{6}(x-1)6x$

$6x+12(x-1)=7x(x-1)$

$6x+12x-12=7x^2-7x$

$7x^2-7x-18x+12=0$

$7x^2-25x+12=0$

$(7x-4)(x-3)=0$; $\{\frac{4}{7}, 3\}$; $x\neq 0, 1$

64. $\dfrac{4}{x}+\dfrac{1}{x-2}=3$

$x(x-2)\left(\dfrac{4}{x}+\dfrac{1}{x-2}\right)=3x(x-2)$

$4(x-2)+x=3x^2-6x$

$4x-8+x=3x^2-6x$

$3x^2-6x-5x+8=0$

$3x^2-11x+8=0$

$(x-1)(3x-8)=0$; $\{1, \frac{8}{3}\}$; $x\neq 0, 2$

65. a. no

b. inversely related

c. 50 lb · 8 ft = 100 lb · 4 ft = 400 lb·ft

d. $\dfrac{400\,lb\cdot ft}{80\,lb}$ = 5 ft from the pivot point

66. a. The equation is the total distance divided by the total hours minus 3 hours not traveling.

b.

Total Trip Time (hours)	Rate (miles per hour)
4	$\dfrac{300}{4-3} = 300$
5	$\dfrac{300}{5-3} = 150$
6	$\dfrac{300}{6-3} = 100$
7	$\dfrac{300}{7-3} = 75$
8	$\dfrac{300}{8-3} = 60$

c. Due to the 3 hour delay, we replace t with $t - 3$, which causes a shift of 3 units to the right.

d. $\dfrac{300}{24-3} = 14.3 \text{ mph}$

e. $\dfrac{300}{18-3} = 20 \text{ mph}$

f. $\dfrac{300}{10-3} = 42.9 \text{ mph}$

g. $\dfrac{300}{3.5-3} = 600 \text{ mph}$

h. $\dfrac{300}{3.1-3} = 3000 \text{ mph}$

i. $\dfrac{300}{3.01-3} = 30{,}000 \text{ mph}$

j. $\dfrac{300}{3-3}$; not possible, division by zero

67. a. d = 3480 km; D = 384,000 km

$$\frac{\alpha}{206{,}265\,\text{sec}} = \frac{3480}{384{,}000}$$

$\alpha \approx 1869.277$ sec ≈ 31.255 min $\approx 0.519°$

b. and c. Answer will vary.

d. 360° 3600 sec/1° = 1,296,000 sec

$$\frac{1{,}296{,}000\,\text{sec}}{2\pi D} = \frac{\alpha\,\text{sec}}{d}$$

$$\frac{206{,}265\,\text{sec}}{D} = \frac{\alpha\,\text{sec}}{d}$$

$$\frac{d}{D} = \frac{\alpha}{206{,}265}$$

Chapter 4 Test

1. $\dfrac{15b^2c^3}{10b^3c} = \dfrac{5 \cdot 3bbccc}{5 \cdot 2bbbc} = \dfrac{3c^2}{2b}$; $b \neq 0,\ c \neq 0$

2. $\dfrac{3x^2 - 7x + 2}{2x^2 - 5x + 2} = \dfrac{(3x-1)(x-2)}{(2x-1)(x-2)} = \dfrac{3x-1}{2x-1}$; $x \neq \frac{1}{2}, 2$

3. $\dfrac{2x^2 + 7x + 6}{x^2 + 4x + 4} = \dfrac{(2x+3)(x+2)}{(x+2)(x+2)} = \dfrac{2x+3}{x+2}$; $x \neq -2$

4. $\dfrac{12xy^2}{7y - y^2} \div \dfrac{6x^2}{7-y} = \dfrac{12xy^2}{y(7-y)} \cdot \dfrac{7-y}{6x^2} = \dfrac{2y}{x}$; $y \neq 0, 7;\ x \neq 0$

5. $\dfrac{x^2 + 8x + 16}{x-2} \cdot \dfrac{x^2 - 4}{x+4} = \dfrac{(x+4)(x+4)(x+2)(x-2)}{(x-2)(x+4)} = (x+4)(x+2)$; $x \neq -4, 2$

6. $\dfrac{30\,\frac{mi}{gal}}{60\,\frac{mi}{hr}} = \dfrac{1}{2}\,\dfrac{mi}{gal}\,\dfrac{hr}{mi} = \dfrac{1}{2}\,\dfrac{hr}{gal}$

7. $\dfrac{x-2}{x+2} - \dfrac{x}{x^2 + 2x} = \dfrac{x(x-2) - x}{x(x+2)} = \dfrac{x[(x-2)-1]}{x(x+2)} = \dfrac{x-3}{x+2}$; $x \neq -2, 0$

8. $x \neq -3, 0, 3$

$$\dfrac{\dfrac{x^2 - 3x}{3}}{\dfrac{9 - x^2}{x}}$$

$$= \dfrac{x^2 - 3x}{3} \cdot \dfrac{x}{9 - x^2}$$

$$= \dfrac{x(x-3)x}{3(3-x)(3+x)}$$

$$= \dfrac{x^2(x-3)}{-3(x-3)(x+3)}$$

$$= -\dfrac{x^2}{3(x+3)}$$

9. $x \neq 0$

$$\dfrac{\dfrac{3}{x} - x}{x + \dfrac{2}{x}}$$

$$= \dfrac{\dfrac{3 - x^2}{x}}{\dfrac{x^2 + 2}{x}}$$

$$= \dfrac{3 - x^2}{x} \cdot \dfrac{x}{x^2 + 2}$$

$$= \dfrac{3 - x^2}{x^2 + 2}$$

10. Defined for $\Re$, $x \neq -2$.
hole at $(-2, 0)$

$$\begin{array}{r} x^2 + 4x + 4 \\ x + 2 \overline{)\, x^3 + 6x^2 + 12x + 8} \\ -(x^3 + 2x^2) \\ 4x^2 + 12x \\ -(4x^2 + 8x) \\ 4x + 8 \\ -(4x + 8) \\ 0 \end{array}$$

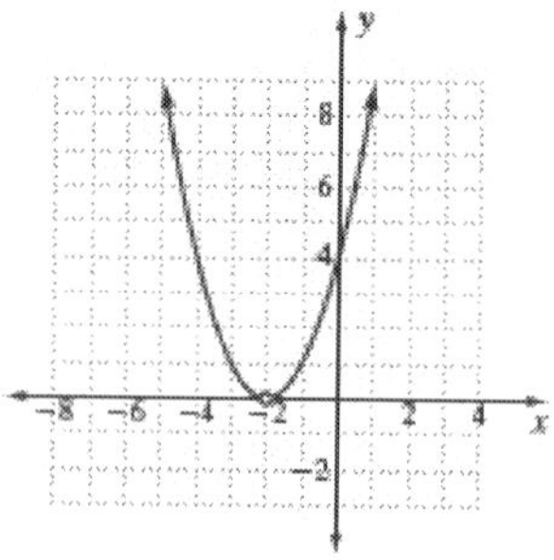

11. Defined for $\Re$, $x \neq -4$.
asymptote is $x = -4$

$$\begin{array}{r} x^2 + 2x + 4 - \frac{8}{x+4} \\ x + 4 \overline{)\, x^3 + 6x^2 + 12x + 8} \\ -(x^3 + 4x^2) \\ 2x^2 + 12x \\ -(2x^2 + 8x) \\ 4x + 8 \\ -(4x + 16) \\ -8 \end{array}$$

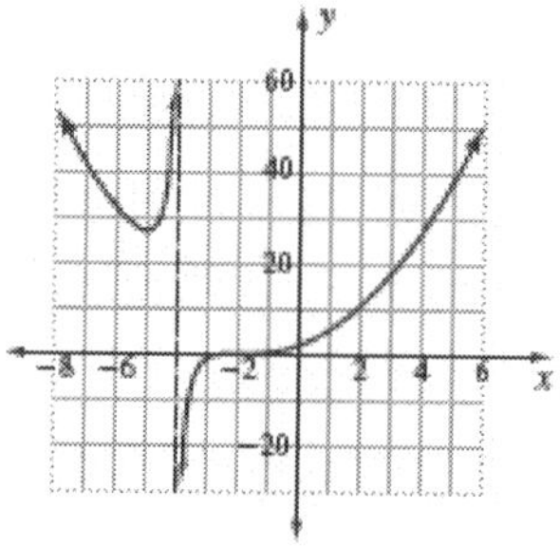

12. $$\frac{1}{3} - \frac{1}{x} = \frac{2x}{3}$$

$$3x\left(\frac{1}{3} - \frac{1}{x}\right) = \frac{2x}{3}3x$$

$$x - 3 = 2x^2$$

$$2x^2 - x + 3 = 0$$

$$x = \frac{-(-1) \pm \sqrt{(-1)^2 - 4(2)(3)}}{2(2)}$$

$$x = \frac{1 \pm \sqrt{-23}}{4}$$

$$\left\{\frac{1}{4} \pm i\left(\frac{\sqrt{23}}{4}\right)\right\};\ x \neq 0$$

13. $$\frac{1}{x-4} + \frac{1}{x} = \frac{10}{3x}$$

$$3x(x-4)\left(\frac{1}{x-4} + \frac{1}{x}\right) = \frac{10}{3x}3x(x-4)$$

$$3x + 3(x-4) = 10(x-4)$$

$$3x + 3x - 12 = 10x - 40$$

$$-12 + 40 = 10x - 6x$$

$$28 = 4x$$

$$x = 7;\ x \neq 0, 4$$

14. $\frac{3}{x}+\frac{2}{x+1}=4$

$x(x+1)\left(\frac{3}{x}+\frac{2}{x+1}\right)=4x(x+1)$

$3(x+1)+2x=4x^2+4x$

$3x+3+2x=4x^2+4x$

$4x^2+4x-5x-3=0$

$4x^2-x-3=0$

$(4x+3)(x-1)=0 \quad \Rightarrow \{-\frac{3}{4}, 1\};\ x\neq -1, 0$

15. $\frac{7\ lb}{9\ mo}\cdot\frac{16\ oz}{1\ lb}\cdot\frac{1\ mo}{30\ day}\cdot\frac{1\ day}{24\ hr}\approx 0.0173\ {}^{oz}/_{hr}$

16. $\frac{133\ lb}{18\ yr}\cdot\frac{16\ oz}{1\ lb}\cdot\frac{1\ yr}{365\ day}\cdot\frac{1\ day}{24\ hr}\approx 0.0135\ {}^{oz}/_{hr}$

17. $s=45, f=0.6;\ d=\frac{45^2}{30\cdot 0.6}=112.5$ ft

18. $f=0.4, d=208;\ 208=\frac{s^2}{30\cdot 0.4};\ 208\cdot 12=s^2;\ s\approx 50$ mph

19. a. $\frac{1}{5}+\frac{1}{4}=\frac{1}{x};\ 20x\left(\frac{1}{5}+\frac{1}{4}\right)=\frac{1}{x}20x;\ 4x+5x=20;\ 9x=20;\ x=2.2$ minutes

b. $\frac{1}{5}+\frac{1}{4}+\frac{1}{x}=\frac{1}{2};\ 20x\left(\frac{1}{5}+\frac{1}{4}+\frac{1}{x}\right)=\frac{1}{2}20x;\ 4x+5x+20=10x;\ 9x+20=10x;\ x=20$ min

20. a. $AB=\sqrt{6^2+15^2}=3\sqrt{29}\approx 16.155$

$ED=\sqrt{4^2+10^2}=2\sqrt{29}\approx 10.770$

$\frac{EF}{4}=\frac{15}{6};\ EF=\frac{4\cdot 15}{6}=10$

b. $\tan B=\frac{6}{15}\Rightarrow \tan B=\frac{2}{5}$

c. The triangles are similar and angle E has the same measure as angle B.

d. $6^2+15^2=c^2 \quad \Rightarrow \sin A=\frac{15}{3\sqrt{29}}\Rightarrow \sin A=\frac{5}{\sqrt{29}}$ or $\sin A=\frac{5\sqrt{29}}{29}$

$261=c^2$

$\sqrt{261}=c$

$c=\sqrt{9\cdot 29}$

$c=3\sqrt{29}$

Cumulative Review: Chapters 0 to 4

1. Evaluate: $-x^2$ for $x = -3$; $-(-3)^2 = -9$

2. Slope of the line is $-\frac{1}{3}$ and the given point is (6, 3). Use $y - y_1 = m(x - x_1)$.
$y - 3 = -\frac{1}{3}(x - 6)$; $y - 3 = -\frac{1}{3}x + 2$; $y = -\frac{1}{3}x + 5$

3. Factoring: $x^2 + 4x - 12 = 0$
$(x - 2)(x + 6) = 0$
$x = 2$ or $x = -6$

Square Root: $x^2 + 4x = 12$
$x^2 + 4x + 4 = 12 + 4$
$(x + 2)^2 = 16$
$|x + 2| = 4$
$x + 2 = -4$ or $x + 2 = 4$
$x = -6$ or $x = 2$

Quadratic Formula: $a = 1, b = 4, c = -12$

$$x = \frac{-4 \pm \sqrt{4^2 - 4(1)(-12)}}{2(1)} \text{ or } x = \frac{-4 \pm \sqrt{64}}{2} \text{ or } x = \frac{-4-8}{2} = \frac{-12}{2} = -6 \text{ or } x = \frac{-4+8}{2} = \frac{4}{2} = 2$$

4. Factoring: $4x^2 - 19x + 12 = 0$

$(4x - 3)(x - 4) = 0$

$x = \frac{3}{4}$ or $x = 4$

Quadratic Formula: $a = 4, b = -19, c = 12$

$$x = \frac{-(-19) \pm \sqrt{(-19)^2 - 4(4)(12)}}{2(4)}$$

$$x = \frac{19 \pm \sqrt{169}}{8}$$

$$x = \frac{19-13}{8} = \frac{6}{8} = \frac{3}{4} \text{ or } x = \frac{19+13}{8} = \frac{32}{8} = 4$$

5. a. Using the answer to number 3, the x-intercepts are –6 and 2. The line of symmetry is half way between the x-intercepts, which is $x = -2$. The x-value of the vertex is –2.
$f(-2) = (-2)^2 + 4(-2) - 12$; $f(-2) = -16$ The vertex is $(-2, -16)$.

b. Using the answer to number 4, the x-intercepts are $\frac{3}{4}$ and 4. The line of symmetry is half way between the x-intercepts, which is $x = \frac{19}{8}$. The x-value of the vertex is $\frac{19}{8}$.
$f(\frac{19}{8}) = 4(\frac{19}{8})^2 - 19(\frac{19}{8}) + 12$; $f(\frac{19}{8}) = -\frac{169}{16}$ The vertex is $(\frac{19}{8}, -\frac{169}{16})$.

6. The soap is 0.56% impure: $\frac{0.56}{100} = \frac{5600}{1000000}$ or 5600 parts per million.

7. a. $\frac{3.29}{4} = 0.8225$; $\frac{5.29}{8} = 0.6613$; $\frac{8.99}{16} = 0.5619$. This is not a linear variation because the ratios of y to x are not equal. A graph through the points will not pass through the origin.

b. Using LinReg, $y = 0.473x + 1.44$ fits with $r \approx 0.9998$.

c. $y = 0.473x + 1.44$; $y = 0.473(32) + 1.44 = \$16.58$

8. a. $|x - 2| = 3$

$x - 2 = 3$ or $x - 2 = -3$

$x = 5$ or $x = -1 \Rightarrow \{-1, 5\}$

b. $|x + 5| = 2$

$x + 5 = 2$ or $x + 5 = -2$

$x = -3$ or $x = -7 \Rightarrow \{-7, -3\}$

9. $\left(\frac{x}{2}\right)^2 + h^2 = x^2$

$\frac{x^2}{4} + h^2 = x^2$

$h^2 = x^2 - \frac{x^2}{4}$

$h^2 = \frac{3x^2}{4} \Rightarrow h = \frac{x\sqrt{3}}{2}$

$A = \frac{1}{2}(x)\left(\frac{x\sqrt{3}}{2}\right)$

$A = \frac{x^2\sqrt{3}}{4}$

10. $\frac{t_2t_3}{t_1t_2t_3} + \frac{t_1t_3}{t_1t_2t_3} + \frac{t_2t_1}{t_1t_2t_3} = \frac{1}{t}$ $\qquad \frac{t_2t_3 + t_1t_3 + t_2t_1}{t_1t_2t_3} = \frac{1}{t}$

$t = \frac{t_1t_2t_3}{t_2t_3 + t_1t_3 + t_2t_1}$; time required is not the sum of the individual times.

11. a. Inverse variation: $y = \frac{1}{x^2}$

b. Linear variation: $y = 2x$

c. Decreasing function: $y = -x + 2$

12. a. $\sqrt{2500} = \sqrt{25} \cdot \sqrt{100} = 5 \cdot 10 = 50$

b. $\sqrt{0.0025} = \sqrt{\frac{25}{10000}} = \frac{\sqrt{25}}{\sqrt{10000}} = \frac{5}{100} = 0.05$

c. $\sqrt{2.25} = \frac{\sqrt{225}}{\sqrt{100}} = \frac{15}{10} = 1.5$

d. $\sqrt{22500} = \sqrt{225} \cdot \sqrt{100} = 15 \cdot 10 = 150$

13. a.

x	y	1st diff	2nd diff
1	15		
		−7	
2	8		2
		−5	
3	3		2
		−3	
4	0		2
		−1	
5	−1		

Quadratic; $2a = 2$; $a = 1$

When $x = 0$, $y = 24$.; $c = 24$;

$15 = 1(1)^2 + b(1) + 24$; $b = -10$

$y = x^2 - 10x + 24$

b.

x	y	1st diff	2nd diff
1	4		
		5	
2	9		0
		5	
3	14		0
		5	
4	19		0
		5	
5	24		

Linear; $m = 5$. When $x = 0$, $y = -1$.

$y = 5x - 1$

c.

x	y	1st diff	2nd diff
1	−3		
		7	
2	4		0
		7	
3	11		0
		7	
4	18		0
		7	
5	25		

Linear; $m = 7$. When $x = 0$, $y = -10$.

$y = 7x - 10$

d.

x	y	1st diff	2nd diff
1	1		
		−2	
2	−1		4
		2	
3	1		−4
		−2	
4	−1		4
		2	
5	1		

Neither; $(-1)^n$ or $y = 1$ for odd x and $y = -1$ for even x.

14. a.

Factor	3x	+4
2x	$6x^2$	8x
−1	−3x	−4

Diagonal sum = 5x

Diagonal product = $-24x^2$

Factors are 8x and −3x.

$6x^2 + 5x - 4 = (3x + 4)(2x - 1)$

b. $3x^3 + 6x^2 + 3x = 3x(x^2 + 2x + 1)$

Factor	x	+1
x	x^2	x
+1	x	1

Diagonal sum = 2x

Diagonal product = x^2

Factors are x & x.

$3x^3 + 6x^2 + 3x = 3x(x + 1)^2$

15. $y = -\frac{2}{3}x + 5$

$2x + 3y = 15$

$2x + 3(-\frac{2}{3}x + 5) = 15$

$2x - 2x + 15 = 15$

$15 = 15$; identity

Coincident lines

Infinite number of solutions

CHAPTER 5

Exponents and Radicals

Exercises 5.1: Exponents and Their Properties

1. **a.** $x^4x^5 = x^{4+5} = x^9$

b. $y^2y^a = y^{2+a}$

c. $a^na^3 = a^{n+3}$

3. **a.** $(3x)^4 = 3^4x^4 = 81x^4$

b. $(ab)^3 = a^3b^3$

c. $(4n)^2 = 4^2n^2 = 16n^2$

5. **a.** $(x^2)^4 = x^{2\cdot4} = x^8$

b. $(2x^3)^5 = 2^5x^{3\cdot5} = 32x^{15}$

c. $(3a^4b^2)^3 = 3^3a^{4\cdot3}b^{2\cdot3} = 27a^{12}b^6$

7. **a.** $x^5/x^2 = x^{5-2} = x^3$

b. $y^3/y^5 = y^{3-5} = y^{-2} = 1/y^2$

c. $a^5a^3/a^6 = a^{5+3-6} = a^2$

9. **a.** $2x^{-1} = \dfrac{2}{x}$

b. $\left(\dfrac{y}{x}\right)^{-1} = \dfrac{y^{-1}}{x^{-1}} = y^{-1}x^{-(-1)} = \dfrac{x}{y}$

c. $\left(\dfrac{s}{t}\right)^0 = 1$

d. $(x^2y^{-1})^{-1} = x^{2(-1)}y^{-1(-1)} = x^{-2}y^1 = \dfrac{y}{x^2}$

e. $(x^{-1}y^3)(x^3y^{-2}) = x^{-1}x^3y^3y^{-2} = x^{-1+3}y^{3-2} = x^2y$

f. $(x^3y^{-2})^{-1} = x^{3(-1)}y^{(-2)(-1)} = x^{-3}y^2 = \dfrac{y^2}{x^3}$

11. **a.** $\left(\dfrac{3a^2}{c}\right)^{-3} = \left(\dfrac{c}{3a^2}\right)^3 = \dfrac{c^3}{3^3a^{2\cdot3}} = \dfrac{c^3}{27a^6}$

b. $\left(\dfrac{-a}{c^2}\right)^{-2} = \left(\dfrac{c^2}{-a}\right)^2 = \dfrac{c^{2\cdot2}}{(-1)^2a^2} = \dfrac{c^4}{a^2}$

c. $\dfrac{-1}{c^{-3}} = -1\cdot c^3 = -c^3$

d. $\dfrac{2}{a^{-2}} = 2\cdot a^2 = 2a^2$

e. $\dfrac{12a^3b^{-2}}{4a^{-1}b^{-5}} = 3a^{3-(-1)}b^{(-2)-(-5)} = 3a^4b^3$

f. $\dfrac{3}{3^x} = 3\cdot3^{-x} = 3^{1-x}$

13. a. $5^{-1} = \frac{1}{5} = 0.2$

b. $(-2)^{-1} = \frac{1}{-2} = -\frac{1}{2} = -0.5$

c. $0.5^{-2} = \left(\frac{1}{2}\right)^{-2} = \left(\frac{2}{1}\right)^{2} = 2^2 = 4$

d. $-0.25^{-2} = -\left(\frac{1}{4}\right)^{-2} = -\left(\frac{4}{1}\right)^{2} = -4^2 = -16$

e. $0.2^{-1} = \left(\frac{1}{5}\right)^{-1} = \left(\frac{5}{1}\right)^{1} = 5$

f. $(-3)^{-2} = \left(\frac{1}{(-3)}\right)^{2} = \frac{1^2}{(-3)^2} = \frac{1}{9}$

15. a. Adding like terms: $y^3 + 2y^3 = y^3(1 + 2) = 3y^3$

b. Adding like terms: $x^{-1} + x^{-1} = x^{-1}(1 + 1) = 2x^{-1} = \frac{2}{x}$

c. Neither: already simplified

17. a. Neither: $x^2 + x^{-2} = x^2 + \frac{1}{x^2}$

b. Adding like terms: $y^{-3} + y^{-3} = 2y^{-3} = \frac{2}{y^3}$

c. Multiplying like bases: $y^{-4}y^2 = y^{-4+2} = y^{-2} = \frac{1}{y^2}$

19. a. Neither: already simplified

b. Multiplying like bases: $x^3x^{-1} = x^{3-1} = x^2$

c. Adding like terms: $x^2 + 2x^2 = 3x^2$

21. Let r = radius of golf ball: $\dfrac{\frac{4\pi(5.3r)^3}{3}}{\frac{4\pi r^3}{3}} = \frac{4\pi 5.3^3 r^3}{3} \cdot \frac{3}{4\pi r^3} = 5.3^3 \approx 149$ times larger.

23. No output for 0^0; undefined at x = 0.

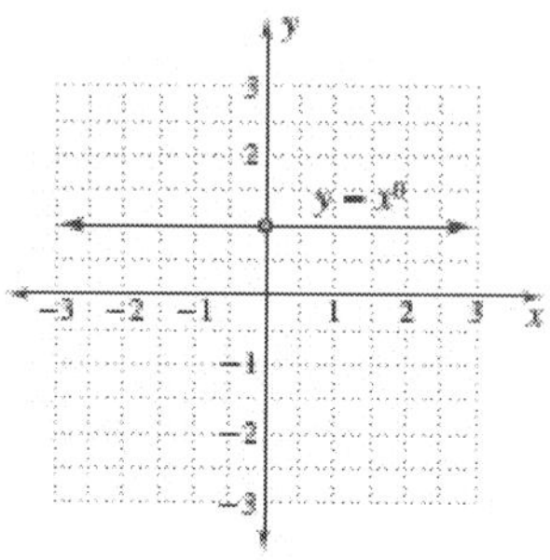

25. $-x^2 \neq (-x)^2$. In $-x^2$, the implied -1 is not included in the base.

27. The exponent does not apply to the 3. Answer is $\frac{3}{x^2}$.

29. $6\left(\frac{5}{6}\right)^4 = 6\cdot\frac{5}{6}\cdot\left(\frac{5}{6}\right)^3 = 5\left(\frac{5}{6}\right)^3$

31. a. $\frac{(x+h)^{-1}-x^{-1}}{h} = \frac{(x+h)^{-1}}{h} - \frac{x^{-1}}{h} = \frac{1}{(x+h)h} - \frac{1}{xh} = \frac{x}{x(x+h)h} - \frac{(x+h)}{x(x+h)h}$

$= \frac{x-x-h}{x(x+h)h} = \frac{-h}{x(x+h)h} = \frac{-1}{x(x+h)}$

b. $\frac{(x+h)^{-2}-x^{-2}}{h} = \frac{(x+h)^{-2}}{h} - \frac{x^{-2}}{h} = \frac{1}{(x+h)^2 h} - \frac{1}{x^2 h} = \frac{x^2}{x^2(x+h)h} - \frac{(x+h)^2}{x^2(x+h)^2 h}$

$= \frac{x^2-(x^2+2xh+h^2)}{x^2(x+h)^2 h} = \frac{x^2-x^2-2xh-h^2}{x^2(x+h)^2 h} = \frac{-h(2x+h)}{x^2(x+h)^2 h} = \frac{-(2x+h)}{x^2(x+h)^2}$

c. $x^2(-1)(x+1)^{-2} + \frac{2x}{x+1} = \frac{-x^2}{(x+1)^2} + \frac{2x}{x+1} = \frac{-x^2}{(x+1)^2} + \frac{2x(x+1)}{(x+1)^2} = \frac{-x^2+2x^2+2x}{(x+1)^2}$

$= \frac{x^2+2x}{(x+1)^2}$

d. $(x-1)^{-1}(2x+2) + (x^2+2x)(-1)(x-1)^{-2} = \frac{2x+2}{(x-1)} - \frac{x^2+2x}{(x-1)^2} = \frac{(2x+2)(x-1)}{(x-1)^2} - \frac{x^2+2x}{(x-1)^2}$

$= \frac{2x^2-2-x^2-2x}{(x-1)^2} = \frac{x^2-2x-2}{(x-1)^2}$

33. a.

American Wire Gauge Size	**Diameter (inches)**	**Resistance at 20° C per 1000 Feet (ohms)**	$k = d^2 \cdot R$
24	0.0201	25.67	$0.0201^2 \cdot 25.67 \approx 0.0104$
22	0.0254	16.14	$0.0254^2 \cdot 16.14 \approx 0.0104$
20	0.0320	10.15	$0.0320^2 \cdot 10.15 \approx 0.0104$
18	0.0403	6.39	$0.0403^2 \cdot 6.39 \approx 0.0104$
16	0.0508	4.02	$0.0508^2 \cdot 4.02 \approx 0.0104$

b. $R \approx 0.01034d^{-2.001}$

c. As the American wire gauge size decreases, the diameter ***increases.***

Exercises 5.2: Scientific Notation

1. **a.** $0.001 = 10^{-3}$

b. $\dfrac{1}{100{,}000} = 0.000\ 01 = 10^{-5}$

c. $10{,}000 = 10^{4}$

d. 1 hundredth $= 0.01 = 10^{-2}$

e. 1 million $= 1{,}000{,}000 = 10^{6}$

f. 1 millionth $= 0.000\ 001 = 10^{-6}$

3. **a.** $3000 = 3 \cdot 1000 = 3 \times 10^{3}$

b. $350 = 3.5 \cdot 100 = 3.5 \times 10^{2}$

c. $350{,}\bar{0}00 = 3.500 \cdot 100{,}000 = 3.5 \times 10^{5}$

d. $0.003\ 50 = 3.50 \cdot 0.001 = 3.5 \times 10^{-3}$

5. **a.** $2.9979 \times 10^{10} = 29{,}979{,}000{,}000$ cm/sec

b. $1.6726 \times 10^{-24} = 0.000\ 000\ 000\ 000\ 000\ 000\ 000\ 001\ 672\ 6$ g

c. $-1.6022 \times 10^{-19} = -0.000\ 000\ 000\ 000\ 000\ 000\ 160\ 22$ C

d. $1.6022 \times 10^{-19} = 0.000\ 000\ 000\ 000\ 000\ 000\ 160\ 22$ J

7. $4.0 \cdot 1.5 = 6;\ 10^{-2} \cdot 10^{6} = 10^{4};\ 6 \times 10^{4}$

9. $6 \cdot 2.5 = 15 = 1.5 \times 10;\ 10^{-3} \cdot 10^{-5} \cdot 10^{1} = 10^{-7};\ 1.5 \times 10^{-7}$

11. $7.5 \div 2.5 = 3.0;\ 10^{-2} \div 10^{4} = 10^{-6};\ 3.0 \times 10^{-6}$

13. $2.4 \div 0.3 = 8;\ 10^{6} \div 10^{-2} = 10^{8};\ 8.0 \times 10^{8}$

15. **a.** $40\bar{0} \times 10^{-3} = 4.00 \times 10^{2} \times 10^{-3} = 4.00 \times 10^{-1}$

b. $40\bar{0} \times 10^{3} = 4.00 \times 10^{2} \times 10^{3} = 4.00 \times 10^{5}$

17. **a.** $0.05\bar{0} \times 10^{-4} = 5.0 \times 10^{-2} \times 10^{-4} = 5.0 \times 10^{-6}$

b. $0.05\bar{0} \times 10^{5} = 5.0 \times 10^{-2} \times 10^{5} = 5.0 \times 10^{3}$

19. **a.** 1 E 10; calculator reads one number

b. 1 E 10; calculator reads two numbers

c. 1 E 11; calculator reads one number

d. 1 E 11; calculator reads two numbers

e. 1 E 52; calculator reads one number

f. 1 E 53; calculator reads one number

g. Tells the calculator to multiply the preceding number by 10 raised to the exponent that follows.

21. a. 2.8253×10^7 acre·feet · 4.3560×10^4 ft^3/acre·ft = $12.31 \times 10^{11} = 1.231 \times 10^{12}$ ft^3

b. $1.231\times10^{12}\ \text{ft}^3 \cdot \left(\dfrac{1\ \text{mi}}{5280\ \text{ft}}\right)^2 \cdot \dfrac{\text{Rhode Island}}{1045\ \text{mi}^2} \approx 42\ \text{ft}$

c. $\dfrac{(7.7304\times10^{10})(0.5)}{28.32} \approx 8.35\times10^8$ cubic feet ; $\dfrac{1.231\times10^{12}}{8.35\times10^8} \approx 1470$ -- The capacity behind the dam is 1470 times that of the water consumed. Take no comfort in this, the reservoir has been well below capacity for years.

23. a. $V_{circ} = \sqrt{\dfrac{(6.67\times10^{-11}\text{N}\cdot\text{m}^2\cdot\text{kg}^{-2})(3.30\times10^{23}\ \text{kg})}{3.24\times10^6\ \text{m}}}$;

To keep units correct, set up two problems, one with numbers, the other with units.

$$\sqrt{\frac{(6.67\times10^{-11})(3.30\times10^{23})}{3.24\times10^6}} = \sqrt{\frac{6.67\cdot3.30\times10^{23-11-6}}{3.24}} \approx 2.61\times10^3$$

$$\sqrt{\frac{(\text{Nm}^2\text{kg}^{-2})(\text{kg})}{\text{m}}} = \sqrt{\frac{(\text{kg}\cdot\text{m}\cdot\text{sec}^{-2})(\text{m}^2\text{kg}^{-2+1})}{\text{m}}} = \sqrt{\text{kg}^{1-1}\text{m}^{1+2-1}\text{sec}^{-2}} = \sqrt{\frac{\text{m}^2}{\text{sec}^2}} = \frac{\text{m}}{\text{sec}}$$

$V_{circ} = 2.61\times10^3$ m/sec

b. Note: the units will be the same as in part a.

$$\sqrt{\frac{(6.67\times10^{-11})(1.90\times10^{27})}{8.14\times10^7}} = \sqrt{\frac{6.67\cdot1.90\times10^{27-11-7}}{8.14}} \approx 3.95\times10^4$$

$V_{circ} = 3.95 \times 10^4$ m/sec

c. Find V_{circ} first: units will be the same as in part a.

$$\sqrt{\frac{(6.67\times10^{-11})(5.98\times10^{24})}{7.18\times10^6}} = \sqrt{\frac{6.67\cdot5.98\times10^{24-11-6}}{7.18}} \approx 7.45\times10^3\ \text{m/sec}$$

$V_{esc} = \sqrt{2} \cdot 7.45 \times 10^3 \approx 1.05 \times 10^4$ m/sec

d. Find V_{circ} first:

$$\sqrt{\frac{(6.67\times10^{-11})(6.44\times10^{23})}{4.19\times10^6}} = \sqrt{\frac{6.67\cdot6.44\times10^{23-11-6}}{4.19}} \approx 3.20\times10^3\ \text{m/sec}$$

$V_{esc} = \sqrt{2} \cdot 3.19 \times 10^3 \approx 4.53 \times 10^3$ m/sec

e. If velocity drops below circular velocity, the vehicle will be pulled back toward the planet.

f. If the velocity is between circular velocity and escape velocity, the orbit becomes oval instead of circular.

25. **a.** $1, i, -1, -i, 1, i, -1, -i, 1$

b. $i^7 = -3E - 13 - i$; $i^8 = 1 - 2E - 13i$.

c. For i^7, the real part is $-3E - 13$ and the imaginary part is $-i$. The real part is approaching zero in scientific notation, so the answer is $-i$.
For i^8, the real part is 1 and $-2E - 13i$ is the imaginary part. The imaginary part in scientific notation is approaching zero. The answer is 1.

d. 16th is $1 - 4E - 13i$ or 1
35th is $5E - 13 - i$ or $-i$
46th is $-1 + 4E - 13i$ or -1
89th is $-3.9E - 12 + i$ or i

e. The scientific notation part, being to the -12 or -13 power of ten, is zero and is discarded. The answer is the real (1 or -1) or imaginary (i or $-i$) that remains.

Exercises 5.3: Rational Exponents

Guess and check is recommended for exercise 1. Only final solutions are shown.

1. **a.** $27^x = 9; x = \frac{2}{3}$ **b.** $9^x = 27; x = \frac{3}{2}$ **c.** $81^x = 27; x = \frac{3}{4}$

3. $3 \cdot 3 \cdot 3 \cdot 3 = (3 \cdot 3) \cdot (3 \cdot 3)$

5. $(2 \cdot 2)(2 \cdot 2)(2 \cdot 2)(2 \cdot 2)(2 \cdot 2) = (2 \cdot 2 \cdot 2 \cdot 2 \cdot 2)(2 \cdot 2 \cdot 2 \cdot 2 \cdot 2)$

7. **a.** $64 = (2 \cdot 2 \cdot 2) \cdot (2 \cdot 2 \cdot 2)$
$64^{1/2} = (2 \cdot 2 \cdot 2) = 8$

b. $16 = 2 \cdot 2 \cdot 2 \cdot 2$
$16^{1/4} = 2$

c. $32 = 2 \cdot 2 \cdot 2 \cdot 2 \cdot 2$
$32^{1/5} = 2$

9. **a.** $32 = 2 \cdot 2 \cdot 2 \cdot 2 \cdot 2$
$32^{2/5} = (32^{1/5})^2 = 2^2 = 4$

b. $16 = (2 \cdot 2) \cdot (2 \cdot 2)$
$16^{3/2} = (16^{1/2})^3 = 4^3 = 64$

c. $36 = 6 \cdot 6$
$36^{3/2} = (36^{1/2})^3 = 6^3 = 216$

d. $32 = 2 \cdot 2 \cdot 2 \cdot 2 \cdot 2$
$32^{6/5} = (32^{1/5})^6 = 2^6 = 64$

11. a. $16 = 2 \cdot 2 \cdot 2 \cdot 2$
$16^{5/4} = (16^{1/4})^5 = 2^5 = 32$

b. $64 = (2 \cdot 2 \cdot 2) \cdot (2 \cdot 2 \cdot 2)$
$64^{3/2} = (64^{1/2})^3 = 8^3 = 512$

c. $25 = 5 \cdot 5$
$25^{3/2} = (25^{1/2})^3 = 5^3 = 125$

d. $81 = 3 \cdot 3 \cdot 3 \cdot 3$
$81^{3/4} = (81^{1/4})^3 = 3^3 = 27$

13. a. $1.5 = \frac{3}{2}$; $25^{3/2} = (25^{1/2})^3 = 5^3 = 125$

b. $2.5 = \frac{5}{2}$; $4^{5/2} = (4^{1/2})^5 = 2^5 = 32$

c. $16^{5/2} = (16^{1/2})^5 = 4^5 = 1024$

d. $0.5 = \frac{1}{2}$; $0.09 = 0.3 \cdot 0.3$; $0.09^{1/2} = 0.3$

15. a. once a year = annual

b. twice a year = semiannual

c. 4 times a year = quarterly

d. 12 times a year = monthly

e. 52 times a year = weekly

f. 365 times a year = daily

17. $\$6.00(1 + 0.05)^{10} = \$9.77$

19. $\$1000(1 + 0.07)^5 = \$1{,}402.55$

21. a. 10,000 * (1 + 0.04 ÷ 12) ^ (12 * 3) = 11,272.72

b. 10,000 (1 + 0.04 ÷ 4) (4 * 3) = 11,268.25

23. $\$6.00(1 + \frac{0.05}{2})^{2 \cdot 10} = \$9.83$

25. $\$200(1 + 0.01)^{30} = \$269.57$

27. 8 months $= \frac{8}{12} = \frac{2}{3}$ yr.; $\$1350(1 + \frac{0.16}{12})^{12 \cdot \frac{2}{3}} = \$1500.91$

29. 2 months $= \frac{2}{12} = \frac{1}{6}$ yr; $\$650(1 + \frac{0.16}{12})^{12 \cdot \frac{1}{6}} = 667.45$

31. $5\frac{1}{3} = \frac{16}{3}$; $\$1000(1 + \frac{0.07}{4})^{4 \cdot \frac{16}{3}} \approx \$1{,}447.88$

33. $\$1000(1 + \frac{0.07}{4})^{4 \cdot 4.75} \approx \$1{,}390.45$

35. Answers could vary. Two possible solutions are $a = b = 2$ and $a = 2$, $b = 4$.

37. This is true for all b > 0.

39. $\frac{r}{4} > \frac{r}{12}$; base is smaller.

41. $1 + \frac{r}{n}$

43. a. Men: $\dfrac{\left(\dfrac{pounds}{2.205}\right)^{1.2}}{(inches\cdot 2.54)^{3.3}}\cdot 3{,}000{,}000$ Women: $\dfrac{\left(\dfrac{pounds}{2.205}\right)^{1.2}}{(inches\cdot 2.54)^{3.3}}\cdot 4{,}000{,}000$

b. $\dfrac{\left(\dfrac{100}{2.205}\right)^{1.2}\cdot 4{,}000{,}000}{(61\cdot 2.54)^{3.3}}=23.04$ $\dfrac{\left(\dfrac{130}{2.205}\right)^{1.2}\cdot 4{,}000{,}000}{(61\cdot 2.54)^{3.3}}=31.56$

c. $\dfrac{\left(\dfrac{132}{2.205}\right)^{1.2}\cdot 4{,}000{,}000}{(70\cdot 2.54)^{3.3}}=20.41$ $\dfrac{\left(\dfrac{165}{2.205}\right)^{1.2}\cdot 4{,}000{,}000}{(70\cdot 2.54)^{3.3}}=26.68$

d. $\dfrac{\left(\dfrac{153}{2.205}\right)^{1.2}\cdot 3{,}000{,}000}{(74\cdot 2.54)^{3.3}}=15.21$ $\dfrac{\left(\dfrac{193}{2.205}\right)^{1.2}\cdot 3{,}000{,}000}{(74\cdot 2.54)^{3.3}}=20.10$

e. $\dfrac{\left(\dfrac{128}{2.205}\right)^{1.2}\cdot 3{,}000{,}000}{(67\cdot 2.54)^{3.3}}=17.05$ $\dfrac{\left(\dfrac{162}{2.205}\right)^{1.2}\cdot 3{,}000{,}000}{(67\cdot 2.54)^{3.3}}=22.62$

Exercises 5.4: Roots and Rational Exponents

1. a. $\sqrt[3]{64}=\sqrt[3]{4^3}=4$ **b.** $\sqrt[5]{32}=\sqrt[5]{2^5}=2$

3. a. $\sqrt[3]{-27}=\sqrt[3]{(-3)^3}=-3$ **b.** $\sqrt[4]{625}=\sqrt[4]{5^4}=5$

5. a. $\sqrt{-4}$, not a real number **b.** $\sqrt[4]{-16}$, not a real number

7. a. $-\sqrt[3]{64}=-\sqrt[3]{4^3}=-4$ **b.** $-\sqrt[3]{-1000}=-\sqrt[3]{(-10)^3}=-(-10)=10$

9. a. $-\sqrt[4]{256}=-\sqrt[4]{4^4}=-4$ **b.** $\sqrt[4]{-625}$, not a real number

11. a. $\dfrac{\sqrt[4]{243}}{\sqrt[4]{3}}=\sqrt[4]{\dfrac{243}{3}}=\sqrt[4]{81}=3$ **b.** $\sqrt[2]{\dfrac{27}{3}}=\sqrt[2]{9}=3$

c. $\sqrt[3]{16}\sqrt[3]{4}=\sqrt[3]{64}=4$ **d.** $\dfrac{\sqrt[4]{64}}{\sqrt[4]{4}}=\sqrt[4]{\dfrac{64}{4}}=\sqrt[4]{16}=2$

13. a. $\sqrt{100n} = \sqrt{100}\sqrt{n} = 10\sqrt{n}$ **b.** $\sqrt{0.01n} = \sqrt{0.01}\sqrt{n} = 0.1\sqrt{n}$

c. $\sqrt{\dfrac{n}{100}} = \dfrac{\sqrt{n}}{\sqrt{100}} = \dfrac{\sqrt{n}}{10}$ **d.** $\sqrt[3]{\dfrac{x}{1000}} = \dfrac{\sqrt[3]{x}}{\sqrt[3]{1000}} = \dfrac{\sqrt[3]{x}}{10}$

e. $\sqrt[3]{8n} = \sqrt[3]{8}\sqrt[3]{n} = 2\sqrt[3]{n}$

15. a. $(\sqrt[3]{-8})^5 = (-8)^{5/3} = [(-2)^3]^{5/3} = (-2)^5 = -32$ **b.** $\sqrt[3]{27^2} = 27^{2/3} = (3^3)^{2/3} = 3^2 = 9$

17. a. $\sqrt[4]{16^2} = 16^{2/4} = 16^{1/2} = 4$ **b.** $(\sqrt[4]{-64})^2$, not a real number

19. a. $x^{1/6} = \sqrt[6]{x}$ **b.** $\sqrt[3]{x} = x^{1/3}$ **c.** $x^{0.5} = x^{1/2} = \sqrt{x}$

21. a. $x^{3/2} = (\sqrt{x})^3 \; or \sqrt{x^3}$ **b.** $\sqrt[4]{x^3} = (x^3)^{1/4} = x^{3/4}$ **c.** $x^{0.8} = x^{4/5} = (\sqrt[5]{x})^4 \; or \; \sqrt[5]{x^4}$

23. a. $\sqrt[4]{x} \cdot \sqrt[3]{x^2} = x^{1/4} \cdot x^{2/3} = x^{1/4+2/3} = x^{3/12+8/12} = x^{11/12}$

b. $x^{2/3} \cdot x^{1/5} = x^{2/3 + 1/5} = x^{10/15 + 3/15} = x^{13/15}$

25. a. $x^{5/8} \div x^{1/2} = x^{5/8 - 1/2} = x^{5/8 - 4/8} = x^{1/8}$

b. $\sqrt[8]{x^3} \cdot \sqrt{x^4} = x^{3/8} \cdot x^{4/2} = x^{3/8+2} = x^{3/8+16/8} = x^{19/8}$

27. a. $\sqrt[2]{\sqrt[4]{x}} = (x^{1/4})^{1/2} = x^{1/8} = \sqrt[8]{x};\; x \ge 0$ **b.** $\sqrt[4]{\sqrt[3]{x^5}} = (x^{5/3})^{1/4} = x^{5/12} = \sqrt[12]{x^5}\;;\; x \ge 0$

29. a. $\sqrt[3]{x}\sqrt[2]{x^3} = x^{1/3}x^{3/2} = x^{1/3+3/2} = x^{2/6+9/6} = x^{11/6}$ **b.** $\dfrac{x^{3/4}}{x^{1/3}} = x^{3/4-1/3} = x^{9/12-4/12} = x^{5/12}$

31. a. $\sqrt{a}\sqrt{a^3} = \sqrt{a^{1+3}} = \sqrt{a^4} = a^2;\; a \ge 0$ **b.** $\sqrt[3]{b}\sqrt[3]{b^2} = \sqrt[3]{b^{1+2}} = \sqrt[3]{b^3} = b$

c. $\dfrac{\sqrt{x^5}}{\sqrt{x^3}} = \sqrt{\dfrac{x^5}{x^3}} = \sqrt{x^{5-3}} = \sqrt{x^2} = x;\; x > 0$

33. a. $\sqrt[3]{\sqrt[2]{x}} = \left(x^{1/2}\right)^{1/3} = x^{1/6} = \sqrt[6]{x};\; x \ge 0$ **b.** $\sqrt[5]{\sqrt[2]{x^3}} = \left(x^{3/2}\right)^{1/5} = x^{3/10} = \sqrt[10]{x^3}$

35. a. $\sqrt{y^2} = |y|$ **b.** $\sqrt[4]{z^8} = z^{8/4} = z^2$ **c.** $\sqrt{x^6} = |x^3|$

37. a. $(x^4)^{1/2} = x^{4/2} = x^2$ **b.** $(x^{2/3})^3 = x^{6/3} = x^2$ **c.** $(y^6)^{1/6} = y^{6/6} = |y|$

39. $\sqrt[p]{x} \cdot \sqrt[q]{x} = x^{1/p} x^{1/q} = x^{1/p+1/q} = x^{q/pq+p/pq} = x^{(p+q)/pq}$

41. True; Definition of even n^{th} root.

43. False; The odd power of a negative number is still negative.

45. True; $2^3 = 8$ and 2 is a real number.

47. True; Outputs of graphs are positive; Satisfies definition of nth root.

49. False; cube root of -8 is -2.

51. False; definition of even power and property of product of two (or any even number of) negative numbers.

53. True; Definition of nth root.

55. a. Graph is a smooth curve. **b.** $y = 19.0x^{2.49}$

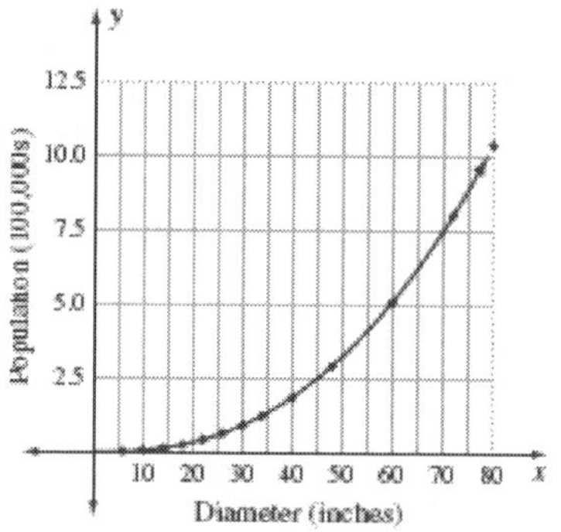

c. $y = 19.0x^{5/2}$

d. No; There is still 0.5 in the exponent not explained by cross sectional area.

e. No; 6000 gallons/4/30 = 50 gallons per day.

f. Answers will vary.

Exercises 5.5: More Operations with Radicals

1. $3\sqrt{2}-4\sqrt{2}+7\sqrt{2}=(3-4+7)\sqrt{2}=6\sqrt{2}$

3. $\sqrt{20}+\sqrt{45}=2\sqrt{5}+3\sqrt{5}=(2+3)\sqrt{5}=5\sqrt{5}$

5. $\sqrt{75}+\sqrt{27}=5\sqrt{3}+3\sqrt{3}=(5+3)\sqrt{3}=8\sqrt{3}$

7. $\sqrt{25x}+\sqrt{4x}+\sqrt{2.25x}=5\sqrt{x}+2\sqrt{x}+1.5\sqrt{x}=8.5\sqrt{x}$

9. $3\sqrt[3]{x}-\sqrt[3]{x}=(3-1)\sqrt[3]{x}=2\sqrt[3]{x}$

11. $3\sqrt[4]{x}-2\sqrt[2]{x}$; not like terms

13. $\sqrt[3]{8x}+\sqrt[3]{27x}=2\sqrt[3]{x}+3\sqrt[3]{x}=(2+3)\sqrt[3]{x}=5\sqrt[3]{x}$

15. $a\sqrt[4]{81ab}-\sqrt[4]{a^5b}=3a\sqrt[4]{ab}-a\sqrt[4]{ab}=(3a-a)\sqrt[4]{ab}=2a\sqrt[4]{ab}$

17. $\sqrt[4]{16x^4y}+x\sqrt[4]{y}=2x\sqrt[4]{y}+x\sqrt[4]{y}=(2x+x)\sqrt[4]{y}=3x\sqrt[4]{y}$

19. a. $(2-\sqrt{3})(2+\sqrt{3})=4-(\sqrt{3})^2=4-3=1$

b. $(5-\sqrt{2})(5-\sqrt{2})=5^2-10\sqrt{2}+(\sqrt{2})^2=25-10\sqrt{2}+2=27-10\sqrt{2}$

21. a. $(x-\sqrt{5})(x+\sqrt{5})=x^2-(\sqrt{5})^2=x^2-5$

b. $(x-\sqrt{7})(x-\sqrt{7})=x^2-2x\sqrt{7}+(\sqrt{7})^2=x^2-2x\sqrt{7}+7$

23. a. $(1-\sqrt{a})^2=1^2-2\sqrt{a}+(\sqrt{a})^2=1-2\sqrt{a}+a$

b. $(\sqrt{a}-\sqrt{b})(\sqrt{a}-\sqrt{b})=(\sqrt{a})^2-2\sqrt{a}\sqrt{b}+(\sqrt{b})^2=a-2\sqrt{ab}+b$

25.

Multiply	x^2	$-x\sqrt[3]{2}$	$+\sqrt[3]{4}$
x	x^3	$-x^2\sqrt[3]{2}$	$x\sqrt[3]{4}$
$+\sqrt[3]{2}$	$+x^2\sqrt[3]{2}$	$-x\sqrt[3]{2}\sqrt[3]{2}=-x\sqrt[3]{4}$	$+\sqrt[3]{2}\sqrt[3]{4}=\sqrt[3]{8}=2$

$(x+\sqrt[3]{2})(x^2-x\sqrt[3]{2}+\sqrt[3]{4})=x^3+2$

27. a. $3-\sqrt{2}$ **b.** $3+\sqrt{a}$

c. $a+\sqrt{b}$ **d.** $\sqrt{2}+\sqrt{3}$

29. $(1+\sqrt{3})^2-2(1+\sqrt{3})-2=0$

$1+2\sqrt{3}+3-2-2\sqrt{3}-2=0$

$4-4+2\sqrt{3}-2\sqrt{3}=0$

$0=0$

Other solution: $x=1-\sqrt{3}$.

31. $(-2-\sqrt{2})^2+4(-2-\sqrt{2})+2=0$

$4+4\sqrt{2}+2-8-4\sqrt{2}+2=0$

$6-6+4\sqrt{2}-4\sqrt{2}=0$

$0=0$

Other Solution: $x=-2+\sqrt{2}$

33. a. $\frac{4}{\sqrt{5}}=\frac{4\sqrt{5}}{\sqrt{5}\sqrt{5}}=\frac{4\sqrt{5}}{5}$ **b.** $\frac{8}{\sqrt{6}}=\frac{8\sqrt{6}}{\sqrt{6}\sqrt{6}}=\frac{8\sqrt{6}}{6}=\frac{4\sqrt{6}}{3}$

35. a. $\frac{a}{\sqrt{c}}=\frac{a\sqrt{c}}{\sqrt{c}\sqrt{c}}=\frac{a\sqrt{c}}{c}$ **b.** $\frac{a}{\sqrt{a}}=\frac{a\sqrt{a}}{\sqrt{a}\sqrt{a}}=\frac{a\sqrt{a}}{a}=\sqrt{a}$

37. $\frac{4}{7-\sqrt{5}}=\frac{4(7+\sqrt{5})}{(7-\sqrt{5})(7+\sqrt{5})}=\frac{4(7+\sqrt{5})}{49-5}=\frac{4(7+\sqrt{5})}{44}=\frac{7+\sqrt{5}}{11}$

39. $\frac{x}{x-\sqrt{y}}=\frac{x(x+\sqrt{y})}{(x-\sqrt{y})(x+\sqrt{y})}=\frac{x^2+x\sqrt{y}}{x^2-y}$

41. a. $\frac{1}{\sqrt[3]{2}}=\frac{\sqrt[3]{4}}{\sqrt[3]{2}\sqrt[3]{4}}=\frac{\sqrt[3]{4}}{\sqrt[3]{8}}=\frac{\sqrt[3]{4}}{2}$ **b.** $\frac{2}{\sqrt[3]{4}}=\frac{2\sqrt[3]{2}}{\sqrt[3]{4}\sqrt[3]{2}}=\frac{2\sqrt[3]{2}}{\sqrt[3]{8}}=\frac{2\sqrt[3]{2}}{2}=\sqrt[3]{2}$

c. $\frac{3}{\sqrt[3]{3}}=\frac{3\sqrt[3]{9}}{\sqrt[3]{3}\sqrt[3]{9}}=\frac{3\sqrt[3]{9}}{\sqrt[3]{27}}=\frac{3\sqrt[3]{9}}{3}=\sqrt[3]{9}$

43. $d=\sqrt{(\Delta x)^2+(\Delta y)^2}=\sqrt{(x_2-x_1)^2+(y_2-y_1)^2}$

45. $\text{m}=\frac{-3-3}{4-(-6)}=\frac{-6}{10}=\frac{-3}{5}$; $\text{d}=\sqrt{10^2+(-6)^2}=\sqrt{100+36}=\sqrt{136}=2\sqrt{34}\approx 11.662$

47. $\text{m}=\frac{b-\frac{b}{2}}{a-\frac{a}{2}}=\frac{\frac{b}{2}}{\frac{a}{2}}=\frac{b}{a}$; $\text{d}=\sqrt{(\frac{a}{2})^2+(\frac{b}{2})^2}=\sqrt{\frac{a^2}{4}+\frac{b^2}{4}}=\sqrt{\frac{a^2+b^2}{4}}=\frac{\sqrt{a^2+b^2}}{2}$

49. $\dfrac{y}{x+r} = \dfrac{\sqrt{r^2-x^2}}{x+r} = \dfrac{\sqrt{r^2-x^2}}{x+r}\cdot\dfrac{\sqrt{r^2-x^2}}{\sqrt{r^2-x^2}} = \dfrac{r^2-x^2}{(x+r)\sqrt{r^2-x^2}} = \dfrac{(r+x)(r-x)}{(x+r)\sqrt{r^2-x^2}} = \dfrac{r-x}{\sqrt{r^2-x^2}} = \dfrac{(-1)(x-r)}{y}$

51. a. AB = $\sqrt{12^2+5^2} = 13$ miles; 13(5280) = 68,640 ft; 68,640 · \$1500 = \$102,960,000

b. 12(5280) = 63,360 ft (farm land); 5(5280) = 26,400 ft (swamp);
63,360(\$1000) + 26,400(\$1500) = \$102,960,000

c. $AC_1 = \sqrt{5^2+1^2} = \sqrt{26}$ miles; $\sqrt{26}$ (5280) ≈ 26,922.8 ft (swamp);
11(5280) = 58,080 ft (farmland); \$1000(58,080) + \$1500(26,922.8) ≈ \$98,464,000

d. $AC_2 = \sqrt{5^2+2^2} = \sqrt{29}$ miles; $\sqrt{29}$ (5280) ≈ 28,433.7 ft (swamp);
10(5280) = 52,800 ft (farmland); \$1000(52,800) + \$1500(28,433.7) ≈ \$95,451,000

e. $AC_n = (5280)\sqrt{25+x^2}$; $BC_n = (5280)(12-x)$;
cost = $\$1500(5280)(\sqrt{25+x^2}) + \$1000(5280)(12-x)$

f. At $x \approx 4.47$, cost is \$92,876,000 (use a spreadsheet or graph to find min. cost).

Exercises 5.6: Solving Root and Power Equations

1. From the graph: x = 14

3. From the graph: { } or ∅

Calculator graphs are not included for exercises 5 through 25.

5. Defined when $2 - x \geq 0$ or $x \leq 2$.

$\sqrt{2-x} = 2;\ \left(\sqrt{2-x}\right)^2 = 2^2$

$2 - x = 4;\ -x = 2;\ x = -2$

$\sqrt{2-(-2)} = \sqrt{4} = 2$

7. Defined for all x.

$\sqrt[3]{2x-5} = 5;\ \left(\sqrt[3]{2x-5}\right)^3 = 5^3$

$2x - 5 = 125;\ x = 65$

$\sqrt[3]{2(65)-5} = 5$

9. Defined when $x \geq 5$ (see #7).

$\sqrt{x-5} - 6 = -2;\ \sqrt{x-5} = 4$

$\left(\sqrt{x-5}\right)^2 = 4^2;\ x - 5 = 16;\ x = 21$

$\sqrt{21-5} - 6 = \sqrt{16} - 6 = 4 - 6 = -2$

11. Defined when $x \leq 15/17$.

$(2x)^2 = 15 - 17x;\ 4x^2 + 17x - 15 = 0$

$(4x\ \ 3)(x + 5) = 0;\ x = 3/4,\ 5$

$2(3/4) = \sqrt{15 - 17(3/4)};\ 3/2 = 3/2$

$2(-5) = \sqrt{15 - 17(-5)};\ -10 = 10$

x = -5 is an extraneous solution. So, x = ¾.

13. Defined when x ≥ −7 (see #11).

$\sqrt{2x+14}=6;\ \left(\sqrt{2x+14}\right)^2=6^2$

$2x+14=36;\ 2x=22,\ x=11$

$\sqrt{2(11)+14}=\sqrt{22+14}=\sqrt{36}=6$

15. Defined when x ≥ 0.

$\sqrt[4]{8x}=4;\ 8x=4^4;\ 8x=256$

$x=32;\ 2\sqrt[4]{8(32)}-1=7;\ 7=7$

17. Defined when x ≥ $\frac{5}{3}$ (see #15).

$\sqrt{3x-5}=-1$

Principal square root must be positive; { } or ∅.

19. Defined when 5x − 1 ≥ 0 or x ≥ $\frac{1}{5}$.

$\sqrt{5x-1}=3;\ \left(\sqrt{5x-1}\right)^2=3^2$

$5x-1=9;\ 5x=10;\ x=2$

$\sqrt{5(2)-1}=\sqrt{10-1}=\sqrt{9}=3$

21. Defined when 2x + 7 ≥ 0 or x ≥ −3.5 **and** x ≥ 0; use x ≥ 0.

$\sqrt{2x+7}=\sqrt{x}+2;\ \left(\sqrt{2x+7}\right)^2=\left(\sqrt{x}+2\right)^2;\ 2x+7=x+4\sqrt{x}+4;\ 2x-x+7-4=4\sqrt{x}$

$x+3=4\sqrt{x};\ (x+3)^2=\left(4\sqrt{x}\right)^2;\ x^2+6x+9=16x;\ x^2-10x+9=0;\ (x-9)(x-1)=0$

$\sqrt{2(1)+7}=\sqrt{1}+2;\ 3=3;\ \ \sqrt{2(9)+7}=\sqrt{9}+2;\ 5=5;\ \ \{1,9\}$

23. Defined when x + 4 ≥ 0 or x ≥ −4 **and** 3x ≥ 0 or x ≥ 0; use x ≥ 0.

$\sqrt{3x}-2=\sqrt{x+4};\ \left(\sqrt{3x}-2\right)^2=\left(\sqrt{x+4}\right)^2;\ 3x-4\sqrt{3x}+4=x+4;\ 2x=4\sqrt{3x}$

$x=2\sqrt{3x};\ (x)^2=\left(2\sqrt{3x}\right)^2=x^2=4(3x);x^2=12x;\ x^2-12x=0;\ x(x-12)=0$

x = 0 or x = 12; $\sqrt{3(0)}-2=\sqrt{0+4};\ -2=\sqrt{4}$; principal square root must be positive; discard answer; $\sqrt{3(12)}-2=\sqrt{12+4};\ 6-2=4;\ \{12\}$.

25. Defined when 3x − 3 ≥ 0 or x ≥ 1 and 4x ≥ 0 or x ≥ 0; use x ≥ 1.

$\sqrt{3x-3}=\sqrt{4x}-1;\ \left(\sqrt{3x-3}\right)^2=\left(\sqrt{4x}-1\right)^2;\ 3x-3=4x-2\sqrt{4x}+1;\ x+4=2\sqrt{4x}$

$(x+4)^2=\left(2\sqrt{4x}\right)^2;\ x^2+8x+16=4(4x);\ x^2+8x+16=16x;\ x^2-8x+16=0;$

$(x-4)^2=0;\ x=4;\ \sqrt{3(4)-3}=\sqrt{4(4)}-1;\ 3=3;\ \{4\}$

27. a. $x^2=1;\ \sqrt{x^2}=\sqrt{1};\ |x|=1;\ x=\pm1$

$x^2-1=0;\ (x+1)(x-1)=0;\ x=\pm1$

b. $x^4=1;\ \sqrt[4]{x^4}=\sqrt[4]{1};\ |x|=1;\ x=\pm1$

$x^4-1=0;\ (x^2+1)(x^2-1)=0;\ x^2+1=0,\ x^2=-1;\ x=\pm i$

$x^2-1=0;\ (x+1)(x-1)=0;\ x=\pm1$

c. $x^3 = 1;\ \sqrt[3]{x^3} = \sqrt[3]{1};\ x = 1$

$x^3 - 1 = 0;\ (x-1)(x^2+x+1) = 0;\ x = 1\ or\ x = \dfrac{-1 \pm \sqrt{1^2 - 4(1)(1)}}{2(1)} = -\dfrac{1}{2} \pm \dfrac{\sqrt{3}}{2}i$

29. $4x^2 = 25;\ 4x^2 - 25 = 0;\ (2x-5)(2x+5) = 0;\ x = \pm 2.5$; no roots with imaginary units.

31. $16x^4 - 25 = 600;\ 16x^4 = 625;\ (16x^4)^{1/4} = (625)^{1/4};\ 2|x| = 5;\ |x| = 2.5;$
$x = \pm 2.5$; 2 solutions with imaginary units (4th power has 4 solutions).

33. $100 = 40(1+x)^5;\ 2.5 = (1+x)^5;\ \sqrt[5]{2.5} = \sqrt[5]{(1+x)^5}\ ;\ 1+x \approx 1.201;\ x \approx 0.201$

35. $10 = 30(1+x)^{15};\ \frac{1}{3} = (1+x)^{15};\ \sqrt[15]{\frac{1}{3}} = \sqrt[15]{(1+x)^{15}}\ ;\ 1+x \approx 0.929;\ x \approx -0.071$

37. a. $r = \sqrt{24L};\ r^2 = 24L;\ L = \dfrac{r^2}{24}$ **b.** $r = \sqrt{12L};\ r^2 = 12L;\ L = \dfrac{r^2}{12}$

c. $r_{dry} = \sqrt{24(50)} \approx 34.64;\ r_{wet} = \sqrt{12(50)} \approx 24.49;$

$\dfrac{r_{wet}}{r_{dry}} = \dfrac{\sqrt{12L}}{\sqrt{24L}} = \sqrt{\dfrac{12L}{24L}} = \sqrt{\dfrac{1}{2}} = \dfrac{1}{\sqrt{2}} = \dfrac{\sqrt{2}}{2}$; slow down when the road is wet.

39. $v = \sqrt{2gs};\ v^2 = 2gs;\ s = \dfrac{v^2}{2g}$ **41.** $Q_b = \sigma A T^4;\ \dfrac{Q_b}{\sigma A} = T^4;\ T = \sqrt[4]{\dfrac{Q_b}{\sigma A}}$

43. $K = \dfrac{mV^2}{2};\ \dfrac{2K}{m} = V^2;\ V = \sqrt{\dfrac{2K}{m}}$

45. $t = 13;\ 119{,}000 = 55{,}000(1+r)^{13}$ $\dfrac{119{,}000}{55{,}000} = (1+r)^{13}$ $\left(\dfrac{119}{55}\right)^{1/13} = 1 + r$

$r = \left(\dfrac{119}{55}\right)^{1/13} - 1 \approx 6.1\%$ in 2010 $t = 25;\ S = 55{,}000(1+0.061)^{25};\ S \approx \$243{,}000$

47. $t = 25;\ 75{,}000 = 14{,}000(1 + r)^{25}$ $\quad \dfrac{75{,}000}{14{,}000} = (1 + r)^{25}$ $\quad \left(\dfrac{75}{14}\right)^{1/25} = 1 + r$

$r = \left(\dfrac{75}{14}\right)^{1/25} - 1 \approx 6.94\%$ in 2010 $t = 40;\ S = 14{,}000(1 + 0.0694)^{40};\ S \approx \$205{,}315$

49. $t = 8;\ 52.02 = 6.953(1 + r)^{8}$ $\quad \dfrac{52.02}{6.953} = (1 + r)^{8}$ $\quad \left(\dfrac{52.02}{6.953}\right)^{1/8} = 1 + r$

$r = \left(\dfrac{52.02}{6.953}\right)^{1/8} - 1 \approx 28.6\%$

51. $t = 32;\ 5 = 96(1 + r)^{32}$ $\quad \dfrac{5}{96} = (1 + r)^{32};\ \left(\dfrac{5}{96}\right)^{1/32} = 1 + r;\ r = \left(\dfrac{5}{96}\right)^{1/32} - 1 \approx -8.8\%$

53. See exercise for prices in 1994 and 1998. $t = 8$ from 1994.

$120 = 470(1 + r)^{8};\ \dfrac{120}{470} = (1 + r)^{8};\ \left(\dfrac{120}{470}\right)^{1/8} = 1 + r;\ r = \left(\dfrac{120}{470}\right)^{1/8} - 1 \approx -15.7\%$

$t = 4$ from 1998

$120 = 290(1 + r)^{4};\ \dfrac{120}{290} = (1 + r)^{4};\ \left(\dfrac{120}{290}\right)^{1/4} = 1 + r;\ r = \left(\dfrac{120}{290}\right)^{1/4} - 1 \approx -19.8\%$

55. $t = 61;\ 68{,}400 = 3000(1 + r)^{61}$ $\quad \dfrac{68{,}400}{3000} = (1 + r)^{61}$ $\quad \left(\dfrac{684}{30}\right)^{1/61} = 1 + r$

$r = \left(\dfrac{684}{30}\right)^{1/61} - 1 \approx 5.3\%$ in 2010 $t = 73;\ S = 3000(1 + 0.053)^{73};\ S \approx \$127{,}000$

57. $1.25 = 0.25(1 + r)^{25}$ $\quad 5 = (1 + r)^{25}$ $\quad 5^{1/25} = 1 + r$ $\quad r = 5^{1/25} - 1 \approx 6.64\%$

59. $7.25 = 3.80(1 + r)^{19}$ $\quad \dfrac{145}{76} = (1 + r)^{19}$ $\quad \left(\dfrac{145}{76}\right)^{1/19} = 1 + r$ $\quad r = \left(\dfrac{145}{76}\right)^{1/19} - 1 \approx 3.46\%$

61. $A = 1000\left(1 + \dfrac{0.03}{365}\right)^{365t};\ A = 1000\left(1 + \dfrac{0.03}{12}\right)^{12t};\ A = 1000\left(1 + \dfrac{0.03}{1}\right)^{t}$; 74.3 years; 26.1 years

Chapter 5 Review Exercises

1. **a.** $\dfrac{a^0b^{-1}}{c^2} = \dfrac{1}{bc^2}$ **b.** $\dfrac{a^3b^{-2}}{c^{-1}} = \dfrac{a^3c}{b^2}$ **c.** $\left(\dfrac{2x}{y^2}\right)^4 = \dfrac{2^4x^4}{(y^2)^4} = \dfrac{16x^4}{y^8}$

2. **a.** $\dfrac{2}{x^{-1}} = 2x$ **b.** $\left(\dfrac{2x^2}{y}\right)^3 = \dfrac{2^3(x^2)^3}{y^3} = \dfrac{8x^6}{y^3}$

c. $\left(\dfrac{x^{-1}}{2y^2}\right)^{-3} = \left(\dfrac{1}{2xy^2}\right)^{-3} = \left(\dfrac{2xy^2}{1}\right)^3 = 2^3x^3(y^2)^3 = 8x^3y^6$

3. **a.** $\left(x^3y^{-4}\right)^2 = x^6y^{-8} = \dfrac{x^6}{y^8}$

b. $\left(3x^3y^2\right)^2\left(4xy^{-3}\right) = 9x^6y^4(4xy^{-3}) = 36x^{6+1}y^{4-3} = 36x^7y$

4. **a.** $\dfrac{6x^2y^{-3}}{3x^{-2}y^4} = \dfrac{2x^{2+2}}{y^{4+3}} = \dfrac{2x^4}{y^7}$ **b.** $\dfrac{4a^{-3}b^2}{(6a^3b^{-2})^2} = \dfrac{4a^{-3}b^2}{36a^6b^{-4}} = \dfrac{b^{2+4}}{9a^{6+3}} = \dfrac{b^6}{9a^9}$

5. **a.** $3.45 \times 10^{-15} = 0.000\ 000\ 000\ 000\ 003\ 45$ **b.** $6.400 \times 10^{-3} = 0.006\ 400$

c. $4.005 \times 10^5 = 400{,}500$ **d.** $4.7800 \times 10^3 = 4780.0$

6. **a.** $\dfrac{4.5\times10^{-2}}{0.9\times10^5} = 5 \times 10^{-7}$ **b.** $\dfrac{3.6\times10^4}{0.9\times10^{-3}} = 4 \times 10^7$

7. **a.** $64^{1/2} = 8$ **b.** $64^{1/6} = 2$ **c.** $64^{5/6} = (64^{1/6})^5 = 2^5 = 32$

8. **a.** $32^{1/5} = 2$ **b.** $16^{3/4} = (16^{1/4})^3 = 2^3 = 8$ **c.** $32^{4/5} = (32^{1/5})^4 = 2^4 = 16$

9. **a.** $16^{0.75} = 16^{3/4} = (16^{1/4})^3 = 2^3 = 8$ **b.** $25^{3/2} = (25^{1/2})^3 = 5^3 = 125$

c. $27^{4/3} = (27^{1/3})^4 = 3^4 = 81$

10. **a.** $9^{1.5} = 9^{3/2} = (9^{1/2})^3 = 3^3 = 27$ **b.** $8^{4/3} = (8^{1/3})^4 = 2^4 = 16$

c. $4^{5/2} = (4^{1/2})^5 = 2^5 = 32$

11. a. $(a^2)^{3/2} = (a^{2/2})^3 = a^3$

b. $x^{1/3} \cdot x^{2/3} = x^{1/3 + 2/3} = x^1 = x$

c. $a^{1/4}a^{3/2} = a^{1/4 + 3/2} = a^{1/4 + 6/4} = a^{7/4}$

d. $\dfrac{b^{2/3}}{b^{1/2}} = b^{2/3-1/2} = b^{4/6-3/6} = b^{1/6}$

e. $x \cdot x^n = x^{n+1}$

f. $(x^3y^4)^{1/2} = x^{3/2}y^{4/2} = x^{3/2}y^2$

12. a. $(x^{1/2} \cdot y^{1/4})^2 = x^{2/2}y^{2/4} = xy^{1/2}$

b. $\dfrac{x^n}{x} = x^{n-1}$

c. $(x^{1/4})^{2/3} = x^{2/12} = x^{1/6}$

d. $\dfrac{a^{2/3}}{a^{4/3}} = a^{2/3-4/3} = a^{-2/3} = \dfrac{1}{a^{2/3}}$

e. $\dfrac{a}{a^x} = a^{1-x} = a^{-(x-1)} = \dfrac{1}{a^{x-1}}$

f. $\left(\dfrac{x^{1/2}}{y^2}\right)^{2/3} = \dfrac{x^{2/6}}{y^{4/3}} = \dfrac{x^{1/3}}{y^{4/3}}$

13. a. $\sqrt{9n} = 3\sqrt{n};\ n \ge 0$

b. $\sqrt{90n^2} = \sqrt{9n^2}\sqrt{10} = 3|n|\sqrt{10}$

c. $\sqrt{0.09x} = 0.3\sqrt{x};\ x \ge 0$

d. $\sqrt{900x^4} = 30x^2$

14. a. $\sqrt{0.9n^2} = \sqrt{9n^2}\sqrt{0.1} = 3|n|\sqrt{0.1}$

b. $\sqrt{9000x^2} = \sqrt{900x^2}\sqrt{10} = 30|x|\sqrt{10}$

c. $\sqrt{0.009x} = \sqrt{0.09}\sqrt{0.1x} = 0.3\sqrt{0.1x};\ x \ge 0$

d. $\sqrt{90x^3} = \sqrt{9x^2}\sqrt{10x} = 3x\sqrt{10x};\ x \ge 0$

15. a. $\left(\sqrt[5]{32}\right)^6 = 2^6 = 64$ **b.** $\left(\sqrt[3]{64}\right)^4 = 4^4 = 256$ **c.** $\left(\sqrt[4]{81}\right)^3 = 3^3 = 27$

16. a. $\sqrt[3]{-8} = (-8)^{1/3} = -2$

b. $\sqrt[4]{\dfrac{625}{16}} = \left(\dfrac{625}{16}\right)^{1/4} = \dfrac{5}{4}$

c. $\sqrt[3]{8^2} = (8)^{2/3} = (8^{1/3})^2 = 2^2 = 4$

17. a. $\sqrt[3]{\dfrac{27}{8}} = \left(\dfrac{27}{8}\right)^{1/3} = \dfrac{27^{1/3}}{8^{1/3}} = \dfrac{3}{2}$

b. $\sqrt[5]{-32} = (-32)^5 = -2$

c. $\sqrt{4^3} = 4^{3/2} = (4^{1/2})^3 = 2^3 = 8$

18. a. $\dfrac{\sqrt[3]{40}}{\sqrt[3]{5}} = \sqrt[3]{\dfrac{40}{5}} = \sqrt[3]{8} = 8^{1/3} = 2$

b. $\dfrac{-\sqrt[4]{2}}{\sqrt[4]{32}} = -\sqrt[4]{\dfrac{2}{32}} = -\sqrt[4]{\dfrac{1}{16}} = -\left(\dfrac{1}{16}\right)^{1/4} = -\dfrac{1}{2}$

c. $\dfrac{-\sqrt[3]{24}}{\sqrt[3]{81}} = -\sqrt[3]{\dfrac{24}{81}} = -\sqrt[3]{\dfrac{8}{27}} = -\left(\dfrac{8}{27}\right)^{1/3} = -\dfrac{2}{3}$

19. a. $125^{2/3} = \left(\sqrt[3]{125}\right)^2 = 5^2 = 25$

b. $64^{1/3} = \sqrt[3]{64} = 4$

c. $32^{2/5} = \left(\sqrt[5]{32}\right)^2 = 2^2 = 4$

20. a. $64^{3/2} = \left(\sqrt{64}\right)^3 = 8^3 = 512$

b. $32^{3/5} = \left(\sqrt[5]{32}\right)^3 = 2^3 = 8$

c. $27^{2/3} = \left(\sqrt[3]{27}\right)^2 = 3^2 = 9$

21. a. $(-64)^{1/3} = \sqrt[3]{-64} = -4$

b. $-64^{1/2} = -\sqrt{64} = -8$

c. $(-16)^{1/4}$ is not a real number.

22. a. $-16^{0.75} = -16^{3/4} = -\left(\sqrt[4]{16}\right)^3 = -2^3 = -8$

b. $(-64)^{1/2}$ is not a real number.

c. $-27^{1/3} = -\sqrt[3]{27} = -3$

23. a. $x^{2/3} = \left(\sqrt[3]{x}\right)^2$

b. $x^{1.5} = x^{3/2} = \left(\sqrt{x}\right)^3$

c. $a^{0.75} = a^{3/4} = \left(\sqrt[4]{a}\right)^3$

24. a. $a^{3/2} = \left(\sqrt{a}\right)^3$

b. $a^{1.25} = a^{5/4} = \left(\sqrt[4]{a}\right)^5$

c. $x^{0.8} = x^{4/5} = \left(\sqrt[5]{x}\right)^4$

25. a. $\sqrt{z}\sqrt{z^5} = \sqrt{z^6} = z^3$

b. $\sqrt[3]{x^2} \cdot \sqrt[3]{x} = \sqrt[3]{x^3} = x$

c. $\sqrt[3]{x^5} \cdot \sqrt[3]{x^1} = \sqrt[3]{x^6} = x^2$

26. a. $\sqrt[4]{32} = \sqrt[4]{16}\sqrt[4]{2} = 2\sqrt[4]{2}$

b. $\sqrt{72b^3} = \sqrt{36b^2}\sqrt{2b} = 6b\sqrt{2b}$

c. $\sqrt[4]{32x^6y^5z^4} = \sqrt[4]{16x^4y^4z^4}\sqrt[4]{2x^2y} = 2xyz\sqrt[4]{2x^2y}$

27. a. $\sqrt[3]{x} \cdot \sqrt[2]{x} = x^{1/3}x^{1/2} = x^{1/3+1/2} = x^{2/6+3/6} = x^{5/6}$

b. $\sqrt{x^3} \cdot \sqrt{x^2} = \sqrt{x^5} = x^{5/2}$

c. $\sqrt[2]{\sqrt[3]{x}} = \sqrt[2\cdot 3]{x} = \sqrt[6]{x} = x^{1/6}$

28. a. $\sqrt{x^3} \cdot \sqrt{x^4} = \sqrt{x^7} = x^{7/2}$

b. $\sqrt[3]{x} \cdot \sqrt[4]{x} = x^{1/3}x^{1/4} = x^{4/12+3/12} = x^{7/12}$

c. $\sqrt[3]{\sqrt[4]{x}} = \sqrt[3\cdot 4]{x} = \sqrt[12]{x} = x^{1/12}$

29. a. $\dfrac{\sqrt{x}}{\sqrt{x^3}} = \sqrt{\dfrac{x}{x^3}} = \sqrt{\dfrac{1}{x^2}} = \dfrac{1}{x}$

b. $\dfrac{\sqrt[3]{x^5}}{\sqrt[3]{x^2}} = \sqrt[3]{\dfrac{x^5}{x^2}} = \sqrt[3]{x^3} = x$

c. $\dfrac{\sqrt[3]{x}}{\sqrt[4]{x}} = \dfrac{x^{1/3}}{x^{1/4}} = x^{1/3-1/4} = x^{4/12-3/12} = x^{1/12}$

30. a. $\dfrac{\sqrt[4]{x}}{\sqrt[3]{x^2}} = \dfrac{x^{1/4}}{x^{2/3}} = \dfrac{1}{x^{2/3-1/4}} = \dfrac{1}{x^{8/12-3/12}} = \dfrac{1}{x^{5/12}}$

b. $\dfrac{\sqrt[3]{x^4}}{\sqrt[3]{x^7}} = \sqrt[3]{\dfrac{x^4}{x^7}} = \sqrt[3]{\dfrac{1}{x^3}} = \dfrac{1}{x}$ **c.** $\dfrac{\sqrt{x^5}}{\sqrt{x}} = \sqrt{\dfrac{x^5}{x}} = \sqrt{x^4} = x^2$

31. a. $\sqrt{3} + 2\sqrt{12} = \sqrt{3} + 2\sqrt{4}\sqrt{3} = \sqrt{3} + 4\sqrt{3} = 5\sqrt{3}$

b. $\sqrt{9x} + \sqrt{x} = 3\sqrt{x} + \sqrt{x} = 4\sqrt{x},\ x \ge 0$

32. a. $\sqrt{x} + \sqrt{x^3} = \sqrt{x} + x\sqrt{x} = (1+x)\sqrt{x};\ x \ge 0$ **b.** $\sqrt{16x} - \sqrt{x} = 4\sqrt{x} - \sqrt{x} = 3\sqrt{x};\ x \ge 0$

33. a. $(3 + \sqrt{6})(3 - \sqrt{6}) = 9 - 6 = 3$ **b.** $(3 - \sqrt{x})(3 - \sqrt{x}) = 9 - 6\sqrt{x} + x$

34. a. $(2 - \sqrt{2})(2 - \sqrt{2}) = 4 - 4\sqrt{2} + 2 = 6 - 4\sqrt{2}$ **b.** $(5 - \sqrt{2x})(5 + \sqrt{2x}) = 25 - 2x$

35. a. $\dfrac{1}{b-\sqrt{a}} \cdot \dfrac{b+\sqrt{a}}{b+\sqrt{a}} = \dfrac{b+\sqrt{a}}{b^2 - a}$ **b.** $\dfrac{\sqrt{x}-3}{\sqrt{x}+4} \cdot \dfrac{\sqrt{x}-4}{\sqrt{x}-4} = \dfrac{x - 7\sqrt{x} + 12}{x - 16}$

c. $\dfrac{a}{\sqrt[4]{a^2 b}} \cdot \dfrac{\sqrt[4]{a^2 b^3}}{\sqrt[4]{a^2 b^3}} = \dfrac{a\sqrt[4]{a^2 b^3}}{ab} = \dfrac{\sqrt[4]{a^2 b^3}}{b}$

36. a. $\dfrac{1}{\sqrt[3]{9x}} \cdot \dfrac{\sqrt[3]{3x^2}}{\sqrt[3]{3x^2}} = \dfrac{\sqrt[3]{3x^2}}{\sqrt[3]{27x^3}} = \dfrac{\sqrt[3]{3x^2}}{3x}$ **b.** $\dfrac{3x}{\sqrt[5]{x^3}} \cdot \dfrac{\sqrt[5]{x^2}}{\sqrt[5]{x^2}} = \dfrac{3x\sqrt[5]{x^2}}{\sqrt[5]{x^5}} = \dfrac{3x\sqrt[5]{x^2}}{x} = 3\sqrt[5]{x^2}$

c. $\dfrac{1}{\sqrt{a}+\sqrt{b}} \cdot \dfrac{\sqrt{a}-\sqrt{b}}{\sqrt{a}-\sqrt{b}} = \dfrac{\sqrt{a}-\sqrt{b}}{a-b}$

37. $(1-\sqrt{2})^2 - 2(1-\sqrt{2}) - 1 = 1 - 2\sqrt{2} + 2 - 2 + 2\sqrt{2} - 1 = 0$

To find the other, solution use quadratic formula:

$$x = \frac{-(-2) \pm \sqrt{(-2)^2 - 4(1)(-1)}}{2(1)} = \frac{2 \pm \sqrt{8}}{2} = \frac{2 \pm 2\sqrt{2}}{2} = 1 \pm \sqrt{2}$$

The other solution is $1 + \sqrt{2}$.

38. $(2-\sqrt{6})^2-4(2-\sqrt{6})-2=4-4\sqrt{6}+6-8+4\sqrt{6}-2=0$

To find the other solution, use quadratic formula:

$$x=\frac{-(-4)\pm\sqrt{(-4)^2-4(1)(-2)}}{2(1)}=\frac{4\pm\sqrt{24}}{2}=\frac{4\pm2\sqrt{6}}{2}=2\pm\sqrt{6}$$

The other solution is $2+\sqrt{6}$.

39. $x=\frac{1}{2}+\frac{\sqrt{7}}{2}$

40. a. $d=\sqrt{(-1-4)^2+(-6-6)^2}=\sqrt{25+144}=\sqrt{169}=13$

b. $d=\sqrt{(1-(-3))^2+(-1-2)^2}=\sqrt{16+9}=\sqrt{25}=5$

c. $d=\sqrt{(3-\sqrt{2})^2+(3-(-\sqrt{2}))^2}=\sqrt{9-6\sqrt{2}+2+9+6\sqrt{2}+2}=\sqrt{22}$

d. $d=\sqrt{(-\sqrt{3}-\sqrt{3})^2+(-\sqrt{5}-\sqrt{5})^2}=\sqrt{(-2\sqrt{3})^2+(-2\sqrt{5})^2}=\sqrt{12+20}=\sqrt{32}=4\sqrt{2}$

41. $\$7(1+0.04)^{10}=\$10.36$; $\$7(1+\frac{0.04}{2})^{2\cdot 10}=\$10.40$

42. $\$1200(1+\frac{0.06}{365})^{365\cdot 7}=\$1826.29$

43. $\$1600(1+\frac{0.05}{12})^{12\cdot 4}=\$1953.43$

44. $\$1200(1+\frac{0.06}{365})^{365\cdot 2.5}=\$1394.18$

45. $\$1600(1+\frac{0.05}{12})^{12\cdot 3.75}=\$1929.21$

46. $9.00=5.50(1+r)^{10}$ $\quad \frac{9.00}{5.50}=(1+r)^{10}$ $\quad \left(\frac{9.00}{5.50}\right)^{1/10}-1=r$ $\quad r\approx 5.05\%$

47. $8.00=5.50(1+r)^{10}$ $\quad \frac{8.00}{5.50}=(1+r)^{10}$ $\quad \left(\frac{8.00}{5.50}\right)^{1/10}-1=r$ $\quad r\approx 3.82\%$

48. $15{,}000=10{,}000(1+r)^5$ $\quad \frac{15{,}000}{10{,}000}=(1+r)^5$ $\quad (1.5)^{1/5}-1=r$ $\quad r\approx 8.45\%$

49. $17{,}000=7000(1+r)^{22}$ $\quad \frac{17{,}000}{7{,}000}=(1+r)^{22}$ $\quad \left(\frac{17}{7}\right)^{1/22}-1=r$ $\quad r\approx 7.67\%$

50. $r=\sqrt[3]{\frac{3V}{4\pi}}$; $r^3=\frac{3V}{4\pi}$; $V=\frac{4\pi r^3}{3}$

51. a. $x^3 + y^3 = a^3$; $y^3 = a^3 - x^3$; $y = \sqrt[3]{a^3 - x^3}$

52. $F = k\dfrac{Q_1Q_2}{r^2}$; $r^2 = k\dfrac{Q_1Q_2}{F}$; $r = \sqrt{k\dfrac{Q_1Q_2}{F}}$

53. $F = k\dfrac{M_1M_2}{r^2}$ $\quad M_1 = \dfrac{Fr^2}{kM_2}$

54. $r^3 = a^2p$; $r = \sqrt[3]{a^2p}$

55. $x^4 = a^2(x^2 - y^2)$ $\quad \dfrac{x^4}{a^2} = x^2 - y^2$ $\quad y^2 = x^2 - \dfrac{x^4}{a^2} = x^2(1 - \dfrac{x^2}{a^2})$

$y = \pm\sqrt{x^2\dfrac{(a^2 - x^2)}{a^2}} = \pm\dfrac{x}{a}\sqrt{a^2 - x^2}$

56. $\dfrac{27kx^2}{8} = (y-k)^3$; $y - k = \sqrt[3]{\dfrac{27kx^2}{8}}$; $y = \dfrac{3}{2}\sqrt[3]{kx^2} + k$

57. a. From the graph: x = 12

With symbols: $\sqrt{3x} + 2 = \sqrt{6x-8}$; $\left(\sqrt{3x} + 2\right)^2 = \left(\sqrt{6x-8}\right)^2$; $3x + 4\sqrt{3x} + 4 = 6x - 8$

$4\sqrt{3x} = 3x - 12$; $\left(4\sqrt{3x}\right)^2 = (3x-12)^2$; $16(3x) = 9x^2 - 72x + 144$; $48x = 9x^2 - 72x + 144$

$9x^2 - 120x + 144 = 0$; $(9x-12)(x-12) = 0$; $x = 12$; $x = \frac{4}{3}$ is an extraneous solution.

b. $\sqrt{6x-8} = 4$; $6x - 8 = 16$; $6x = 24$; $x = 4$

c. $\sqrt{3x} + 2 = -x + 8$; $\sqrt{3x} = -x + 6$; $\left(\sqrt{3x}\right) = (-x+6)^2$; $3x = x^2 - 12x + 36$;

$x^2 - 15x + 36 = 0$; $(x-3)(x-12) = 0$; $x = 3$; $x = 12$, with 12 being extraneous .

d. $3x - 8 = \sqrt{6x-8}$; $(3x-8)^2 = (\sqrt{6x-8})^2$; $9x^2 - 48x + 64 = 6x - 8$

$9x^2 - 54x + 72 = 0$; $(x-2)(x-4) = 0$; $x = 2$; $x = 4$, with 4 being extraneous.

58. a. From the graph: $x \approx 0.25$ or $x = 2$

With symbols: $\sqrt{8x} - 1 = \sqrt{5x-1}$; $\left(\sqrt{8x} - 1\right)^2 = \left(\sqrt{5x-1}\right)^2$; $8x - 2\sqrt{8x} + 1 = 5x - 1$;

$3x + 2 = 2\sqrt{8x}$; $(3x+2)^2 = \left(2\sqrt{8x}\right)^2$; $9x^2 + 12x + 4 = 4(8x)$; $9x^2 + 12x + 4 = 32x$;

$9x^2 - 20x + 4 = 0$; $(9x-2)(x-2) = 0$; $x = \frac{2}{9}$; $x = 2$

b. $2 = \sqrt{5x-1}$; $4 = 5x - 1$; $5x = 5$; $x = 1$

c. $\sqrt{5x-1} = 5x - 7$; $(\sqrt{5x-1})^2 = (5x-7)^2$; $5x - 1 = 25x^2 - 70x + 49$;

$25x^2 - 75x + 50 = 0$; $(x-2)(x-1) = 0$; $x = 1$; $x = 2$, with 1 being extraneous.

d. $\sqrt{8x}-1=-2x+2$; $\left(\sqrt{8x}\right)^2=(-2x+3)^2$; $8x=4x^2-12x+9$; $4x^2-20x+9=0$;

$(2x-1)(2x-9)=0$; $x=\frac{1}{2}$; $x=\frac{9}{2}$, with $\frac{9}{2}$ being extraneous.

59. a. From the graph: $x = 18$

With symbols:

$\sqrt{2x}+1=\sqrt{3x-5}$; $\left(\sqrt{2x}+1\right)^2=\left(\sqrt{3x-5}\right)^2$; $2x+2\sqrt{2x}+1=3x-5$; $2\sqrt{2x}=x-6$;

$\left(2\sqrt{2x}\right)^2=(x-6)^2$; $4(2x)=x^2-12x+36$; $8x=x^2-12x+36$;

$x^2-20x+36=0$; $(x-2)(x-18)=0$; $x=18$; $x=2$ is an extraneous solution.

b. $\sqrt{2x}+1=5$; $\sqrt{2x}=4$; $2x=16$; $x=8$

c. $\sqrt{3x-5}=x-3$; $\left(\sqrt{3x-5}\right)^2=(x-3)^2$; $3x-5=x^2-6x+9$;

$x^2-9x+14=0$; $(x-2)(x-7)=0$; $x=2$; $x=7$; $x=2$ is an extraneous solution.

d. $\sqrt{2x}+1=-x+5$; $\sqrt{2x}=-x+4$; $(\sqrt{2x})^2=(-x+4)^2$; $2x=x^2-8x+16$

$x^2-10x+16=0$; $(x-8)(x-2)=0$; $x=2$; $x=8$ is an extraneous solution.

60. a. Defined for $x \geq 5/3$; $\left(\sqrt{3x-5}\right)^2=(-3x+7)^2$;$3x-5=9x^2-42x+49$;$9x^2-45x+54=0$;

$(9x-18)(x-3)=0$;$x=2,3$;$\sqrt{3(2)-5}=-3(2)+7$;$1=1$;$\sqrt{3(3)-5}=-3(3)+7$;$2=-2$

$x = 3$ is extraneous. So, $x = 2$.

b. Defined for all x; $2x+14=27$;$x=\frac{13}{2}=6.5$;$\sqrt[3]{2(6.5)+14}=3$;$3=3$;$x=6.5$

c. Defined for $x \geq 1/3$; $3x+1=\left(\frac{7-3}{2}\right)^4$;$3x=15$;$x=5$;$2\sqrt[4]{3(5)+1}+3=7$;$7=7$;$x=5$

d. Defined for all x; $11-4x=3^3$;$x=-4$;$\sqrt[3]{11-4(-4)}-10=-7$;$-7=-7$;$x=-4$

61. a. $AP=\sqrt{(20-x)^2+10^2}$; $PB=\sqrt{x^2+20^2}$; total length $=\sqrt{x^2+20^2}+\sqrt{(20-x)^2+10^2}$

b. $x \approx 13.33$ km (Use a spreadsheet or graph to find min. total length.)

c. If the villages are equal distances from the river, the pump would be placed halfway between them, at $x = 10$ km.

Chapter 5 Test

1. a. $\dfrac{a^0b^2}{b^{-1}} = 1 \cdot b^{2-(-1)} = b^3$

b. $\dfrac{a^{-2}b^3}{(bc)^2} = \dfrac{b^3}{a^2b^2c^2} = \dfrac{b}{a^2c^2}$

c. $\left(\dfrac{x^2y}{3x}\right)^{-2} = \left(\dfrac{3x}{x^2y}\right)^2 = \dfrac{9x^2}{x^4y^2} = \dfrac{9}{x^2y^2}$

2. a. $(x^{-1}y^3)^2 = x^{-2}y^6 = \dfrac{y^6}{x^2}$

b. $a^6 \cdot a^3 = a^{6+3} = a^9$

c. $\dfrac{x}{x^n} = x^{1-n} = x^{-(n-1)} = \dfrac{1}{x^{n-1}}$

3. a. $\dfrac{b^{3/4}}{b^2} = \dfrac{1}{b^{2-0.75}} = \dfrac{1}{b^{1.25}}$

b. $\left(\dfrac{a^{1/3}}{b^{1/2}}\right)^6 = \dfrac{a^{6/3}}{b^{6/2}} = \dfrac{a^2}{b^3}$

c. $\left(x^{3/4}y^{2/3}\right)^6 = x^{18/4}y^{12/3} = x^{4.5}y^4$

4. a. $(b^2)^{3/2} = b^{6/2} = b^3$

b. $b^{2/3}b^{1/3} = b^{3/3} = b$

c. $3 \cdot 3^n = 3^{1+n}$

5. $a\left(\dfrac{b}{a}\right)^4 = a\dfrac{b^4}{a^4} = \dfrac{b^4}{a^3}$; $b\left(\dfrac{b}{a}\right)^3 = b\dfrac{b^3}{a^3} = \dfrac{b^4}{a^3}$; the expressions are equal.

6. a. 0.000000000000000380

b. 423000000000000

c. The division of 6.3 by 7 give 0.9 or 9×10^{-1} and the division of the tens give 10^{-6}. The result is adds the exponents on the tens to give 9×10^{-7}.

7. a. $\sqrt{36x} = 6\sqrt{x};\ x \geq 0$

b. $\sqrt{0.36x^2} = 0.6|x|$

c. $\sqrt{360x^4} = \sqrt{36x^4}\sqrt{10} = 6x^2\sqrt{10}$

8. a. $64^{2/3} = \left(\sqrt[3]{64}\right)^2 = 4^2 = 16$

b. $32^{4/5} = \left(\sqrt[5]{32}\right)^4 = 2^4 = 16$

c. $81^{3/4} = \left(\sqrt[4]{81}\right)^3 = 3^3 = 27$

9. a. $\sqrt[4]{\dfrac{16}{81}} = \dfrac{\sqrt[4]{16}}{\sqrt[4]{81}} = \dfrac{2}{3}$

b. $\sqrt[3]{-27} = -3$

c. $-\sqrt[3]{-64} = -(-4) = 4$

10. a. $\sqrt[3]{\sqrt[5]{x^{10}}} = \sqrt[15]{x^{10}} = x^{10/15} = x^{2/3}$

b. $\sqrt[3]{x^2}\sqrt[3]{x^4} = \sqrt[3]{x^{2+4}} = \sqrt[3]{x^6} = x^{6/3} = x^2$

c. $\dfrac{\sqrt[3]{54x^2}}{\sqrt[3]{128x^{-1}}} = \sqrt[3]{\dfrac{54x^{2-(-1)}}{128}} = \sqrt[3]{\dfrac{27x^3}{64}} = \dfrac{\sqrt[3]{27x^3}}{\sqrt[3]{64}} = \dfrac{3x}{4}$

11. a. $(2 - \sqrt{x})(2 + \sqrt{x}) = 4 - x$

b. $(8 - \sqrt{3})(3 + \sqrt{8}) = 24 + 8\sqrt{8} - 3\sqrt{3} - \sqrt{3}\sqrt{8} = 24 + 16\sqrt{2} - 3\sqrt{3} - 2\sqrt{6}$

12. a. $\dfrac{1}{x+\sqrt{y}} \cdot \dfrac{x-\sqrt{y}}{x-\sqrt{y}} = \dfrac{x-\sqrt{y}}{x^2-y}$

b. $\dfrac{1}{\sqrt[3]{x}} \cdot \dfrac{\sqrt[3]{x^2}}{\sqrt[3]{x^2}} = \dfrac{\sqrt[3]{x^2}}{\sqrt[3]{x^3}} = \dfrac{\sqrt[3]{x^2}}{x}$

13. Yes; $(4-\sqrt{10})^2 - 8(4-\sqrt{10}) + 6 = 16 - 8\sqrt{10} + 10 - 32 + 8\sqrt{10} + 6 = 0$

14. $2000\left(1+\dfrac{0.08}{4}\right)^{4\cdot 3} = \$2536.48$

15. $18{,}000 = 12{,}000(1 + \mathrm{r})^5$

$$\left(\frac{18{,}000}{12{,}000}\right) = (1+r)^5$$

$$\left(\frac{18{,}000}{12{,}000}\right)^{1/5} = 1+r$$

$$r = \left(\frac{18{,}000}{12{,}000}\right)^{1/5} - 1 \approx 8.45\%$$

16. $M = \dfrac{mgl}{\pi r^2 s}$

$$r^2 = \frac{mgl}{M\pi s}$$

$$r = \sqrt{\frac{mgl}{M\pi s}}$$

17. $d = \dfrac{Wl^3}{3EI}$; $l^3 = \dfrac{3EId}{W}$; $l = \sqrt[3]{\dfrac{3EId}{W}}$

18. a. From the graph, x = 8.

With symbols: $\sqrt{x+1}=\sqrt{2x}-1$; $\left(\sqrt{x+1}\right)^2=\left(\sqrt{2x}-1\right)^2$; $x+1=2x-2\sqrt{2x}+1$; $2\sqrt{2x}=x$

$\left(2\sqrt{2x}\right)^2=x^2$; $4(2x)=x^2$; $x^2-8x=0$; $x(x-8)=0$; $x=8$; $x=0$ is an extraneous solution.

b. $\sqrt{x+1}=2$; $\left(\sqrt{x+1}\right)^2=2^2$; $x+1=4$; $x=3$

19. a. $\sqrt{2x}-1=x-5$; $\sqrt{2x}=x-4$; $\left(\sqrt{2x}\right)^2=(x-4)^2$; $2x=x^2-8x+16$;

$x^2-10x+16=0$; $(x-2)(x-8)=0$; $x=8$; $x=2$ is an extraneous solution.

b. $\sqrt{x+1}=-\frac{1}{5}x-\frac{7}{5}$; $\left(\sqrt{x+1}\right)^2=\left(-\frac{1}{5}x-\frac{7}{5}\right)^2$; $x+1=\frac{1}{25}x^2+\frac{14}{25}x+\frac{49}{25}$;

$x^2-11x+24=0$; $(x-3)(x-8)=0$; $x=3$, $x=8$; both are extraneous solutions.

20. a. If n is odd, y = x; if n is even, y = |x|.

b. If n is even, $\sqrt[n]{x}$ is called the principal n^{th} root of x. If n is odd, the odd n^{th} root of x is the number $\sqrt[n]{x}$.

Cumulative Review: Chapters 1 to 5

1. $5 - 3(x + 2) = 32 \qquad 5 - 3x - 6 = 32 \qquad -3x = 33 \quad x = -11$

2. $\frac{6+9x}{15} = \frac{3(2+3x)}{3 \cdot 5} = \frac{2+3x}{5}$

3. $A = \frac{a+b+c}{3} \qquad 3A = a+b+c \qquad b = 3A - a - c$

4. $3 - x < 16 < 10 - x$ *Add x to all sides.*

$3 < 16 + x < 10$ *Subtract 16 from all sides.*

$-13 < x < -6$

5. Let y = total cost in \$; x = kilowatt-hours used; assume a linear equation.

$m = \frac{83.33 - 68.46}{1410 - 1120} \approx 0.051$; $83.33 = 0.051(1410) + b$; $b = 11.42$; $y \approx 0.051x + 11.42$

If calculator linear regression is used, equation is: $y \approx 0.051x + 11.03$, round off error.

6. $f(3) = 3 - (3) = 0$; $f(-2) = 3 - (-2) = 3 + 2 = 5$

7. a. $\frac{-2-1}{4-(-1)} = -\frac{3}{5}$ **b.** Negative reciprocal; $\frac{5}{3}$

8. Solve for y first: $3y = 6x + 3$; $y = 2x + 1$, slope is 2; parallel function through the origin has $b = 0$; $y = 2x$.

9. Inputs: $\Re$. Outputs: 4.

10. $50(2.72) + x(4.00) = (50 + x)(3.00)$

$136 + 4x = 150 + 3x \qquad x = 14$ credits of A

11.

Loan Amount	Interest %	Interest
x (credit card)	20% = 0.20	0.20x
y (credit union)	7.5% = 0.075	0.075y
x + y	10% - 0.10	0.10(x + y)

$0.20x + 0.075y = 0.10(x + y)$; $0.20x - 0.10x = 0.10y - 0.075y$; $0.10x = 0.025y$

Set up ratio of x to y; $\frac{x}{y} = \frac{0.025}{0.10} = \frac{1}{4}$; \$4 at credit union for each \$1 on credit card.

12. a. $f(-2) = 2(-2)^2 - 5(-2) + 6 = 8 + 10 + 6 = 24$; $f(0) = 2(0)^2 - 5(0) + 6 = 6$

b. $f(0)$ is the y-intercept or $f(x)$-intercept on the graph.

c. x coordinate of vertex is $\frac{-(-5)}{2(2)} = \frac{5}{4} = 1\frac{1}{4} = 1.25$;

y coordinate is $f(1.25) = 2(1.25)^2 - 5(1.25) + 6 = 3.125 - 6.25 + 6 = 2.875 = 2\frac{7}{8}$

d. $x = \frac{-b \pm \sqrt{b^2 - 4ac}}{2a}$; $x = \frac{-(-5) \pm \sqrt{(-5)^2 - 4(2)(6)}}{2(2)} = \frac{5 \pm \sqrt{-23}}{4}$; no real x-intercepts.

13. a. $\{-4, 1\}$ **b.** $\{-3, 0\}$ **c.** No real number solution

14. a. $x^2 + 3x - 4 = 0$

$(x + 4)(x - 1) = 0$

$x + 4 = 0$; $x = -4$; or $x - 1 = 0$; $x = 1$

$\{-4, 1\}$

b. $x^2 + 3x = 0$

$x(x + 3) = 0$

$x = 0$ or $x + 3 = 0$; $x = -3$

$\{-3, 0\}$

15. $x < -3$ or $x > 0$

16. $x^2 + 3x + 5 = 0; x = \frac{-3 \pm \sqrt{3^2 - 4(1)(5)}}{2(1)} = \frac{-3 \pm i\sqrt{11}}{2}$

17. Shift $y = x^2$ a units horizontally.

18. $(x + \frac{7}{2})^2 = x^2 + 7x + \frac{49}{4}$

19.

Multiply	x^3	$-3x^2$	$+3x$	-1
x	x^4	$-3x^3$	$+3x^2$	$-x$
−1	$-x^3$	$+3x^2$	$-3x$	$+1$

$(x^3 - 3x^2 + 3x - 1)(x - 1) = x^4 - 4x^3 + 6x^2 - 4x + 1$

20. a.

x	y	Differences	
		1st	2nd
1	2		
		3	
2	5		2
		5	
3	10		2
		7	
4	17		2
		9	
5	26		

$26 + (9 + 2) = 37$; quadratic

$2a = 2,\ a = 1;\ \ c = 2 - (3 - 2) = 1$

$2 = 1(1)^2 + b(1) + 1;\ b = 0$

$y = x^2 + 1$

b.

x	y	Differences	
		1st	2nd
1	−5		
		3	
2	−2		0
		3	
3	1		0
		3	
4	4		0
		3	
5	7		

$7 + 3 = 10$; linear

$m = 3;\ \ b = -5 - 3 = -8$

$y = 3x - 8$

c.

x	y	Differences	
		1st	2nd
1	−3		
		4	
2	1		2
		6	
3	7		2
		8	
4	15		2
		10	
5	25		

$25 + (10 + 2) = 37$; quadratic

$2a = 2;\ a = 1;\ \ c = -3 - (4 - 2) = -5$

$-3 = 1(1)^2 + b(1) - 5;\ b = 1$

$y = x^2 + x - 5$

d.

x	y	Differences	
		1st	2nd
1	−2		
		1	
2	−1		1
		2	
3	1		2
		4	
4	5		4
		8	
5	13		

Notice pattern in 1st and 2nd differences.

Next number is $13 + (8 + 8) = 29$.

Neither quadratic nor linear.

21. a. $(8-2i)(8+2i)=8^2-(2i)^2=64-4(-1)=64+4=68$

b. $(5-4i)(3+2i)=15-2i-8i^2=15-2i+8=23-2i$

c. $\frac{1}{2+i}\cdot\frac{2-i}{2-i}=\frac{2-i}{4-(-1)}=\frac{2-i}{5}$

22. $x^3-1=(x-1)(x^2+x+1)=0$; $(x-1)=0$, $x=1$ or $x=\frac{-1\pm\sqrt{1^2-4(1)(1)}}{2(1)}$

$x=\frac{-1\pm\sqrt{-3}}{2}=-\frac{1}{2}\pm\frac{i\sqrt{3}}{2}$ $\left\{1,\ -\frac{1}{2}\pm i\frac{\sqrt{3}}{2}\right\}$

23. $3.5^2+12^2=12.25+144=156.25=12.5^2$, right triangle

24. $3^2+x^2=12^2; x^2=135; x=11.62$ feet; 76° (by guess and check)

25. $\frac{\$8.99}{\text{package}}\cdot\frac{1\text{ package}}{56\text{ diapers}}\cdot\frac{10\text{ diapers}}{1\text{ day}}\cdot 365\text{ days}\approx\$585.96$

26. a. $1+\frac{b}{a-b}=\frac{a-b}{a-b}+\frac{b}{a-b}=\frac{a}{a-b}$, **b.** $\frac{x}{y-x}+\frac{y}{x-y}=-\frac{x}{x-y}+\frac{y}{x-y}=\frac{y-x}{x-y}=-1$

27. $x^2-3x+9-\frac{19}{x+3}$; an asymptote at x = -3; the value of x^3+8 at x = -3 is -19.

28. $\frac{x^2+3x}{x-1}\cdot\frac{x^2-1}{3x^2}=\frac{x(x+3)(x-1)(x+1)}{(x-1)x(3x)}=\frac{(x+3)(x+1)}{3x}$

29. 0.825 billion $=8.25\times10^8$; $3(8.25\times10^8)=2.475\times10^9=2{,}475{,}000{,}000$

30. $19\le\frac{(150)(704.5)}{x^2}$; $x^2\le\frac{(150)(704.5)}{19}$; $x\le 75$ in. or

$\frac{(150)(704.5)}{x^2}\le 24$; $\frac{(150)(704.5)}{24}\le x^2$; $66\le x$; $\approx 66\text{ in}\le x\le 75\text{ in}$

31. a. $32^{4/5} = (32^{1/5})^4 = 2^4 = 16$

b. $\sqrt[3]{125^2} = 125^{2/3} = (125^{1/3})^2 = 5^2 = 25$

c. $\sqrt[4]{-81}$; no real number solution

d. $-\sqrt{2}\sqrt{8} = -\sqrt{16} = -4$

e. $x^{3/5} \cdot x^{1/3} = x^{3/5 + 1/3} = x^{9/15 + 5/15} = x^{14/15}$

f. $\dfrac{x^{1/2}}{x^{-3/2}} = x^{1/2-(-3/2)} = x^{4/2} = x^2$

g. $\sqrt{18a^3} = \sqrt{9a^2}\sqrt{2a} = 3a\sqrt{2a}$

h. $\sqrt[3]{16x^3y^4z^5} = \sqrt[3]{8x^3y^3z^3}\sqrt[3]{2yz^2} = 2xyz\sqrt[3]{2yz^2}$

i. $(x^{2/3})^{1/2} = x^{2/6} = x^{1/3}$

j. $\sqrt{8x} + \sqrt{2x} + \sqrt{50x} = 2\sqrt{2x} + \sqrt{2x} + 5\sqrt{2x} = 8\sqrt{2x}$

k. $(2-\sqrt{2})(2+\sqrt{2}) = 4-2 = 2$

32. a. $x + 2 = 3\sqrt{x}$

$(x+2)^2 = (3\sqrt{x})^2$

$x^2 + 4x + 4 = 9x$

$x^2 - 5x + 4 = 0$

$(x - 4)(x - 1) = 0$

$x = 4$ or $x = 1$

Use calculator graphs to check.

Both solutions are found on graph.

$\{1, 4\}$

b. $x = \sqrt{6-x}$

$x^2 = (\sqrt{6-x})^2$

$x^2 = 6 - x$

$x^2 + x - 6 = 0$

$(x + 3)(x - 2) = 0$

$x = -3$ or $x = 2$

Use calculator graphs to check.

Only $x = 2$ is found on the graph.

$\{2\}$

33. $(-2, 2)$ & $(4, 1)$ $d = \sqrt{(4-(-2))^2 + (1-2)^2} = \sqrt{36+1} = \sqrt{37}$

$(-2, 2)$ & $(3, -5)$ $d = \sqrt{(3-(-2))^2 + (-5-2)^2} = \sqrt{25+49} = \sqrt{74}$

$(4, 1)$ & $(3, -5)$ $d = \sqrt{(3-4)^2 + (-5-1)^2} = \sqrt{1+36} = \sqrt{37}$

$\left(\sqrt{37}\right)^2 + \left(\sqrt{37}\right)^2 = \left(\sqrt{74}\right)^2$; figure is an isosceles right triangle.

34. $W = nB^{1.6}$; $W = nB^{8/5}$; $B^{8/5} = \dfrac{W}{n}$; $B = \left(\dfrac{W}{n}\right)^{5/8}$

CHAPTER 6

Exponential and Logarithmic Functions

Exercises 6.1: Exponential Functions

1. a. a^x; base is a, exponent is x.

b. $4x^3$; base is x, exponent is 3.

c. 2^{-x}; base is 2, exponent is –x.

d. x^a; base is x, exponent is a.

3. $2^{x+2} = 2^x \cdot 2^2 = 2^x \cdot 4 = 4 \cdot 2^x$

5. $2^{x-1} = 2^x \cdot 2^{-1} = 2^x \cdot \frac{1}{2} = \dfrac{2^x}{2}$

7. $(\frac{1}{2})^{-x} = (\frac{2}{1})^x = 2^x$

9. $4 \cdot 2^n = 2^2 \cdot 2^n = 2^{n+2}$

11. $\dfrac{3^n}{3} = 3^n \cdot 3^{-1} = 3^{n-1}$

13. $9 \cdot 3^n = 3^2 \cdot 3^n = 3^{n-1}$

15. No common ratio; 1st differences = 6, 10, 14; 2nd differences = 4, 4; quadratic function.
$2a = 4$, $a = 2$; $c = 2 - (6 - 4) = 0$; $2(1)^2 + b(1) = 2$, $b = 0$; $y = 2x^2$

17. Common ratio is 3; exponential function; $a_n = 0.5 \cdot 3^{n-1}$; $y = 0.167 \cdot 3^x$

19. No common ratio; 1st differences = 4, 5, 6; 2nd differences = 1, 1; quadratic function.
$2a = 1$, $a = 0.5$; $c = 3 - (4 - 1) = 0$; $0.5(1)^2 + b(1) = 3$, $b = 2.5$; $y = 0.5x^2 + 2.5x$

21. No common ratio; 1st differences = 8, 8, 8, 8; linear function.
$m = 8$; $30 = 8(1) + b$, $b = 22$; $y = 8x + 22$

23. Common ratio is 2; exponential function; $a_n = 5 \cdot 2^{n-1}$; $y = 2.5 \cdot 2^x$.

25. a. $13.5 \div 4.5 = 3$; $4.5 \div 1.5 = 3$; $1.5 \div 0.5 = 3$; common ratio is 3.

b. $a_1 = 0.5$, $r = 3$; $a_n = 0.5 \cdot 3^{n-1}$

c. $y = 0.167 \cdot 3^x$

d. $0.5 \cdot 3^{n-1} = 0.5 \cdot 3^n \cdot 3^{-1} = \frac{0.5}{3} \cdot 3^n = 0.167 \cdot 3^n$

27. a. $80 \div 40 = 2$; $40 \div 20 = 2$; $20 \div 10 = 2$; common ratio is 2

b. $a_1 = 10$, $r = 2$; $a_n = 10 \cdot 2^{n-1}$

c. $y = 5 \cdot 2^x$

d. $10 \cdot 2^{n-1} = 10 \cdot 2^n \cdot 2^{-1} = \frac{10}{2} \cdot 2^n = 5 \cdot 2^n$

29. a. $\frac{1}{27} \div \frac{1}{9} = \frac{1}{27} \cdot \frac{9}{1} = \frac{1}{3}$; $\frac{1}{9} \div \frac{1}{3} = \frac{1}{3}$; $\frac{1}{3} \div 1 = \frac{1}{3}$; $1 \div 3 = \frac{1}{3}$; common ratio is $\frac{1}{3}$.

b. $a_1 = 3, r = \frac{1}{3}; a_n = 3 \cdot (\frac{1}{3})^{n-1}$

c. $y = 9 \cdot 0.333^x$

d. $3 \cdot (\frac{1}{3})^{n-1} = 3 \cdot (\frac{1}{3})^n \cdot (\frac{1}{3})^{-1} = 3 \cdot 3 \cdot (\frac{1}{3})^n = 9 \cdot (\frac{1}{3})^n = 9 \cdot 0.333^n$

31. a. $2 \div 4 = \frac{1}{2}$; $4 \div 8 = \frac{1}{2}$; $8 \div 16 = \frac{1}{2}$; $16 \div 32 = \frac{1}{2}$; common ratio is $\frac{1}{2}$.

b. $a_1 = 32, r = \frac{1}{2}; a_n = 32 \cdot (\frac{1}{2})^{n-1}$

c. $y = 64 \cdot 0.5^x$

d. $32 \cdot (\frac{1}{2})^{n-1} = 32 \cdot (\frac{1}{2})^n \cdot (\frac{1}{2})^{-1} = 32 \cdot 2 \cdot (\frac{1}{2})^n = 64 \cdot (\frac{1}{2})^n = 64 \cdot 0.5^n$

33. $-6 \div 2 = -3$; $18 \div -6 = -3$; $r = -3$; $a_n = 2 \cdot (-3)^{n-1}$; base is negative, not exponential .

35. $-32 \div 64 = -\frac{1}{2}$; $16 \div -32 = -\frac{1}{2}$; $r = -\frac{1}{2}$; $a_n = 64 \cdot (-\frac{1}{2})^{n-1}$; base is negative, not exponential.

37. a. 1.25 or (1 + 0.25); 25%

b. 0.75 or (1 – 0.25); -25%

39. a. 2.5 or (1 + 1.5); 150%

b. 0.125 or (1 – 0.875); -87.5%

41. a. Linear; $y \approx 6400x + 2600$; rate of growth = 6400 per year; $6400(80)+2600 \approx 514{,}600$; actual: 515,000

b. Exponential; $y \approx 16900(1.03)^x$; rate of growth = 3% per year; $16900(1.03)^{110} \approx 436{,}500$; actual: 436,000,

c. Quadratic; $y \approx 2.89x^2 - 253x + 43{,}400$; $2.89(80)^2 - 253(80) + 43400 \approx 41{,}700$; actual: 41,700

d. Quadratic; $y \approx 6.93x^2 - 161x + 40000$; $6.93(80)^2 - 161(80) + 40000 \approx 71{,}500$; actual: 71,500

43. a. 2 inch ÷ 500 = 0.004 in.

b. 10 folds is 2^{10} sheets; $2^{10} \cdot 0.004 = 4.096$ in.

c. 5280 ft · 12 in/ft = 63360 in.; 63360 ÷ 0.004 = 15,840,000 sheets, or $\approx 2^{24}$ or 24 folds

d. 1 mile = 15,840,000 sheets; 93 million miles = 1.47312×10^{15} sheets, which is between 2^{50} and 2^{51}, so 51 folds would be needed to reach the sun.

45. a. An appropriate window would be X[0, 110] Scl 10; Y[10^{10}, 1.5 x 10^{14}] Scl 10^{12}.

b. The data appears like the power of 2 curve; $y = (1.99 \times 10^{10})(1.07)^x$.

c. The base is 1.07, or 7% per year.

d. Answers will vary depending on the current year. Go to http://www.treasurydirect.gov/NP/BPDLogin?application=np to look up the current year.

Exercises 6.2: Exponential Equations and Graphs

1. a. $2^x = 256$

$2^x = 2^8$

$x = 8$

b. $16^x = 256$

$(2^4)^x = 2^8$

$2^{4x} = 2^8$

$4x = 8$

$x = 2$

3. a. $4^n = 8$

$(2^2)^n = 2^3$

$2^{2n} = 2^3$

$2n = 3$

$n = \frac{3}{2}$

b. $8^n = 16$

$(2^3)^n = 2^4$

$2^{3n} = 2^4$

$3n = 4$

$n = \frac{4}{3}$

5. a. $16^n = 4$

$(2^4)^n = 2^2$

$2^{4n} = 2^2$

$4n = 2$

$n = \frac{1}{2}$

b. $100^x = 1000$

$(10^2)^x = 10^3$

$10^{2x} = 10^3$

$2x = 3$

$x = \frac{3}{2}$

7. a. $64^x = \frac{1}{8}$

$(2^6)^x = 2^{-3}$

$2^{6x} = 2^{-3}$

$6x = -3$

$x = -\frac{1}{2}$

b. $125^x = \frac{1}{5}$

$(5^3)^x = 5^{-1}$

$5^{3x} = 5^{-1}$

$3x = -1$

$x = -\frac{1}{3}$

9. a. $8^n = \frac{1}{2}$

$(2^3)^n = 2^{-1}$

$2^{3n} = 2^{-1}$

$3n = -1$

$n = -\frac{1}{3}$

b. $81^x = \frac{1}{3}$

$(3^4)^x = 3^{-1}$

$3^{4x} = 3^{-1}$

$4x = -1$

$x = -\frac{1}{4}$

11. a. $(\frac{1}{2})^{x+5} = 32$

$(2^{-1})^{x+5} = 2^5$

$2^{-x-5} = 2^5$

$-x - 5 = 5$

$-x = 10$

$x = -10$

b. $(\frac{1}{8})^{x+3} = 8$

$(2^{-3})^{x+3} = 2^3$

$2^{-3x-9} = 2^3$

$-3x - 9 = 3$

$-3x = 12$

$x = -4$

13. a. $2^{x-1} = 2^{2x+5}$

$x - 1 = 2x + 5$

$-6 = x$

b. $3^{2x-1} = 3^{x+3}$

$2x - 1 = x + 3$

$x = 4$

15. a. $2^{x+5} = 4^{x-1}$

$2^{x+5} = (2^2)^{x-1}$

$2^{x+5} = 2^{2x-2}$

$x + 5 = 2x - 2$

$7 = x$

b. $5^{x+1} = 125^{x-3}$

$5^{x+1} = (5^3)^{x-3}$

$5^{x+1} = 5^{3x-9}$

$x + 1 = 3x - 9$

$10 = 2x$

$x = 5$

17. a. $b > 1$, increasing

b. $b < 1$, decreasing

19. a. $b > 1$, increasing

b. $b < 1$, decreasing

21. a. $b > 1$, increasing

b. $b > 1$, increasing

23. a. $b > 1$, decreasing

b. $b > 1$, decreasing

25. a. $b < 1$, increasing

b. $b < 1$, increasing

27. a. $y = 2^x \cdot 2^{-4}$; $y = \frac{1}{16} \cdot 2^x$; y-intercept is $\left(0, \frac{1}{16}\right)$.

b. $y = 2^x \cdot 2^1$; $y = 2 \cdot 2^x$; y-intercept is (0, 2).

29. Use $x = 1$ to identify graphs; $a = 10^x$, $b = 2.72^x$, $c = 2^x$.

31.

x	$y = 2^{x-1}$	$y = 2^x$	$y = 2^{x+1}$
−2	$2^{-2-1} = 2^{-3} = \frac{1}{8}$	$2^{-2} = \frac{1}{4}$	$2^{-2+1} = 2^{-1} = \frac{1}{2}$
−1	$2^{-1-1} = 2^{-2} = \frac{1}{4}$	$2^{-1} = \frac{1}{2}$	$2^{-1+1} = 2^0 = 1$
0	$2^{0-1} = 2^{-1} = \frac{1}{2}$	$2^0 = 1$	$2^{0+1} = 2^1 = 2$
1	$2^{1-1} = 2^0 = 1$	$2^1 = 2$	$2^{1+1} = 2^2 = 4$
2	$2^{2-1} = 2^1 = 2$	$2^2 = 4$	$2^{2+1} = 2^3 = 8$
graph	c	b	a

Adding 1 shifts the graph 1 unit to the left. Subtracting 1 shifts graph 1 unit to the right.

33.

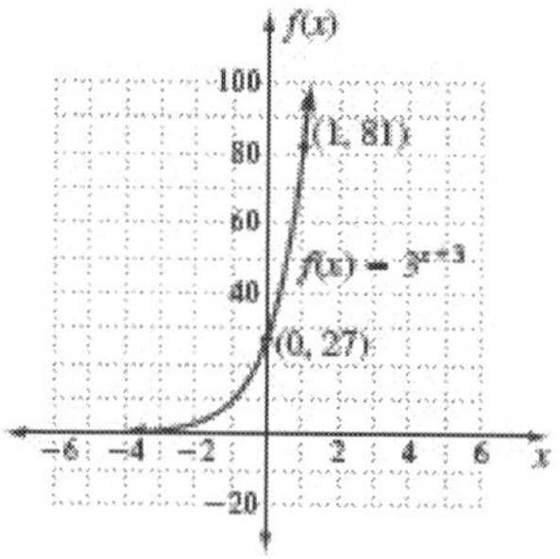

35.

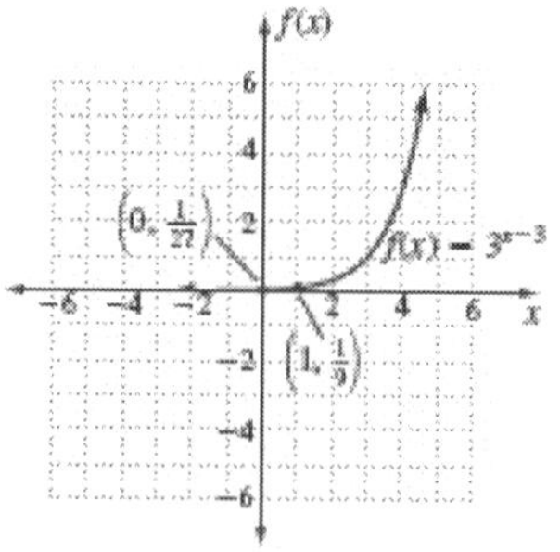

37. Graph will never be below the x-axis. { } or ∅.

39. $4^{6t} = (2^2)^{6t} = 2^{12t}$; $8^{4t} = (2^3)^{4t} = 2^{12t}$; rules are the same.

41. When $x = 1$, $b^x = b$.

43. They do not intersect. $3^x < 3^{x+1}$ for all x.

45. a. y-intercept: $(0, -1)$ **b.** $y = -2^x$ and $y = -2^{-x}$ or $y = -(\frac{1}{2})^x$

c. A ***negative*** coefficient causes the exponential graph to reflect over the x-axis.

47. $a = 2$; $y = a \cdot b^x$; $y = 2 \cdot b^x$; $12 = 2 \cdot (b)^1$; $6 = b$; $y = 2 \cdot 6^x$

49. $a = 3$; $y = a \cdot b^x$; $y = 3 \cdot b^x$; $4.5 = 3 \cdot (b)^1$; $1.5 = b$; $y = 3 \cdot (1.5)^x$

51. a. $x \approx 2.5$ **b.** $x \approx 4.8$ **c.** $y \approx 1.8$

d. $y \approx 4.5$ **e.** $a \approx 1.34$

For exercises 53 and 55 a calculator is used for graphs. Answers are given for reference only.

53. a. $y = 1$ **b.** Graph on calculator. **c.** Horizontal line at $y = 1$

d. yes **e.** $y = 1$

55. a. Graph on calculator.

b. Two sets of points, increasing above x-axis and decreasing below x-axis.

c. no **d.** no; negative base

57. For each graph, make sure to clear out any equations after pressing Y=. Then, enter the left side of the given equation into Y1 and the right side of the equation into Y2. [2nd][TRACE][5][ENTER][ENTER] move your curser close to a point of intersection (adjust window as needed)[ENTER]. Repeat as necessary to find all points of intersection of the two curves.

a. {-0.58, 1, 6.32} **c.** {-1.57, 0}

b. {-0.85, 0, 2, 3.22} **d.** {-1.86, 0.77}

Exercises 6.3: Finding Inverse Functions

1. $(\frac{1}{2}, 1), (\frac{1}{4}, 2), (\frac{1}{8}, 3), (\frac{1}{16}, 4)$

3. $(2, 1), (5, 2), (8, 3), (11, 4)$

5.

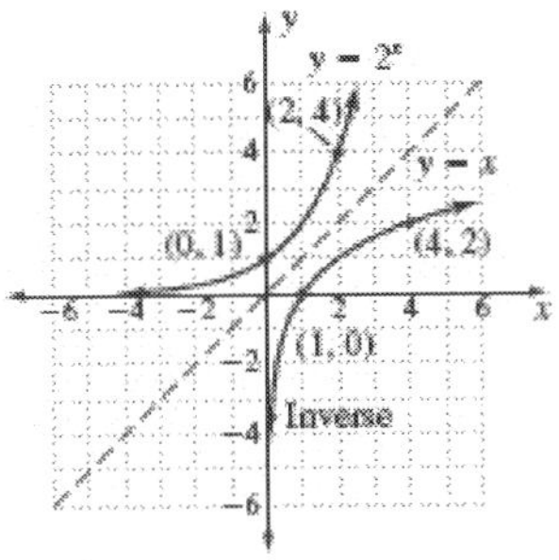

7.

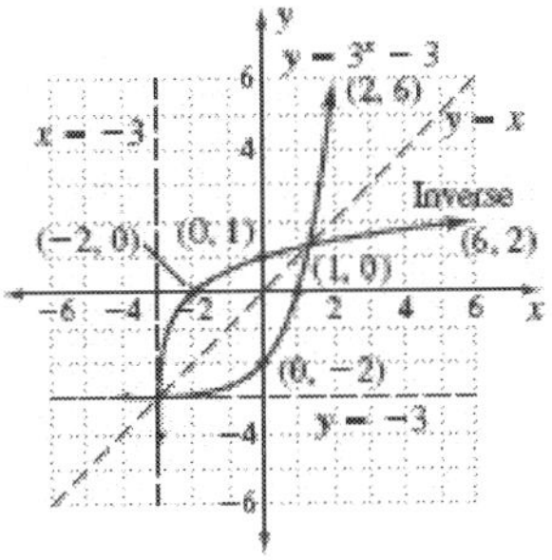

9.

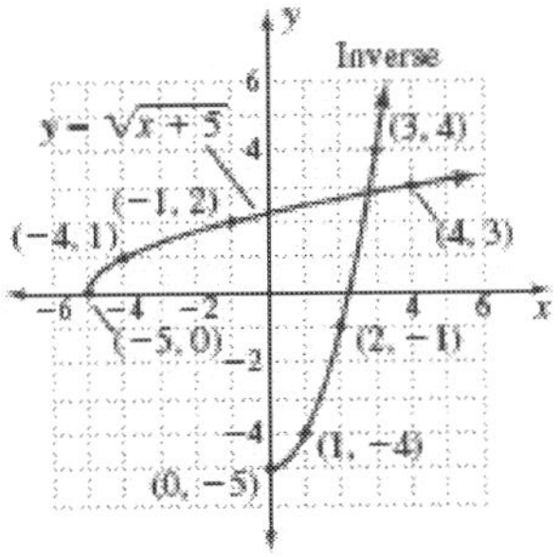

11.

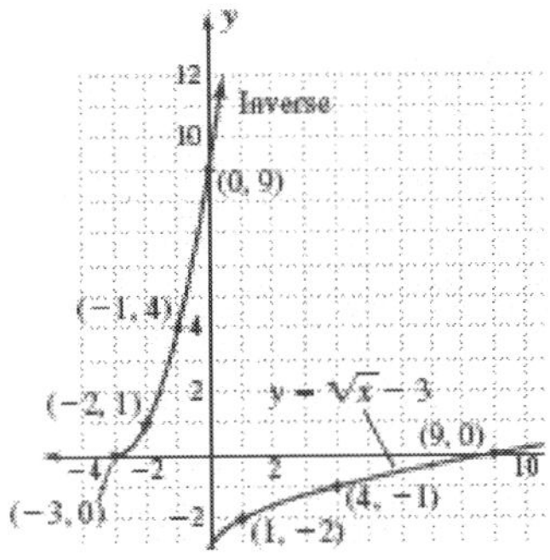

13. $x = 3y;$

$y = \dfrac{x}{3}$

15. $x = 4y - 2; \quad x + 2 = 4y$

$\dfrac{x+2}{4} = y; \quad y = \frac{1}{4}x + \frac{1}{2}$

17. $x = y + 3; \quad y = x - 3$

19. $x = \sqrt{y}; \quad y = x^2; x \geq 0, y \geq 0$

21. $x = \sqrt[5]{y}; \quad y = x^5$

23. $x = y^4; \quad y = \sqrt[4]{x}; x \geq 0, y \geq 0$

25. $x = \sqrt{y-2}; \quad y = x^2 + 2; \quad x \geq 0, \quad y \geq 2$

27. $x = \sqrt{y} + 3; \quad y = (x-3)^2; \quad x \geq 0, \quad y \geq 3$

29. a.

x	$f(x) = 3^x$
−2	$3^{-2} = \frac{1}{9}$
−1	$3^{-1} = \frac{1}{3}$
0	$3^0 = 1$
1	$3^1 = 3$
2	$3^2 = 9$
3	$3^3 = 27$
4	$3^4 = 81$

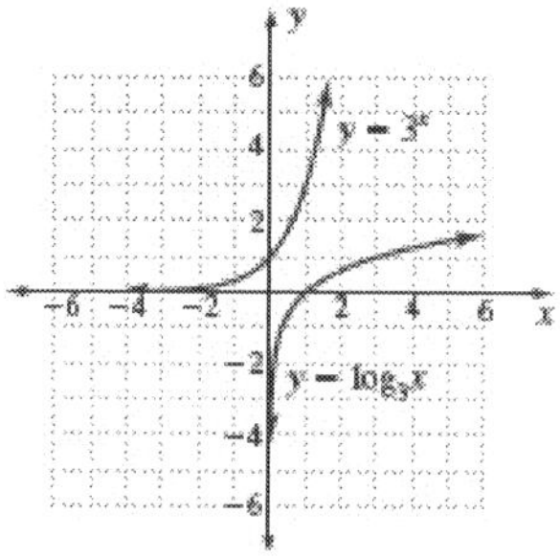

b.

x	y
$\frac{1}{9}$	−2
$\frac{1}{3}$	−1
1	0
3	1
9	2
27	3
81	4

c. Inverse is a function.

d. $y = 3^x$; $x = 3^y$, $f^{-1}(x) = \log_3 x$

31. $f^{-1}(x) = \log_4 x$

33. $h^{-1}(x) = e^x$

35. a. i. $x = y$

ii. $x = \dfrac{1}{y}$; $yx = 1$; $y = \dfrac{1}{x}$

iii. $x = -\dfrac{1}{y}$; $xy = -1$; $y = -\dfrac{1}{x}$

iv. $x = -y$; $y = -x$

v. $x = 1 - y$; $x + y = 1$; $y = 1 - x$

vi. $x = 3 - y$; $x + y = 3$; $y = 3 - x$

b. The inverses functions are the same as the orginal functions.

c. i. 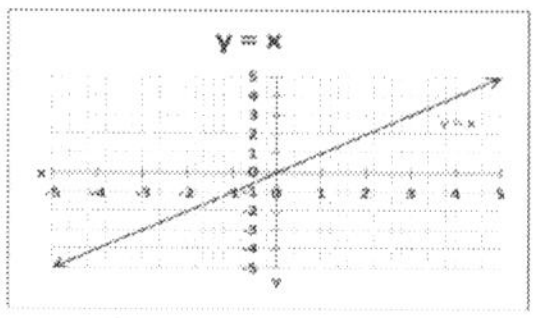

ii. 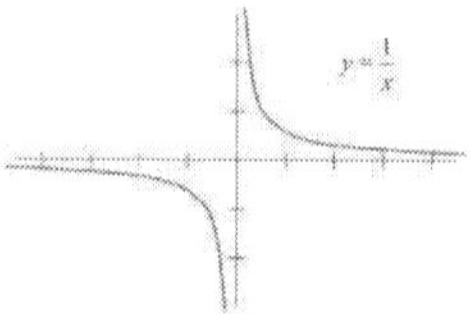

iii. 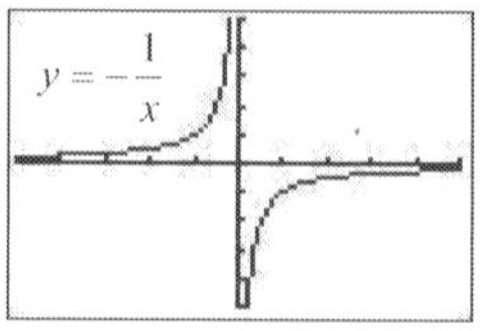

iv. 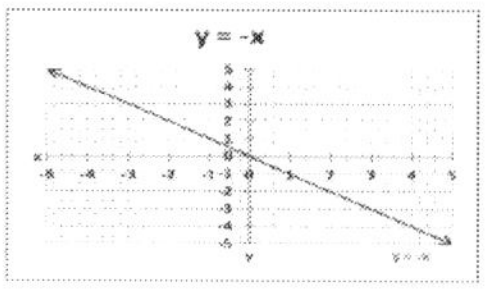

v. 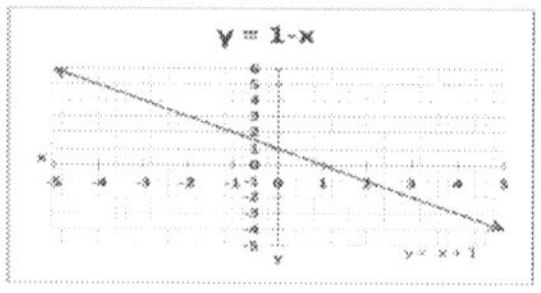

vi. 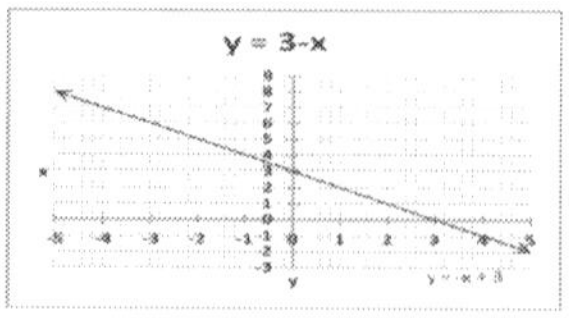

d. A function will be its own inverse when the function has y = x as a line of symmetry.

Exercises 6.4: Solving Exponential Equations: Logarithms

1.

Exponential Equation	Logarithmic Equation
$3^3 = 27$	$\log_3 27 = 3$
$2^3 = 8$	$\log_2 8 = 3$
$10^1 = 10$	$\log_{10} 10 = 1$
$5^3 = 125$	$\log_5 125 = 3$
$3^{-2} = \frac{1}{9}$	$\log_3\left(\frac{1}{9}\right) = -2$

3. **a.** Base = 10; exponent = 3; $10^3 = 1000$ **b.** Base = 10, exponent = 0; $10^0 = 1$
c. Base = 3; exponent = 4; $3^4 = 81$ **d.** Base = 10; exponent = 2; $10^2 = 100$

5. **a.** $\log_2 32 = 5$ **b.** $\log_2 2 = 1$ **c.** $\log_2 1 = 0$ **d.** $\log_{10} 100 = 2$

7. **a.** $\log_{10} 1 = 0$ **b.** $\log_{10} 6 = 0.77815$ **c.** $\log_{10} 10 = 1$
d. $\log_{10} 60 = 1.77815$ **e.** $\log_{10} 100 = 2$ **f.** $\log_{10} 600 = 2.77815$
g. The decimal portion is the same; all are positive.

9. **a.** $\log_{10} 0.6 = -0.22185$ **b.** $\log_{10} 0.1 = -1$
c. $\log_{10} 0.01 = -2$ **d.** $\log_{10} 0.06 = -1.22185$
e. $\log_{10} 0.006 = -2.22185$ **f.** $\log_{10} 0.000\,006 = -5.22185$
g. The decimal portion is the same; all are negative.

11. **a.** $\log_{10} 17 = x$; $x \approx 1.23045$ **b.** $\log_{10} 125 = x$; $x \approx 2.09691$
c. $\log_{10} 400 = x$; $x \approx 2.60206$ **d.** $\log_{10} 0.05 = x$; $x \approx -1.30103$

13.

Exponential Equation	Logarithmic Equation	Solve for x
$5^0 = x$	$\log_5 x = 0$	$x = 1$
$x^3 = 64$	$\log_x 64 = 3$	$64 = 4^3$; $x = 4$
$10^x = -10$	$\log_{10}(-10) = x$	{ } or $\varnothing$
$10^x = 0$	$\log_{10} 0 = x$	{ } or $\varnothing$
$10^{-1} = x$	$\log_{10} x = -1$	$x = \frac{1}{10}$

15. a. $7^2 = x$
$x = 49$

b. $3^x = 3$
$x = 1$

c. $10^x = 0.01$
$x = -2$

d. $10^0 = x$
$x = 1$

e. $x^3 = 64$
$x^3 = 4^3$
$x = 4$

f. $a^1 = x$
$a = x$

17. a. $2^8 = x$
$x = 256$

b. $10^{-2} = x$
$x = \frac{1}{100} = 0.01$

c. $2^x = \frac{1}{4}$
$x = -2$

d. $10^x = 1$
$x = 0$

e. $x^2 = 100$
$x = 10$

f. $a^x = a$
$x = 1$

19. Because we swap x and y, the domain of a function is the ***range*** of its inverse.

21. The logarithm is the ***exponent*** to which we raise a base to obtain a given number.

23. The ***base*** of a logarithmic equation must be a positive number not equal to 1.

25. The set of outputs (range) to a logarithmic function, $y = \log_b x$, $\mathfrak{R}$.

27. The set of outputs (range) to an exponential function, $y = b^x$, is ***y > 0.***

29. a. 72/9 = 8 years.

b. 72/30 = 2.4 years.

c. $\log_{(1+r)} 2 = t$; $t = \dfrac{\log 2}{\log(1+r)}$

d.

Interest, %	1	2	5	8	10	12	20
Rule of 72	72/1=72	72/2=36	72/5=14.4	72/8=9	72/10=7.2	72/12=6	72/20=3.6
Formula	69.6	35	14.2	9	7.3	6.11	3.8
Difference	3.4%	2.9%	1.4%	0%	1.4%	1.8%	5.3%

e. See part d.

f. See part d.

g. According to the difference percentages, from 5% to 12% look best.

Exercises 6.5: Applications of Exponential and Logarithmic Functions

1. a. Base = 3

b. Base = 4

c. Base = 10

3. a. $2^x = 1$

$2^x = 2^0$

$x = 0$

b. $2^x = 3$

$\log_2 3 = x$

$x = \dfrac{\log 3}{\log 2} \approx 1.58$

c. $2^x = 4$

$2^x = 2^2$

$x = 2$

d. $2^x = 5$

$\log_2 5 = x$

$x = \dfrac{\log 5}{\log 2} \approx 2.32$

e. $2^x = 6$

$\log_2 6 = x$

$x = \dfrac{\log 6}{\log 2} \approx 2.58$

f. $2^x = 10$

$\log_2 10 = x$

$x = \dfrac{\log 10}{\log 2} \approx 3.32$

g. $\mathbf{x} \approx 3.58$

5. a. $5^x = 2$

$\log_5 2 = x$

$x = \dfrac{\log 2}{\log 5} \approx 0.43$

b. $5^x = 5$

$5^x = 5^1$

$x = 1$

c. $5^x = 6$

$\log_5 6 = x$

$x = \dfrac{\log 6}{\log 5} \approx 1.11$

d. $5^x = 10$

$\log_5 10 = x$

$x = \dfrac{\log 10}{\log 5} \approx 1.43$

e. $5^x = 25$

$5^x = 5^2$

$x = 2$

f. $5^x = 30$

$\log_5 30 = x$

$x = \dfrac{\log 30}{\log 5} \approx 2.11$

g. $x \approx 2.43$

7. a. $3^x = 10$

$\log_3 10 = x$

$x = \dfrac{\log 10}{\log 3} \approx 2.09590$

b. $x = \log_4 6$

$x = \dfrac{\log 6}{\log 4} \approx 1.29248$

c. $20 = 10(1.07)^x$

$2 = 1.07^x$

$\log_{1.07} 2 = x$

$x = \dfrac{\log 2}{\log 1.07} \approx 10.24477$

d. $15 = 30(0.95)^x$

$0.5 = 0.95^x$

$\log_{0.95} 0.5 = x$

$x = \dfrac{\log 0.5}{\log 0.95} \approx 13.51341$

e. $30 = 60(0.5)^x$

$0.5 = 0.5^x$

$0.5^1 = 0.5^x$

$x = 1$

9. $x \approx 1.47, y \approx 1.47$; $x \approx 7.857, y \approx 7.857$

11. $x \approx 0.361, y \approx 0.361$; $x \approx 0.5432, y \approx 0.2169$; $x \approx 0.2169, y \approx 0.5432$

13. $x \approx 1.258, y \approx 1.258$; $x \approx 14.767, y \approx 14.767$

15. Inverse functions; intersected on the line of symmetry, y = x, or where in pairs of the form (a, b) and (b, a).

17. $9.2 = \left(\frac{2}{3}\right)\log M_0 - 10.7$; $29.85 = \log M_0$; $M_0 \approx 7.1\times10^{29}$ dyne-centimeters.

19. $M_w = \left(\frac{2}{3}\right)\log\left(1.26\times10^{26}\right) - 10.7 \approx 6.7$

21. $\left(3.5\times10^{29}\right)/\left(1.26\times10^{26}\right) \approx 2800$. Japan's was approximately 2800 times that of Northridge.

23. a. $\text{pH} = -\log_{10}(1.0 \times 10^{-14}) = 14$; base

b. $\text{pH} = -\log_{10}0.20 = 0.7$; acid

c. $\text{pH} = -\log_{10}(5.01 \times 10^{-3}) = 2.3$; acid

d. $\text{pH} = -\log_{10}(1.0 \times 10^{-10}) = 10$; base

e. $\text{pH} = -\log_{10}(1.26 \times 10^{-3}) = 2.9$; acid

25. Corrosive acid, 0 to 1; hydrogen peroxide, 3 to 4; shampoo, 4.5 to 5.5; shampoo, 7 to 10; oil bleach, 8 to 9.5; perm solution 8.5 to 9.5; tints/toners, 9.5 to 10.5; straighteners, 11.5 to 14.

27. a. $4.1 = -\log_{10}[H^+]$

$-4.1 = \log_{10}[H^+]$

$10^{-4.1} = [H^+] \approx 7.94 \times 10^{-5}$ M

b. $-0.5 = -\log_{10}[H^+]$

$0.5 = \log_{10}[H^+]$

$10^{0.5} = [H^+] \approx 3.16$ M

c. $12 = -\log_{10}[H^+]$

$-12 = \log_{10}[H^+]$

$10^{-12} = [H^+] = 1.0 \times 10^{-12}$ M

d. $3.5 = -\log_{10}[H^+]$

$-3.5 = \log_{10}[H^+]$

$10^{-3.5} = [H^+] \approx 3.16 \times 10^{-4}$ M

29. pH is negative when $[H^+] > 1$; pH is positive when $[H^+] < 1$.

31. $N = 10\log\left(\frac{I}{I_0}\right)$; $100 = 10\log\left(\frac{I_1}{10^{-12}}\right)$; $10 = \log\left(\frac{I_1}{10^{-12}}\right)$; $10^{10} = \frac{I_1}{10^{-12}}$; $I_1 = 10^{-2}$

$85 = 10\log\left(\frac{I_2}{10^{-12}}\right)$; $8.5 = \log\left(\frac{I_2}{10^{-12}}\right)$; $10^{8.5} = \frac{I_2}{10^{-12}}$; $I_2 = 10^{-3.5}$

$\frac{10^{-2}}{10^{-3.5}} = 10^{1.5} \approx 31.6$

33. $1{,}000{,}000 = 1{,}000(1 + 0.02)^t$

$\frac{1{,}000{,}000}{1{,}000} = (1.02)^t$

$\log_{1.02} 1{,}000 = t$

$t = \frac{\log 1{,}000}{\log 1.02} \approx 349$ yr.

35. $1{,}000{,}000 = 1{,}000(1 + 0.10)^t$

$\frac{1{,}000{,}000}{1{,}000} = (1.10)^t$

$\log_{1.10} 1{,}000 = t$

$t = \frac{\log 1{,}000}{\log 1.10} \approx 73$ yr

37. a. $r = 8\%$

b. $r = 6\%$

39. a. $r = -3\%$

b. $1 + r = 0.95$; $r = -0.05 = -5\%$

41. $2 = (1.08)^t$

$\log_{1.08} 2 = t$

$t = \frac{\log 2}{\log 1.08} \approx 9$ yr.

43. $2 = (1.06)^t$

$\log_{1.06} 2 = t$

$t = \frac{\log 2}{\log 1.06} \approx 11.9$ yr.

45. $0.5 = (1 - 0.10)^t$

$0.5 = (0.90)^t$

$\log_{0.9} 0.5 = t$

$t = \frac{\log 0.5}{\log 0.9} \approx 6.58$ yr

47. $2P = P(1 + r)^t$

$2 = (1 + r)^t$

$\log_{(1+r)} 2 = t$

$t = \frac{\log 2}{\log(1+r)}$

P does not appear in the formula; doubling time is not dependent on P.

49. a. $\frac{1 \text{ half} - \text{life}}{26.3 \text{ hr}} \cdot \frac{24 \text{ hr}}{1 \text{ day}} \cdot \frac{365 \text{ day}}{1 \text{ yr}} \approx 333$ half-lives

b. $\frac{1 \text{ half} - \text{life}}{2.4 \text{ day}} \cdot \frac{365 \text{ day}}{1 \text{ yr}} \approx 152$ half-lives

c. $\frac{1 \text{ half} - \text{life}}{14.3 \text{ day}} \cdot \frac{365 \text{ day}}{1 \text{ yr}} \approx 26$ half-lives

d. S = 1.7 curies; half-life = 24,000 yrs; $\frac{1}{2}$S = 0.85 curies; $y = 1.7(0.99997)^x$; x in years

e. S = 12,000,000 curies; half-life = 15 hrs; $\frac{1}{2}$S = 6,000,000 curies;
$y = 12{,}000{,}000(0.955)^x$; x in hrs.

f. S = 490,000 curies; half-life = 245 days; $\frac{1}{2}$S = 245,000 curies;
$y = 490{,}000(0.997)^x$; x in days

51. Let t = years after 2003 for all data:

a. $y = -2.15t + 37.2$; $y = 37.44(0.932)^t$; 1-0.932 = 0.068 = 6.8% decrease.

b. $y = 3.72t + 15.1$; $y = 15.81(1.159)^t$; 1.159 – 1 = 0.159 = 15.9% increase.

c. $y = 1.72t + 9.021$; $y = 9.16(1.138)^t$; 1.138 – 1 = 0.138 = 13.8% increase.

d. $y = 0.9t + 0.6$; $y = 1.815(1.221)^t$ 1.221 – 1 = 0.221 = 22.1% increase.

The use of checks fell dramatically between 2003 and 2009, consumers are using debit cards and prepaid cards; businesses and government entities are using electronic transfers with ACH instead of issuing checks.

53. a.

Yr.	Option 1	Option 2	Option 3
1	5.25	4.25	4.20
2	5.50	4.55	4.41
3	5.75	4.90	4.63
4	6.00	5.30	4.86
5	6.25	5.75	5.11

b. Option 1: linear; Option 2: quadratic; Option 3: exponential

c. Option 1: $y = 0.25x + 5.00$; Option 2: $y = 0.025x^2 + 0.225x + 4$; Option 3: $y = 4(1.05)^x$

e. Option 1: 20 yr.; Option 2: 9 yr; Option 3: 15 yr.

Exercises 6.6 : Properties of Logarithms and the Logarithmic Scale

1. **a.** $\log 2 + \log x = \log 2x$

b. $\log x^2 = 2\log x$

c. $\log\left(\dfrac{2}{x}\right) = \log 2 - \log x$

d. $\log 2^x = x\log 2$

e. $\log x - \log 2 = \log\left(\dfrac{x}{2}\right)$

3. log of a product property; $\log 2 + \log 3 = \log 6$

5. log of a power property: $\log 3^2 = 2\log 3$

7. **a.** $\log_b xz^3 = \log_b x + \log_b z^3 = \log_b x + 3\log_b z$

b. $\log_b\left(\dfrac{x^2}{y}\right) = \log_b x^2 - \log_b y = 2\log_b x - \log_b y$

9. **a.** $\log_b (xy)^2 = 2\log_b xy = 2(\log_b x + \log_b y)$

b. $\log_b\left(\dfrac{5}{x^2}\right) = \log_b 5 - \log_b x^2 = \log_b 5 - 2\log_b x$

11. **a.** $\log_b\left(\dfrac{\sqrt{x}}{y}\right) = \log_b \sqrt{x} - \log_b y = \log_b x^{\frac{1}{2}} - \log_b y = \frac{1}{2}\log_b x - \log_b y$

b. $\log_b (x-z)^2 = 2\log_b (x-z)$

13. **a.** $\log_b\left(\dfrac{xz^3}{y^2}\right) = \log_b (xz^3) - \log_b y^2 = \log_b x + \log_b z^3 - 2\log_b y$

$= \log_b x + 3\log_b z - 2\log_b y$

b. $\log_b\left(\dfrac{y^3z^2}{x}\right) = \log_b (y^3z^2) - \log_b x = \log_b y^3 + \log_b z^2 - \log_b x$

$= 3\log_b y + 2\log_b z - \log_b x$

15. **a.** $\log_b x + 2\log_b y = \log_b x + \log_b y^2 = \log(xy^2)$

b. $\log_b y - 2\log_b x = \log_b y - \log_b x^2 = \log_b\left(\dfrac{y}{x^2}\right)$

17. a. $\frac{1}{2}(\log_b y + 3\log_b z) = \frac{1}{2}(\log y + \log_b z^3) = \frac{1}{2}\log_b(yz^3) = \log_b(yz^3)^{\frac{1}{2}} = \log_b \sqrt{yz^3}$

b. $\frac{1}{2}\log_b x - \log_b 3 = \log_b x^{\frac{1}{2}} - \log_b 3 = \log_b \sqrt{x} - \log_b 3 = \log_b\left(\frac{\sqrt{x}}{3}\right)$

19. a. $\log(x+1) + \log(x-1) = \log[(x+1)(x-1)] = \log(x^2 - 1)$

b. $\log(x^2 - 1) - \log(x-1) = \log\left(\frac{x^2-1}{x-1}\right) = \log\left(\frac{(x-1)(x+1)}{(x-1)}\right) = \log(x+1)$

21. a. $\log(x^2 + 3x + 2) - \log(x+1) = \log\left(\frac{x^2+3x+2}{x+1}\right) = \log\left(\frac{(x+2)(x+1)}{(x+1)}\right) = \log(x+2)$

b. $\log(x+3) + \log(x-2) = \log[(x+3)(x-2)] = \log(x^2 + x - 6)$

23. $\log\sqrt{x} = \log x^{\frac{1}{2}} = \frac{1}{2}\log x$

25. $\log\left(\frac{1}{x}\right) = \log x^{-1} = -1\log x = -\log x$

27.

$x = \log_b a$

$b^x = a$

$\log b^x = \log a$

$x\log b = \log a$

$x = \frac{\log a}{\log b}$

$\log_b a = \frac{\log a}{\log b}$

29.

$\log_2 x^2 + \log_2 x = \log_2 x^3 = 6$

$3\log_2 x = 6$

$\log_2 x = 2$

$x = 2^2 = 4$

31.

$\log x + \log x = \log x^2 = 2$

$2\log x = 2$

$\log x = 1$

$\text{x} = 10^1 = 10$

33.

$\log(x^2 + x) - \log(x+1)$

$= \log\left(\frac{x^2+x}{x+1}\right)$

$= \log\left(\frac{x(x+1)}{(x+1)}\right)$

$= \log x = 1;\ \text{x} = 10^1 = 10$

35. $\log_3 x^3 - \log_3 x = \log_3\left(\frac{x^3}{x}\right) = \log_3 x^2 = 2\log_3 x = 4;\ \log_3 x = 2;\ \text{x} = 3^2 = 9$

37. a. $\log 3^{x+2} = \log 48$

$(x+2)\log 3 = \log 48$

$x+2 = \dfrac{\log 48}{\log 3}$

$x = \dfrac{\log 48}{\log 3} - 2 \approx 1.5237$

b. $\log_3 48 = x+2$

$x+2 = \dfrac{\log 48}{\log 3}$

$x = \dfrac{\log 48}{\log 3} - 2 \approx 1.5237$

39. a. $\log 4^{x-1} = \log 28$

$(x-1)\log 4 = \log 28$

$x-1 = \dfrac{\log 28}{\log 4}$

$x = \dfrac{\log 28}{\log 4} + 1 \approx 3.4037$

b. $\log_4 28 = x-1$

$x-1 = \dfrac{\log 28}{\log 4}$

$x = \dfrac{\log 28}{\log 4} + 1 \approx 3.4037$

41. $15(1.07)^t = 45;\ 1.07^t = 3$

a. $\log 1.07^t = \log 3$

$t \log 1.07 = \log 3$

$t = \dfrac{\log 3}{\log 1.07} \approx 16.2376$

b. $\log_{1.07} 3 = t$

$t = \dfrac{\log 3}{\log 1.07} \approx 16.2376$

43. $1000(1 - 0.08)^t = 500;\ (0.92)^t = 0.5$

a. $\log 0.92^t = \log 0.5$

$t \log 0.92 = \log 0.5$

$t = \dfrac{\log 0.5}{\log 0.92} \approx 8.3130$

b. $\log_{0.92} 0.5 = t$

$t = \dfrac{\log 0.5}{\log 0.92} \approx 8.3130$

45. A straight line

47.

x	y=4^x
−1	$4^{-1} = \frac{1}{4}$
0	$4^0 = 1$
1	$4^1 = 4$
2	$4^2 = 16$
3	$4^3 = 64$
4	$4^4 = 256$
5	$4^5 = 1024$
6	$4^6 = 4096$

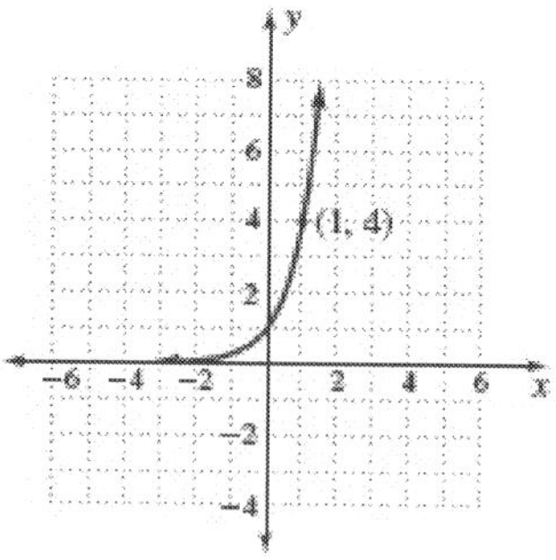

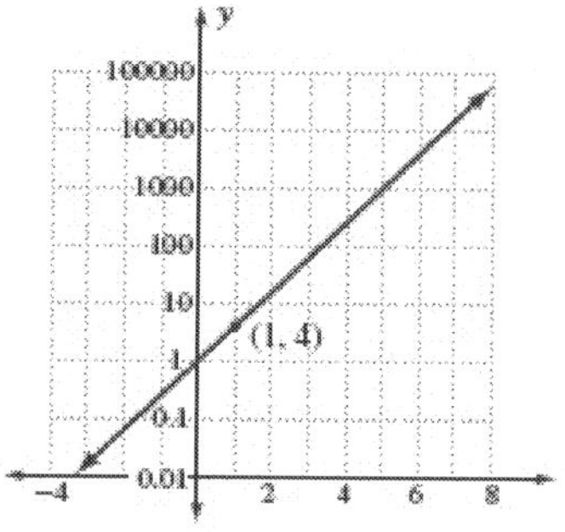

49.

$$\frac{227}{1483}(100) = 15.3\%; \frac{384}{1483}(100) = 25.9\%; \frac{516}{1483}(100) = 34.8\%; \frac{604}{1483}(100) = 40.7\%;$$
$$\frac{698}{1483}(100) = 47.1\%; \frac{937}{1483}(100) = 63.2\%; \frac{1082}{1483}(100) = 73.0\%; \frac{1171}{1483}(100) = 79.0\%;$$
$$\frac{1231}{1483}(100) = 83.0\%; \frac{1271}{1483}(100) = 85.7\%; \frac{1300}{1483}(100) = 87.7\%; \frac{1319}{1483}(100) = 88.9\%;$$
$$\frac{1340}{1483}(100) = 90.4\%; \frac{1355}{1483}(100) = 91.4\%; \frac{1483}{1483}(100) = 100\%$$

51. a. - c.

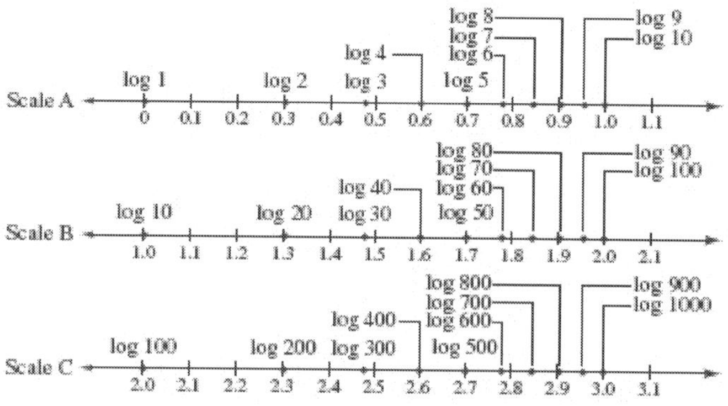

d. Dots are in the same relative positions because log 20 = 1 + log 2, etc.

Exercises 6.7: The Natural Number *e* in Exponential and Logarithmic Equations

1. **a.** 0.066 **b.** 1.445 **c.** 0.368

3. $\$1000e^{0.08 \cdot 1} = \$1083.28$

5. $\$1000e^{0.10 \cdot 1} = \$1105.17$

7. $\$1000e^{0.045 \cdot 1} = \$1046.02$

9. $\$500{,}000 = Pe^{0.10 \cdot 12}$;

$$P = \frac{\$500{,}000}{e^{1.2}} \approx \$150{,}597.11$$

11. $\$500{,}000 = Pe^{0.10 \cdot 20} \Rightarrow P = \dfrac{\$500{,}000}{e^{2.0}} \approx \$67{,}667.65$ \\

13.

Logarithmic Equation	Show base *e* in logarithm	Exponential Equation	y = ?
$\ln y = 1$	$\log_e y = 1$	$y = e^1$	$y = e$
$\ln y = 0$	$\log_e y = 0$	$y = e^0$	$y = 1$
$y = \ln(-1)$	$y = \log_e(-1)$	$e^y = -1$	{ } or $\varnothing$
$y = \ln e^2$	$y = \log_e e^2$	$e^y = e^2$	$y = 2$
$y = \ln e^e$	$y = \log_e e^e$	$e^y = e^e$	$y = e$

15. **a.** $e^x = 4$; $x = \ln 4 = 1.3863$ **b.** $\ln x = -2$; $x = e^{-2} = 0.1353$

17. **a.** $\ln x = 1.5$; $x = e^{1.5} = 4.4817$ **b.** $e^x = 2$; $x = \ln 2 = 0.6931$

19.

$$2 = e^{8r}$$
$$\ln 2 = 8r$$
$$r = \frac{\ln 2}{8} \approx 8.66\%$$

21.

$$2 = e^{rt}$$
$$\ln 2 = rt$$
$$r = \frac{\ln 2}{t}$$

23. $0.5 = e^{0.08t} \Rightarrow \ln 0.5 = 0.08t \Rightarrow t = \dfrac{\ln 0.5}{0.08} \approx 8.7$ yr

25. **a.** $69 \div r = 8$; $r = 69 \div 8 = 8.625\%$ **b.** $r = 69 \div 11 \approx 6.27\%$

c. $t = \dfrac{69}{r}$; $r = \dfrac{69}{t}$

d. Rule of 69 was accurate to within 0.1%.

27. $2P = Pe^{rt};\ 2 = e^{rt};\ \ln 2 = rt;\ t = \dfrac{\ln 2}{r} \approx \dfrac{0.693}{r};\ t \approx \dfrac{0.693(100)}{r(100)} = \dfrac{69.3}{100r}$

29. $K = e^{\left(-\frac{Ea}{RT}+C\right)}$

31. $\ln[H^+] = -\dfrac{EnF}{RT};\ [H^+] = e^{\left(-\frac{EnF}{RT}\right)}$

33. $E = E^0 - \dfrac{RT}{nF}\ln Q; -\dfrac{nF}{RT}\left(E - E^0\right) = \ln Q; Q = e^{-\frac{nF}{RT}\left(E-E^0\right)}$

35. a. $4! = 4 \cdot 3 \cdot 2 \cdot 1 = 24$

b. $10! = 10 \cdot 9 \cdot 8 \cdot 7 \cdot 6 \cdot 5 \cdot 4 \cdot 3 \cdot 2 \cdot 1 = 3{,}628{,}800$

c. $6! = 6 \cdot 5 \cdot 4 \cdot 3 \cdot 2 \cdot 1 = 720$

d. $9! = 9 \cdot 8 \cdot 7 \cdot 6 \cdot 5 \cdot 4 \cdot 3 \cdot 2 \cdot 1 = 362,\ 880$

e. 9 terms shown ≈ 7.3873; $e^2 \approx 7.3891$

37. a.

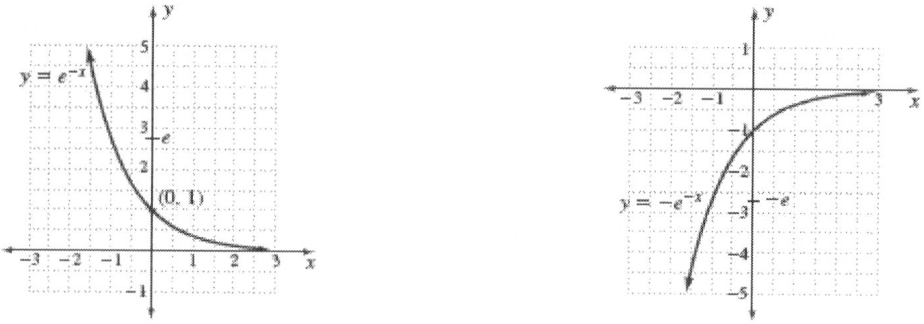

b.

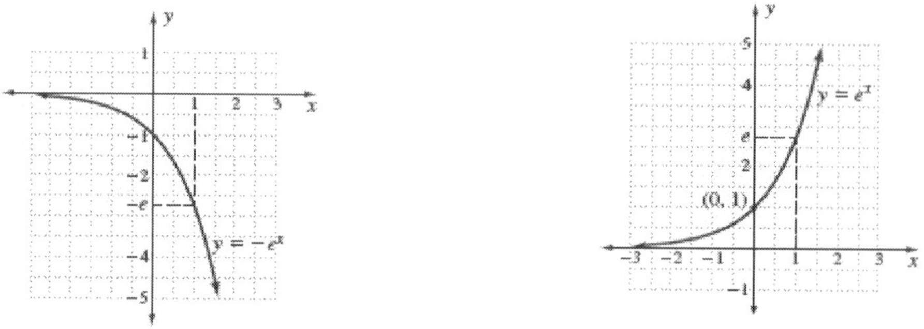

c.

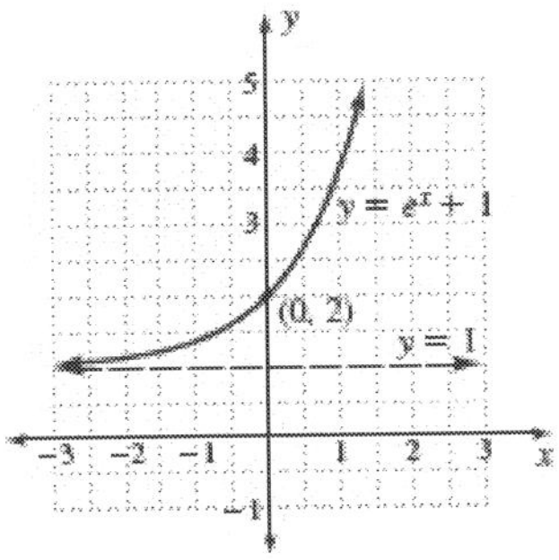

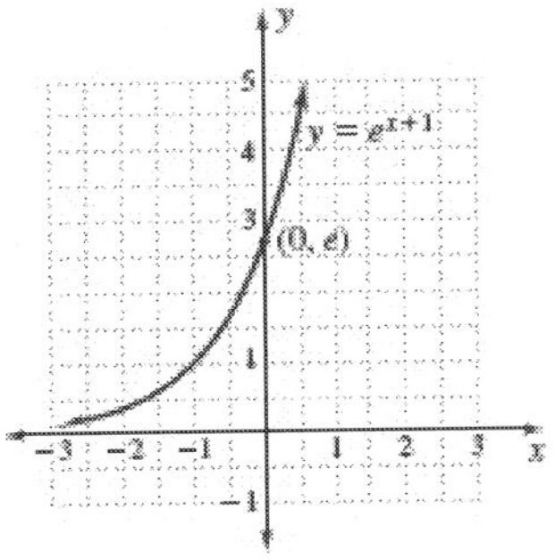

d.

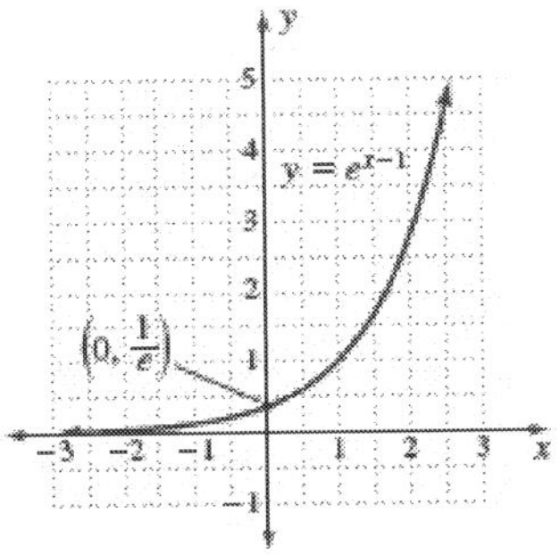

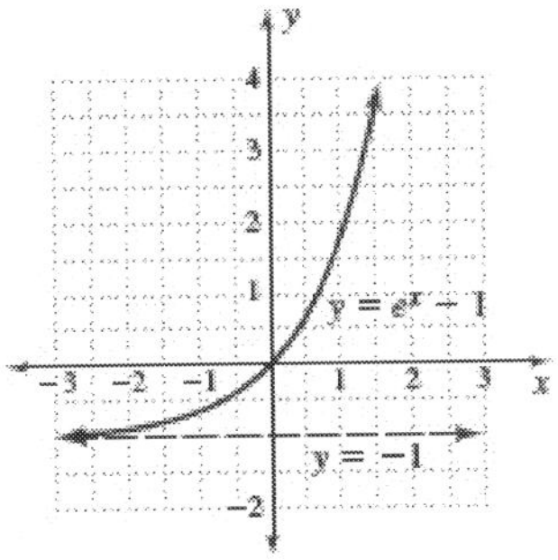

e.

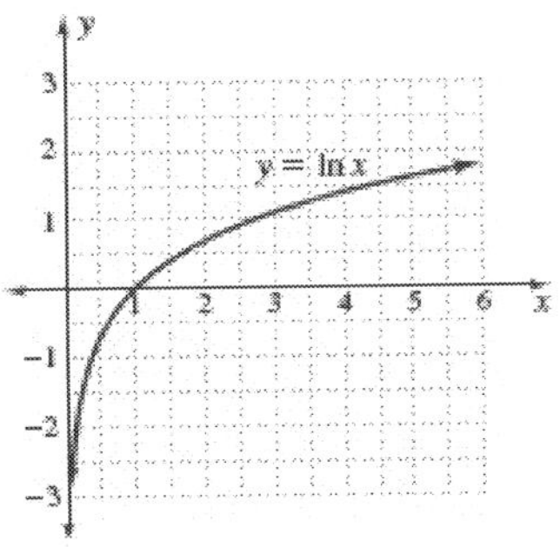

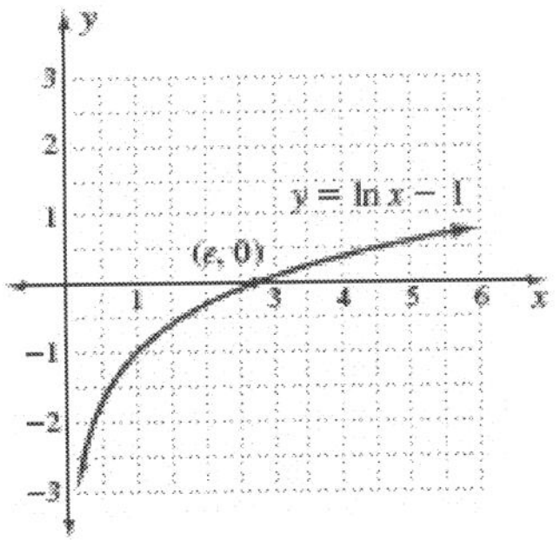

f.

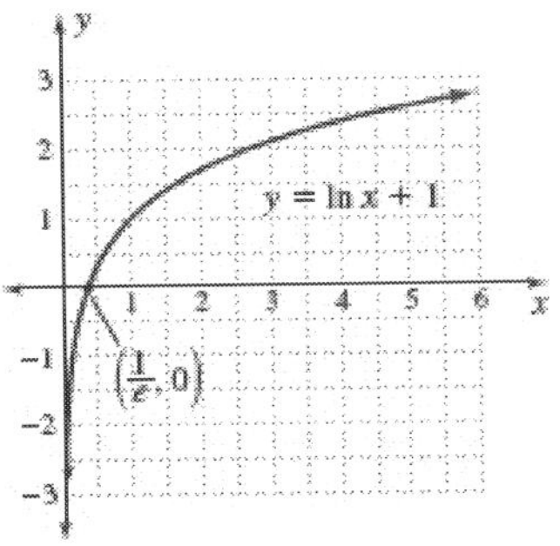

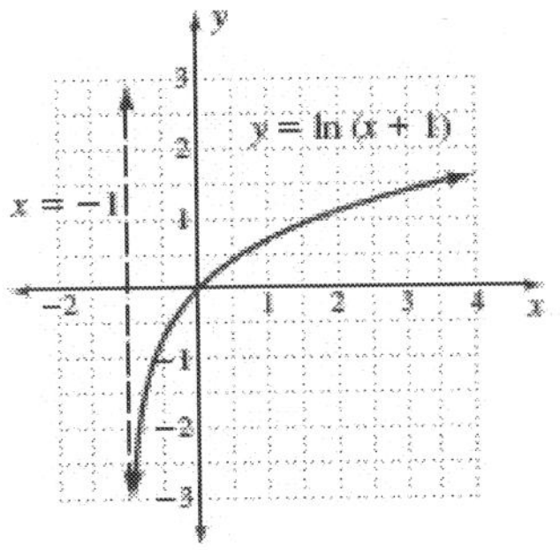

g. Graphs are symmetric over y = x.

Chapter 6 Review Exercises

1. Common ratio is 3; next number = $243 \cdot 3 = 729$; geometric sequence; $y = 9 \cdot 3^{x-1}$ or $y = 3 \cdot 3^x$.

2. No common ratio; first differences = 3, 3, 3, 3; next number is $22 + 3 = 25$; arithmetic sequence; $m = 3$; $10 = 3(1) + b$, $b = 7$; $y = 3x + 7$.

3. No common ratio; first differences = 2, 4, 6, 8; second differences = 2, 2, 2; next number is $21 + (8 + 2) = 31$; quadratic sequence; $2a = 2$, $a = 1$; $c = 1 - (2 - 2) = 1$; $1 = 1(1)^2 + b(1) + 1$, $b = -1$; $y = x^2 - x + 1$.

4. N common ratio; first differences = 3, 5, 7, 9; second differences = 2, 2, 2; next number is $17 + (9 + 2) = 28$; quadratic sequence; $2a = 2$, $a = 1$; $c = -7 - (3 - 2) = -8$; $-7 = 1(1)^2 + b(1) - 8$, $b = 0$; $y = x^2 - 8$.

5. No common ratio; first differences = 3, 3, 3, 3; next number is $17 + 3 = 20$; arithmetic sequence; $m = 3$; $5 = 3(1) + b$, $b = 2$; $y = 3x + 2$.

6. Common ratio = 2; $y = \frac{1}{8} \cdot 2^{x-1}$ or $y = \frac{1}{16} \cdot 2^x$.

7. Common ratio = 2; $a_n = \frac{1}{4} \cdot 2^{n-1}$; $f(x) = 0.125 \cdot 2^x$;
$\frac{1}{4} \cdot 2^{n-1} = 2^{-2} \cdot 2^{-1} \cdot 2^n = 2^{-3} \cdot 2^n = \frac{1}{8} \cdot 2^n = 0.125 \cdot 2^n$.

8. Common ratio = 2; $a_n = 8 \cdot 2^{n-1}$; $f(x) = 4 \cdot 2^x$
$8 \cdot 2^{n-1} = 2^3 \cdot 2^{-1} \cdot 2^n = 2^2 \cdot 2^n = 4 \cdot 2^n$.

9. Common ratio = 2; $a_n = \frac{1}{16} \cdot 2^{n-1}$; $f(x) = 0.03125 \cdot 2^x$;
$\frac{1}{16} \cdot 2^{n-1} = 2^{-4} \cdot 2^{-1} \cdot 2^n = 2^{-5} \cdot 2^n = \frac{1}{32} \cdot 2^n = 0.03125 \cdot 2^n$.

10. Common ratio = $\frac{1}{3}$; $a_n = 6(\frac{1}{3})^{n-1}$; $f(x) = 18(\frac{1}{3})^x$
$6(\frac{1}{3})^{n-1} = 6(\frac{1}{3})^{-1}(\frac{1}{3})^n = 6(3)(\frac{1}{3})^n = 18(\frac{1}{3})^n$.

11. a. $2^{x-3} = 2^{-3} \cdot 2^x = \frac{1}{8} \cdot 2^x$; $\left(0, \frac{1}{8}\right)$
b. $3^{x+1} = 3^1 \cdot 3^x = 3 \cdot 3^x$; $(0, 3)$
c. $3^{x-3} = 3^{-3} \cdot 3^x = \frac{1}{27} \cdot 3^x$; $\left(0, \frac{1}{27}\right)$

12. a. $2^{x+4} = 2^4 \cdot 2^x = 16 \cdot 2^x$; $(0, 16)$
b. $2^{x-2} = 2^{-2} \cdot 2^x = \frac{1}{4} \cdot 2^x$; $\left(0, \frac{1}{4}\right)$
c. $3^{x+2} = 3^2 \cdot 3^x = 9 \cdot 3^x$; $(0, 9)$

13. a. y-intercept = 2; $2 \cdot 2^x = 2^1 \cdot 2^x = 2^{x+1}$
b. y-intercept = $\frac{1}{9}$; $\frac{1}{9} \cdot 3^x = 3^{-2} \cdot 3^x = 3^{x-2}$

14. a. y-intercept = 81; $81 \cdot 3^x = 3^4 \cdot 3^x = 3^{x+4}$
b. y-intercept = $\frac{1}{16}$; $\frac{1}{16} \cdot 2^x = 2^{-4} \cdot 2^x = 2^{x-4}$

15. a. $4^x = 64$

$4^x = 4^3$

$x = 3$

b. $2^x = 2$

$2^x = 2^1$

$x = 1$

c. $a^x = \frac{1}{a}$

$a^x = a^{-1}$

$x = -1$

d. $b^n = 1$

$b^n = b^0$

$n = 0$

16. a. $\left(\frac{1}{2}\right)^x = \frac{1}{8}$

$\left(\frac{1}{2}\right)^x = \left(\frac{1}{2}\right)^3$

$x = 3$

b. $16^x = 8$

$(2^4)^x = 2^3$

$4x = 3$

$x = \frac{3}{4}$

c. $a^x = 1$

$a^x = a^0$

$x = 0$

d. $\pi^x = \frac{1}{\pi^2}$

$\pi^x = \pi^{-2}$

$x = -2$

17. a. $\left(\frac{1}{100}\right)^n = 10$

$(10^{-2})^n = 10^1$

$10^{-2n} = 10^1$

$-2n = 1;\ n = -\frac{1}{2}$

b. $100^n = 10$

$(10^2)^n = 10^1$

$10^{2n} = 10^1$

$2n = 1;\ n = \frac{1}{2}$

c. $4^{x-4} = 64$

$4^{x-4} = 4^3$

$x - 4 = 3$

$x = 7$

d. $27^{x+1} = 81$

$(3^3)^{x+1} = 3^4$

$3^{3x+3} = 3^4$

$3x + 3 = 4;\ 3x = 1;\ x = \frac{1}{3}$

18. a. $\left(\frac{1}{10}\right)^n = 100$

$(10^{-1})^n = 10^2$

$-n = 2$

$n = -2$

b. $\left(\frac{1}{100}\right)^n = 100$

$(10^{-2})^n = 10^2$

$-2n = 2$

$n = -1$

c. $4^{0.5x+2} = 1$

$4^{0.5x+2} = 4^0$

$0.5x + 2 = 0$

$0.5x = -2$

$x = -4$

d. $25^{x-2} = \frac{1}{5}$

$(5^2)^{x-2} = 5^{-1}$

$2x - 4 = -1$

$2x = 3$

$x = \frac{3}{2}$

19. a. $x = 2y - 1;\ x + 1 = 2y;\ y = \frac{x}{2} + \frac{1}{2}$

b. $x = 3y;\ y = \frac{x}{3}$

c. $x = y^3;\ y = \sqrt[3]{x}$

d. $x = \sqrt{y};\ y = x^2,\ x \ge 0,\ y \ge 0$

e. $x = -y;\ y = -x$

f. $x = \frac{1}{y};\ y = \frac{1}{x};\ x \ne 0,\ y \ne 0$

20. a. $x = \sqrt{y+2};\ x^2 = y + 2;\ y = x^2 - 2;\ x \ge 0,\ y \ge -2$

b. $x = \sqrt{y} - 4;\ \sqrt{y} = x + 4;\ y = (x+4)^2;\ x \ge -4,\ y \ge 0$

21. a. $x \ge 0$, $y \ge 1$ for function to have an inverse.
$x = \sqrt{y} + 1;\ x - 1 = \sqrt{y};\ y = (x-1)^2;\ y \ge 0;\ x \ge 1$

b. $x \ge 1$, $y \ge 0$ for function to have an inverse.
$x = \sqrt{y-1};\ x^2 = y - 1;\ y = x^2 + 1;\ x \ge 0,\ y \ge 1$

22. $f^{-1}(x)$ is the inverse function of $f(x)$; x^{-1} is the reciprocal of x.

23.

Exponential Equation	**Logarithmic Equation**	**Solve for x**
$2^x = 16$	$\log_2 16 = x$	$2^x = 2^4;\ x = 4$
$x^2 = 25$	$\log_x 25 = 2$	$x^2 = 5^2;\ x = 5$
$3^x = 81$	$\log_3 81 = x$	$3^x = 3^4;\ x = 4$
$10^{1/2} = x$	$\log_{10} x = \frac{1}{2}$	$x = 10^{1/2} = \sqrt{10} \approx 3.162$
$10^x = 19$	$\log_{10} 19 = x$	$x \approx 1.2788$
$4^0 = x$	$\log_4 x = 0$	$x = 1$

24.

Exponential Equation	**Logarithmic Equation**	**Solve for x**
$10^3 = x$	$\log x = 3$	$x = 1000$
$x^4 = 16$	$\log_x 16 = 4$	$x^4 = 2^4;\ x = 2$
$5^x = 1$	$\log_5 1 = x$	$5^x = 5^0;\ x = 0$
$10^x = 10$	$\log 10 = x$	$x = 1$
$4^x = 9$	$\log_4 9 = x$	$x = \frac{\log 9}{\log 4} \approx 1.585$
$4^x = 4$	$\log_4 4 = x$	$x = 1$

25. $36 = 10^x$

$\log 36 = x$

$x = 1.556$

26. $15 = 10^x$

$\log 16 = x$

$x = 1.176$

27. $10^x = 0.75$

$\log 0.75 = x$

$x = -0.125$

28. $10^{1.5} = x$

$x = 31.623$

29. $4^{x+1} = 32$

$(2^2)^{x+1} = 2^5$

$2^{2x+2} = 2^5$

$2x + 2 = 5$

$2x = 3$

$x = \frac{3}{2}$

30. $4^{x+2} = 1$

$4^{x+2} = 4^0$

$x + 2 = 0$

$x = -2$

31. $3^{2x} = 6$

$\log_3 6 = 2x$

$$2x = \frac{\log 6}{\log 3}$$

$$x = \frac{\log 6}{2\log 3} = 0.815$$

32. $4^{2x} = 0.2$

$\log_4 0.2 = 2x$

$$2x = \frac{\log 0.2}{\log 4}$$

$$x = \frac{\log 0.2}{2\log 4} = -0.580$$

33. $\log_{10} x = -1$

$10^{-1} = x$

$x = 0.1$

34. $\log 0.0001 = x$

$x = -4$

35. $\log_2 2 = x$

$2^x = 2$

$2^x = 2^1$

$x = 1$

36. $\log_3 x = -2$

$3^{-2} = x$

$x = \frac{1}{9}$

37. $\log x = 2$

$10^2 = x$

$x = 100$

38. $\log_2 x = 3$

$2^3 = x$

$x = 8$

39.
$\log_{27} 9 = x$
$27^x = 9$
$(3^3)^x = 3^2$
$3^{3x} = 3^2$
$3x = 2$
$x = \frac{2}{3}$

40.
$\log_2 x = 0$
$2^0 = x$
$x = 1$

41. $\log_b xyz = \log_b x + \log_b y + \log_b z$

42. $\log_b \dfrac{xy^2}{z} = \log_b xy^2 - \log_b z = \log_b x + \log_b y^2 - \log_b z = \log_b x + 2\log_b y - \log_b z$

43. $\log_b \dfrac{x}{y^2} = \log_b x - \log_b y^2 = \log_b x - 2\log_b y$

44. $\log_b \dfrac{\sqrt{x}}{z} = \log_b x^{1/2} - \log_b z = \frac{1}{2}\log_b x - \log_b z$

45. $\log_b \sqrt{\dfrac{z}{x}} = \log_b \dfrac{\sqrt{z}}{\sqrt{x}} = \log_b \sqrt{z} - \log_b \sqrt{x} = \log_b z^{1/2} - \log_b x^{1/2} = \frac{1}{2}\log_b z - \frac{1}{2}\log_b x$

46. $\log_b x^{1/3} y^3 = \log_b x^{1/3} + \log_b y^3 = \frac{1}{3}\log_b x + 3\log_b y$

47. $\log_b x + \frac{1}{2}\log_b y = \log_b x + \log_b y^{1/2} = \log_b xy^{1/2} = \log_b x\sqrt{y}$

48. $\frac{1}{2}\log_b y - 2\log_b z = \log_b y^{1/2} - \log_b z^2 = \log_b \sqrt{y} - \log_b z^2 = \log_b \dfrac{\sqrt{y}}{z^2}$

49. $\log(x-1) + \log(x-2) = \log[(x-1)(x-2)] = \log(x^2 - 3x + 2)$

50. $\log(x-1) + \log(x-2) = \log[(x-1)(x-2)] = \log(x^2 - 3x + 2)$

51. $\log(x^2 + x) - \log x = \log[x(x+1)] - \log x = \log\left(\dfrac{x(x+1)}{x}\right) = \log(x+1)$

52. $\log(x^2 + x + 2) + \log x = \log[(x^2 + x + 2)x] = \log(x^3 + x^2 + 2x)$

53. $\log_3 x^2 - \log_3 x = 2$

$\log_3\left(\dfrac{x^2}{x}\right) = 2$

$\log_3 x = 2$

$3^2 = x;\ x = 9$

54. $\log(x^2 - 4) - \log(x + 2) = 2$

$\log\dfrac{x^2 - 4}{x+2} = \log\dfrac{(x+2)(x-2)}{(x+2)} = 2$

$\log(x - 2) = 2$

$10^2 = x - 2;\ 100 + 2 = x = 102$

55. a. $e^e \approx 15.2$ **b.** $3^e \approx 19.8$ **c.** $e^3 \approx 20.1$

56. a. $\ln 3 = 1.10$ **b.** $\ln \pi = 1.14$ **c.** $\ln e = 1$

57. $e^x = \dfrac{1}{e^2}$

$e^x = e^{-2}$

$x = -2$

58. $e^x = 1$

$e^x = e^0$

$x = 0$

59. $\left(\dfrac{1}{e}\right)^x = e$

$(e^{-1})^x = e^1$

$-1x = 1;\ x = -1$

60. $e^x = e$

$e^x = e^1$

$x = 1$

61. $\ln x = 3$

$e^3 = x$

62. $\ln x = -2$

$e^{-2} = x$

63. $\ln x = \frac{1}{e}$

$e^{1/e} = x$

74. $\ln x = e$

$e^e = x$

65.

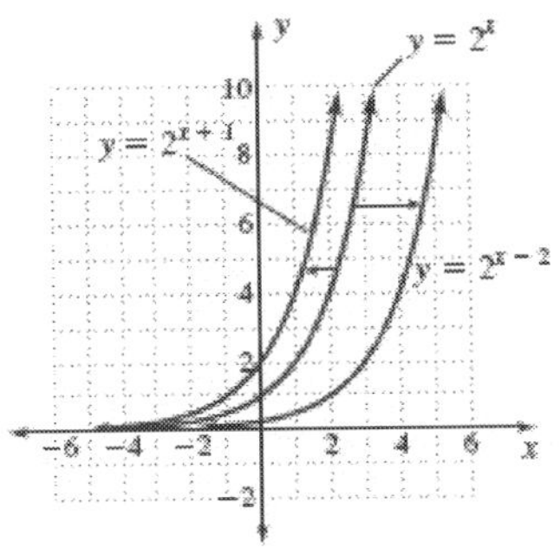

a. Shift 1 unit left.

b. Shift 2 units right.

66.

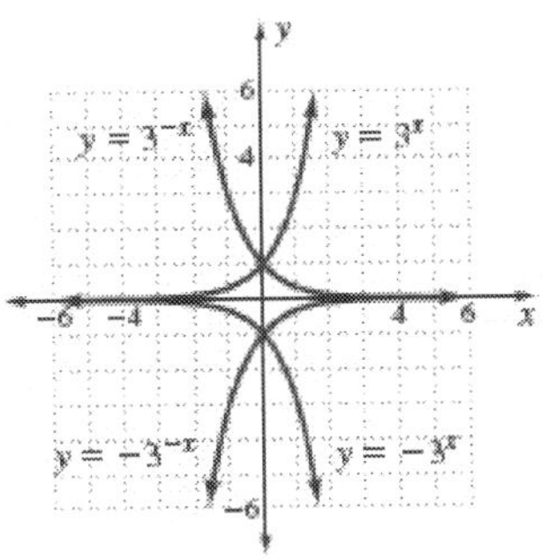

67. a. $3^0 = 1$ and $\log_3 1 = 0$ **b.** $3^1 = 3$ and $\log_3 3 = 1$

c. Graphs mirror each other across y = x.

68.

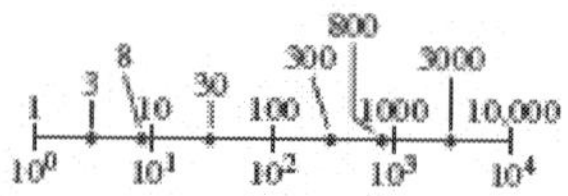

69. Straight line

70. $\log(30) = \log\left(\frac{3000}{100}\right) = \log 3000 - \log 100 = \log 3000 - 2 \approx 1.47712$

71. $\log(500) = \log(5 \bullet 100) = \log 5 + \log 100 = \log 5 + 2 \approx 2.69897$

72. $\log\left(\frac{x}{10}\right) = \log x - \log 10 = \log x - 1$. $\log x$ is 1 greater than $\log\left(\frac{x}{10}\right)$.

73. $\log(1000x)$ is greater than $\log x + 3$

74. $\log(x^2 + 2x + 1) = \log(x+1)^2 = 2\log(x+1)$

75. $\log(8x^3) = \log(2^3x^3) = \log(2x)^3 = 3\log(2x)$

76. $\log \sqrt{x} = \log x^{1/2} = \frac{1}{2}\log x$

77. $\frac{1}{3}\log x = \log x^{1/3} = \log \sqrt[3]{x}$

78. $1000(1 + \frac{0.06}{4})^{4 \cdot 2} \approx \$1126.49$; $2 = (1 + \frac{0.06}{4})^{4t}$; $t = \dfrac{\log 2}{4\log(1 + \frac{0.06}{4})} \approx 11.64$ yr

79. $1000(1 + \frac{0.06}{12})^{12 \cdot 2} \approx \$1127.15$; $2 = (1 + \frac{0.06}{12})^{12t}$; $t = \dfrac{\log 2}{12\log(1 + \frac{0.06}{12})} \approx 11.58$ yr

90. $1000(1 + \frac{0.06}{365})^{365 \cdot 2} \approx 1127.48$; $2 = (1 + \frac{0.06}{365})^{365t}$; $t = \dfrac{\log 2}{365\log(1 + \frac{0.06}{365})} \approx 11.55$ yr

81. $72 \div 6 = 12$ yr; $72 \div 8 = 9$ yr; $72 \div 12 = 6$ yr

82. a. $1000e^{0.06 \cdot 5} \approx \$1349.85$ **b.** $1000e^{0.08 \cdot 5} \approx \$1491.82$

c. $1000e^{0.12 \cdot 5} \approx \$1822.11$

83. a. $1000 = Pe^{0.06 \cdot 5}$

$P = \dfrac{1000}{e^{0.06 \cdot 5}} \approx \$740.82$

b. $1000 = Pe^{0.08 \cdot 5}$

$P = \dfrac{1000}{e^{0.08 \cdot 5}} \approx \$670.32$

c. $1000 = Pe^{0.12 \cdot 5}$

$P = \dfrac{1000}{e^{0.12 \cdot 5}} \approx \$548.81$

84. $69 \div 5 \approx 13.8\%$; $2 = e^{5r}$, $\ln 2 = 5r$, $r = \ln 2 \div 5 \approx 0.1386$

85. a. $y = 25.98(1.055)^t$

b. $1.055 - 1 = 0.055 = 5.5\%$

c. $\dfrac{72}{5.5} \approx 13$ or the year 2012

d. $51.96 = 25.98(1.055)^t$

$2 = 1.055^t$

$\ln 2 = t \ln 1.055$

$t = \dfrac{\ln 2}{\ln 1.055} \approx 12.8$

86. a. \$200

b. $f(n) = 200(2^{n-1})$

87. 40 yr $\div$ 12 $\approx$ 3.33; \$1 $\cdot$ $2^{3.33} \approx$ \$10; doubles in 12 years, find r: $2 = (1 + r)^{12}$.
$1 + r = \sqrt[12]{2}$; $r \approx 0.0595$; $C \approx 1.0595^x$

88. $P = 59{,}509(0.9936)^x$; rate of decay = $0.9936 - 1 = -0.64\%$; $59{,}509(0.9936)^{30} \approx 49{,}011$ in 2010, which seems reasonable since preliminary 2010 census was 48,920.

89. $P = 13{,}447(1.0175)^x$; rate of growth = 1.75%; $13{,}447(1.0175)^{30} \approx 22{,}638$, preliminary 2010 census was 20,249 which was lower than expected.

90. a. about 22 feet

b. about 12 years

c. 60 to 70 years

d. over 200 years

e. 2 and 12; wind breakage

f. about 16 feet

91. a.

Yr.	Option 1	Option 2	Option 3
1	1	1	1
2	2	1.10	1.10
3	3	1.21	1.30
4	4	1.33	1.60
5	5	1.46	2.00

b. Option 1: linear; it has a constant slope. Option 2: exponential; the ratio of consecutive terms is 1.1 to 1. Option 3: quadratic; the second differences are constant.

c. Let x = years after birth and y = birthday gift amount.
Option 1: y = x; Option 2: $y = 1(1 + 0.10)^{(x-1)}$; Option 3: $y = 0.05x^2 - 0.05x + 1$

d. Option 1: 1 year; Option 2: 9 yrs.; Option 3: 5 yrs. (time to double again varies)

e. Option 1 for ages 1 to 20; Options 1 and 3 are tied at $20 at age 20; Option 3 for ages 21 to 52; Option 2 for ages 53 and above.

f. Option 1: 5 yr total = 15; Option 2: 5 yr total = 6.10; Option 3: 5 yr. total = 7.11

92.

a. $400 = 4 \times 10^2 = 10^{0.60206} \times 10^2 = 10^{2.60206}$; n = 2.60206

b. $5000 = 5 \times 10^3 = 10^{0.69897} \times 10^3 = 10^{3.69897}$; n = 3.69897

c. $5{,}000{,}000 = 5 \times 10^6$; n = 6.69897

d. $500{,}000 = 5 \times 10^5$; n = 5.69897

e. $4000 = 4 \times 10^3$; n = 3.60206

f. $500 = 5 \times 10^2$; n = 2.69897

g. $40{,}000 = 4 \times 10^4$; n = 4.60206

h. $4{,}000{,}000 = 4 \times 10^6$; n = 6.60206

i. $0.4 = 4 \times 10^{-1}$; n = −1 + 0.60206 = −0.39794

j. $0.005 = 5 \times 10^{-3}$; n = −3 + 0.69897 = −2.30103

k. $0.0004 = 4 \times 10^{-4}$; n = −4 + 0.60206 = −3.39794

l. $0.5 = 5 \times 10^{-1}$; n = −1 + 0.69897 = −0.30103

m. $x^a \cdot x^b = x^{a+b}$

n. A logarithm is the exponent to which we raise the base to obtain a given number.

o. The exponent on 10 is the first number.

p. *a* is always between 1 and 10; $1 = 10^0$ and $10 = 10^1$.

q. Numbers between 0 and 1 have negative logarithms; their logarithms are the sum of a number between 0 and 1 and the negative exponent on 10.

93. a. $$S = 1000\left(1 + \frac{0.05}{8760}\right)^{8760(1)} = \$1051.270947$$

b. $$S = 1000\left(1 + \frac{0.05}{525600}\right)^{525600(1)} = \$1051.271107$$

c. $$S = 1000e^{0.05} = \$1051.271096$$

d. $$1051.271096 - 1051.271107 = 0.000011 = 1.1 \times 10^{-5}$$

e. They go above and below the maximum limit value found in part c.

f. Sawtooth pattern for Y_1

Chapter 6 Test

1. **a.** Common ratio = 2; geometric sequence

b. $a_n = \frac{1}{8} \cdot 2^{n-1}$; $f(x) = 0.0625 \cdot 2^x$; $2^{-3} \cdot 2^{-1} \cdot 2^n = 2^{-4} \cdot 2^n = \frac{1}{16} \cdot 2^n = 0.0625 \cdot 2^n$

2. **a.** No common ratio; 1st differences = 9, 15, 21, 27; 2nd differences = 6, 6, 6; quadratic

c. $f(x) = 3x^2$

3. **a.** $3^x = 27$
$3^x = 3^3$
$x = 3$

b. $\left(\frac{1}{16}\right)^n = 64$
$(4^{-2})^n = 4^3$
$-2n = 3$
$n = -\frac{3}{2}$

c. $\frac{1}{1000} = 0.1^x$
$10^{-3} = (10^{-1})^x$
$-3 = -x$
$x = 3$

d. $4^{0.5x} = 16$
$4^{0.5x} = 4^2$
$0.5x = 2$
$x = 4$

e. $4^x = \frac{1}{64}$
$4^x = 4^{-3}$
$x = -3$

f. $100^n = 10$
$(10^2)^n = 10^1$
$2n = 1$
$n = \frac{1}{2}$

g. $27^x = 9$
$(3^3)^x = x^2$
$3x = 2$
$x = \frac{2}{3}$

h. $4^{x+1} = 8$
$(2^2)^{x+1} = 2^3$
$2x + 2 = 3$
$2x = 1;\ x = \frac{1}{2}$

4. **a.** $10^x = 3$
$\log 3 = x$
$x \approx 0.477$

b. $5^{2x} = 0.25$
$\log_5 0.25 = 2x$
$x = \dfrac{\log 0.25}{2 \log 5} \approx -0.431$

c. $10^x = -1$
{ } or $\varnothing$

d. $e^x = \frac{1}{e}$
$e^x = e^{-1}$
$x = -1$

e. $5^{x-2} = \frac{1}{125}$
$5^{x-2} = 5^{-3}$
$x - 2 = -3$
$x = -1$

f. $\log_2 x = -2$
$2^{-2} = x$
$x = \frac{1}{4}$

g. $\log_5 x = 3$
$5^3 = x$
$x = 125$

h. $\log_x 32 = 5$
$x^5 = 32$
$x^5 = 2^5$
$x = 2$

i. $\log x = 1$
$10^1 = x$
$x = 10$

j. $\ln x = 1.5$
$e^{1.5} = x$
$x \approx 4.482$

5. **a.** $\ln e^2 = 2\ln e = 2(1) = 2$ **b.** $\log 1 = 0$

c. $\log_2 x + \log_2 (x+1) = \log_2[x(x+1)] = \log_2 (x^2 + x)$

d. $\log_3 (x^2 - 9) - \log_3 (x+3) = \log_3\left(\dfrac{x^2-9}{x+3}\right) = \log_3\left(\dfrac{(x+3)(x-3)}{(x+3)}\right) = \log_3 (x-3)$

6. **a.** $\log_b\left(\dfrac{z^2}{y}\right) = \log_b z^2 - \log_b y = 2\log_b z - \log_b y$

b. $\log_b\left(\dfrac{\sqrt{x}}{y}\right) = \log_b \sqrt{x} - \log_b y = \log_b x^{1/2} - \log_b y = \frac{1}{2}\log_b x - \log_b y$

7. **a.** (–1, –1), (–3, –2), (–5, –3); yes, it is a function.

b. (–1, –1), (–1, –2), (–1, –3); no, it is not a function (same input has several outputs).

8.

x	$y = (\frac{1}{2})^x$
0	1
1	0.5
2	0.25
3	0.125

y	$x = (\frac{1}{2})^y$
0	1
1	0.5
2	0.25
3	0.125

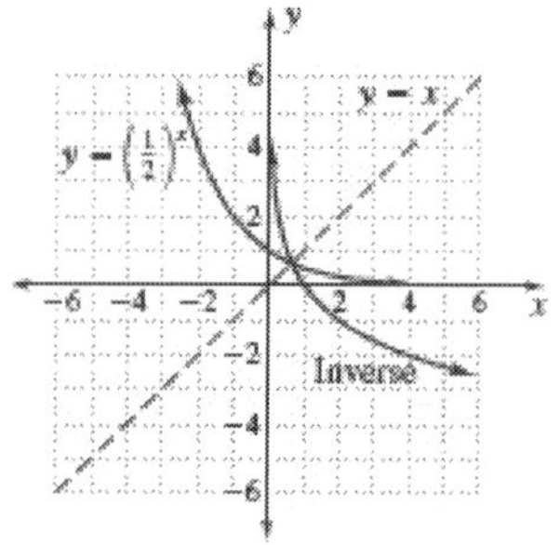

9. $3y - 2x = 6$; or $y = \frac{2}{3}x + 2$

10.

$1360 = 1000(1 + 0.075)^t$
$1.360 = (1.075)^t$
$\log 1.360 = \log 1.075^t$
$\log 1.360 = t\log 1.075$
$t = \dfrac{\log 1.360}{\log 1.075} \approx 4.25$ yr

11.

$1906 = 1000e^{0.075t}$
$1.906 = e^{0.075t}$
$\ln 1.906 = 0.075t$
$t = \dfrac{\ln 1.906}{0.075} \approx 8.6$ yr

12. Compound annually: $2P = P\left(1 + \dfrac{r}{1}\right)^{1(2)}$

Solve by using a calculator: $2 = (1 + r)^2$; $r \approx 0.414$ or 41.4%

Compound monthly: $2P = P\left(1 + \dfrac{r}{12}\right)^{12(2)}$

Solve by using a calculator: $2 = \left(1 + \dfrac{r}{12}\right)^{24}$; $r \approx 0.352$ or 35.2%

Compound continuously: $2P = Pe^{2r}$
$2 = e^{2r}$
$\ln 2 = 2r$; $r = \dfrac{\ln 2}{2} \approx 0.347$ or 34.7%

13. a. $y = 0.97x + 0.07$

b. $y = 0.29(1.05)^x$

c. $y = 0.0013x^2 + 0.0066x + 0.227$

d. Quadratic or linear.

e. $y = 0.25(1 + 0.07)^{71} = \$30.49$

14. $f(x) = \log_2 x$

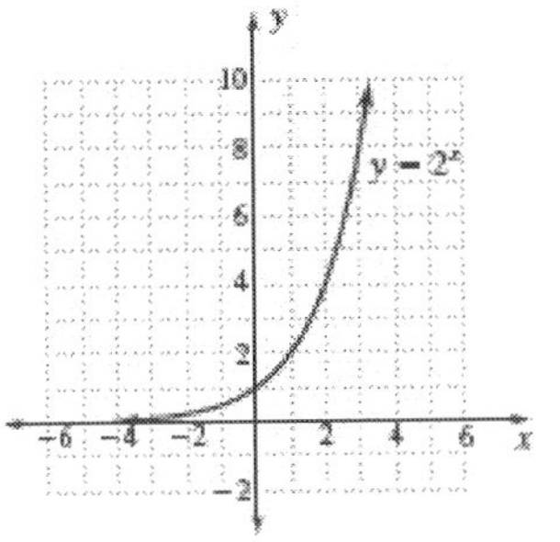

15.

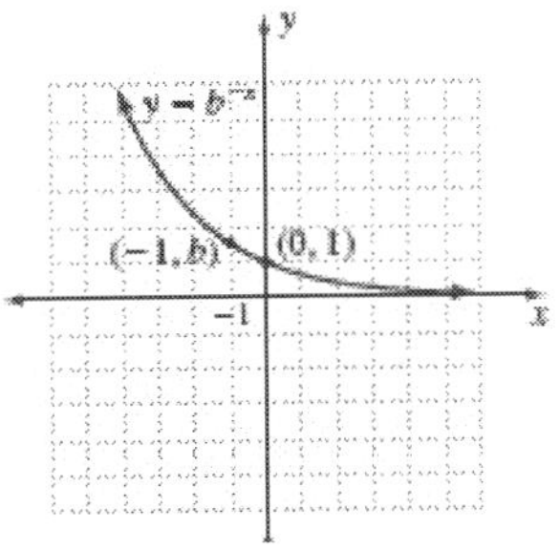

16. 1 + r = 0.93; r = 0.93 – 1 = –0.07 (to find % multiply by 100)

17. log (100x) = log 100 + log x = 2 + log x

18. a. pH = –log $(5.01 \times 10^{-9}) = 8.3$

b. 11.9 = –log $[H^+]$; $-11.9 = \log[H^+]$; $10^{-11.9} = [H^+] \approx 1.26 \times 10^{-12}$ M

19. Extend the logarithm of a product property to *n* factors of x. Write out the *n* factors using the logarithm of a product:

$\log_b x \cdot x \cdot x \cdot \ldots \cdot x$ (*n* factors of x) = $\log_b x + \log_b x + \log_b x + \ldots + \log_b x = n \log_b x$

20. a. An increase of 3 decibels doubles the intensity.

b.

dB	Time	dB	Time
85	8 hrs	115	28.1 sec
88	4 hrs	118	14.1 sec
91	2 hrs	121	7 sec
94	1 hr	127	3.5 sec
97	30 min	130	1.8 sec
100	15 min	133	1 sec
103	7.5 min	136	0.5 sec
106	3.75 min or 225 sec	139	0.1 sec
109	112.5 sec	142	0.05 sec
112	56.3 sec	145	0.027 sec

Cumulative Review: Chapters 1 to 6

1. **a.** $\frac{6}{\sqrt{3}} \cdot \frac{\sqrt{3}}{\sqrt{3}} = \frac{6\sqrt{3}}{3} = 2\sqrt{3}$

b. $\frac{6}{\sqrt{2}} \cdot \frac{\sqrt{2}}{\sqrt{2}} = \frac{6\sqrt{2}}{2} = 3\sqrt{2}$

c. $\frac{x}{\sqrt{y}} \cdot \frac{\sqrt{y}}{\sqrt{y}} = \frac{x\sqrt{y}}{y}$

2. $\frac{186282 \text{ mi}}{\text{sec}} \cdot \frac{60 \text{ sec}}{1 \text{ min}} \cdot \frac{60 \text{ min}}{1 \text{ hr}} = 670615200 \frac{\text{mi}}{\text{hr}}$

$\frac{670615200 \text{ mi}}{\text{hr}} \div \frac{760.98 \text{ mi}}{1 \text{ hr}} = \frac{670615200 \text{ mi}}{1 \text{ hr}} \cdot \frac{1 \text{ hr}}{760.98 \text{ mi}} \approx 881{,}000$

The speed of light is 881,000 times faster than the speed of sound. This is why lightning is seen before the thunder is heard.

3. The –4 shifts the graph down 4 units, so that the vertex and y-intercept are (0, –4) instead of (0, 0) and the x-intercepts are (–2, 0) and (2, 0) instead of (0, 0).

4. **a.** $\sqrt{-16} = 0 + 4i$

b. $4 = 4 + 0i$

c. $3 + \sqrt{-4} = 3 + 2i$

d. $\sqrt{-56} = 0 + 2i\sqrt{14}$

5. **a.** $(2x+5)(x-3) = 0; x = -\frac{5}{2}, 3$

b. $(3x-1)(2x-3) = 0; x = \frac{1}{3}, \frac{3}{2}$

6. **a.** $\frac{4-\sqrt{-48}}{2} = \frac{4-4i\sqrt{3}}{2} = 2-2i\sqrt{3}$

b. $\frac{9+\sqrt{-18}}{6} = \frac{9+3i\sqrt{2}}{6} = \frac{3+i\sqrt{2}}{2}$

7. **a.** $(3 + 2i) - (6 + 2i) =$

$3 + 2i - 6 - 2i = -3$

b. $(4 - 2i) - 2(3 - 4i) =$

$4 - 2i - 6 + 8i = -2 + 6i$

8. **a.** $(3 - 2i)(3 + 2i) =$

$9 - 4(-1) = 13$

b. $(2 - i)(2 + i) =$

$4 - (-1) = 5$

c. $(2 - 3i)(2 + 3i) =$

$4 - 9(-1) = 13$

9. $x = \frac{-1 \pm \sqrt{1^2 - 4(1)(1)}}{2(1)}$

$x = \frac{-1 \pm i\sqrt{3}}{2}$

$\{-\frac{1}{2} \pm \frac{i\sqrt{3}}{2}\}$ or $\{-0.5 \pm 0.87i\}$

10. $x^2(1 - 9x^2) = 0;\ x^2 = 0$ or $1 - 9x^2 = 0$

$x = 0$ (dbl root) or $(1 + 3x)(1 - 3x) = 0$

$x = \pm \frac{1}{3};\ \{0 \text{ (dbl root)}, \pm \frac{1}{3}\}$

11. $(2x-1)(3x+4)=0; 6x^2+5x-4=0$ or any equation equivalent: $6nx^2+5nx-4n=0$ for any real number n except zero.

12. **a.** Until a^2 and b^2 are added, say to c^2, $c \geq 0$, no simplification is possible.
b. If both a and b are positive, the expression equals ab. If either could be negative, then absolute values are needed, as in $|ab|$.
c. Division by zero is not permitted; otherwise, same as part b with $\frac{a}{b}$ or $\left|\frac{a}{b}\right|$.
d. Same as part b with $a+b$ or $|a+b|$.

13. **a.** $x=\frac{-b}{2a}=\frac{1/25}{2(1/80000)}=1600; y=\frac{1600^2}{80000}-\frac{1600}{25}+1224=1160.02; (1600,1160.02)$; where the drain needs to be located.
b. $y=\frac{0^2}{80000}-\frac{0}{25}+1224=1224$
c. $y=\frac{2400^2}{80000}-\frac{2400}{25}+1224=1200$

14. $y=2(1)^x$. Because 1 raised to any power is still 1, the equation is the linear equation, $y=2$.

15. $f(x)=3^x$

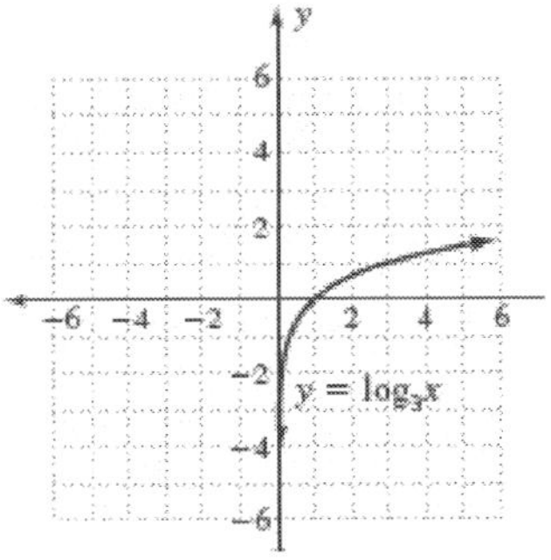

16. **a.**

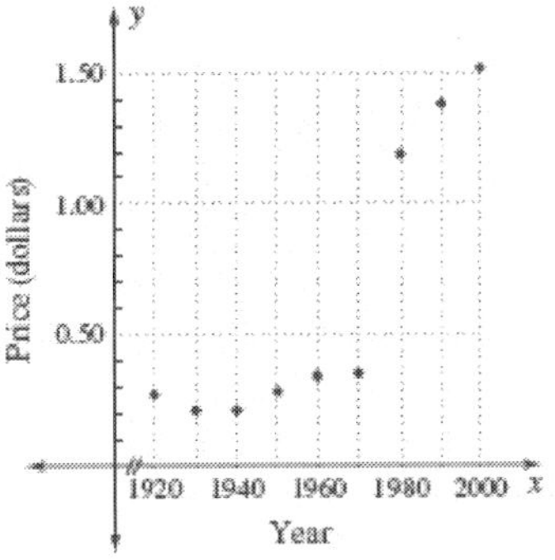

b. Possible answer: exponential. Data start off low and increase rapidly
c. $f(x) \approx 0.156(1.0278)^x$

17. **a.** Turn the 11-minute timer twice.
b. 11-minute and 7-minute both go 7 minutes; Turn the 7 minute over. Stop the 7-minute timer when the 11-minute timer runs out. Run the 11 minute timer by itself for 11 minutes followed by the 4 minutes that are separted out in the 7-minute timer.

18. The answer is self-checking so none is provided.

CHAPTER 7

Conic Sections, Systems of Nonlinear and Linear Equations and Inequalities, Sequences and Series

Exercises 7.1: Conic Sections

1. (±3, 0), (0, ±3)

For exercises 3 to 19, refer to text for sample equations.

3. $x^2 + y^2 = 5^2$; $x^2 + y^2 = 25$

5. $\frac{x^2}{5^2} + \frac{y^2}{2^2} = 1$; $\frac{x^2}{25} + \frac{y^2}{4} = 1$

7. No x-intercept indicates y^2 term comes first; $\frac{y^2}{4^2} - \frac{x^2}{6^2} = 1$; $\frac{y^2}{16} - \frac{x^2}{36} = 1$.

9. Opening to the right indicates squared term is y^2 and a is positive.
$3 = a(-1)^2$; $a = 3$; $x = 3y^2$

11. Opening downward indicates squared term is x^2 and a is negative.
$-1 = a(3)^2$; $a = -\frac{1}{9}$; $y = -\frac{1}{9}x^2$

13. $x^2 + y^2 = 4^2$; $x^2 + y^2 = 16$

15. $\frac{x^2}{1^2} + \frac{y^2}{5^2} = 1$; $x^2 + \frac{y^2}{25} = 1$

17. No x-intercept indicates y^2 term comes first: $\frac{y^2}{3^2} - \frac{x^2}{5^2} = 1$; $\frac{y^2}{9} - \frac{x^2}{25} = 1$.

19. Opening upward indicates squared term is x^2 and a is positive.
$1 = a(-2)^2$; $a = \frac{1}{4}$; $y = \frac{1}{4}x^2$

21. $r < 0$ has no meaning; circle becomes a point.

For exercises 23 to 43, solving for y detail is not shown; graphs are combined.

23. Parabola; y-intercept (0, 2)

25. Circle; intercepts $(\pm 3, 0)$, (0 ± 3)

Calculator graph: $y = \pm\sqrt{9 - x^2}$

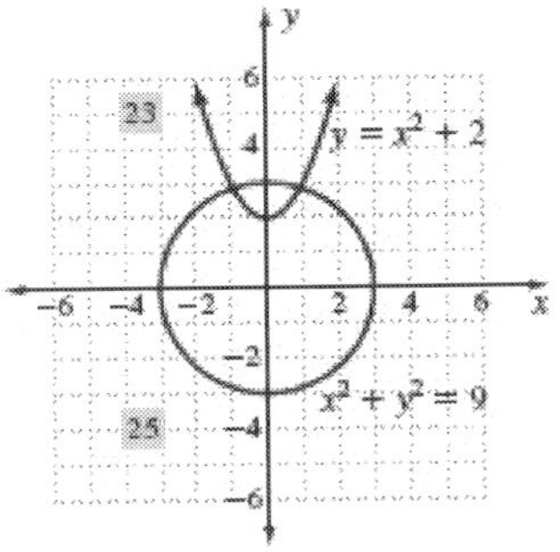

27. Hyperbola; x-intercepts $(\pm 2, 0)$

Calculator: $y = \pm\sqrt{x^2 - 4}$

29. Ellipse: intercepts $(\pm 2, 0)$, $(0, \pm 1)$

Calculator: $y = \pm\sqrt{1 - \dfrac{x^2}{4}}$

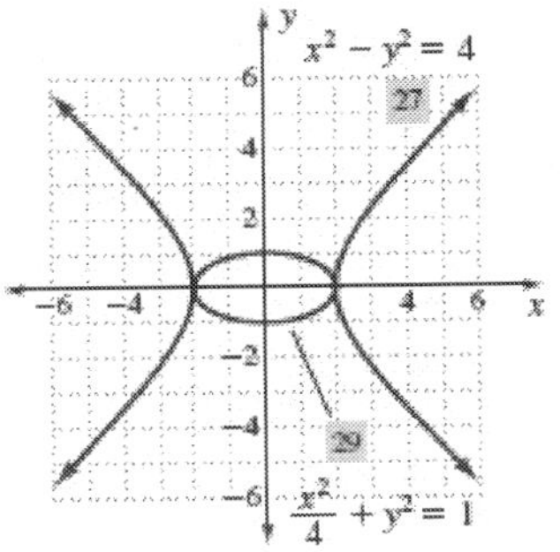

31. Divide both sides by 100:

$\dfrac{x^2}{25} + \dfrac{y^2}{100} = 1$; ellipse,

intercepts: $(\pm 5, 0)$, $(0, \pm 10)$

Calculator: $y = \pm\sqrt{100 - 4x^2}$

33. Add 8 to both sides then divide by 2:

$x^2 + y^2 = 4$; circle

intercepts: $(\pm 2, 0)$, $(0, \pm 2)$

Calculator: $y = \pm\sqrt{4 - x^2}$

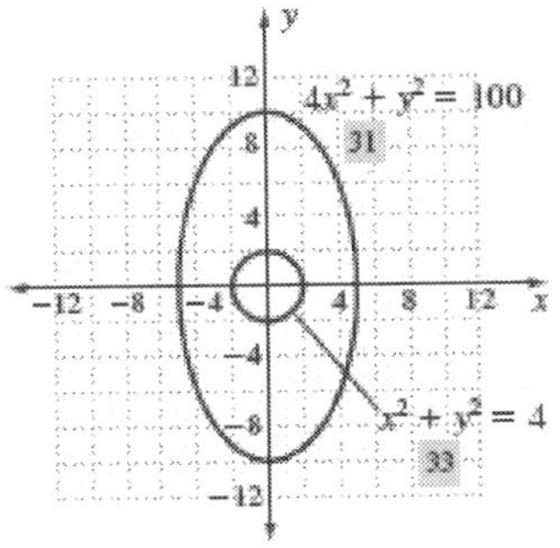

35. y = x; straight line

37. Hyperbola; x-intercepts (±1, 0);

Calculator: y = $\pm\frac{1}{2}\sqrt{x^2-1}$

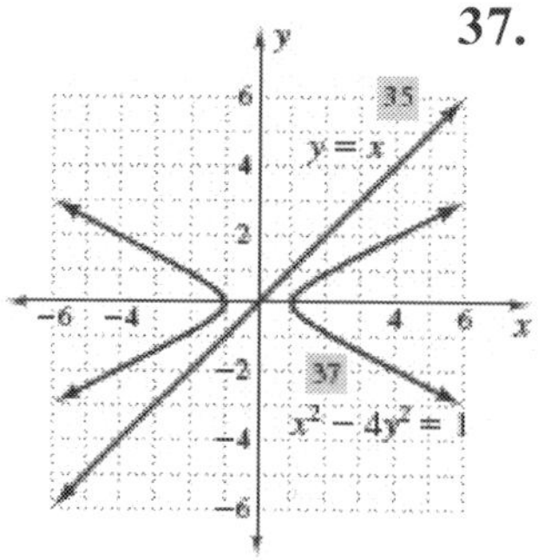

39. Parabola; vertex at (0, 0)

Calculator: y = $\pm\frac{1}{2}\sqrt{x}$

41. Hyperbola; y-intercepts; (0, ±1)

Calculator: y = $\pm\sqrt{1+\dfrac{x^2}{4}}$

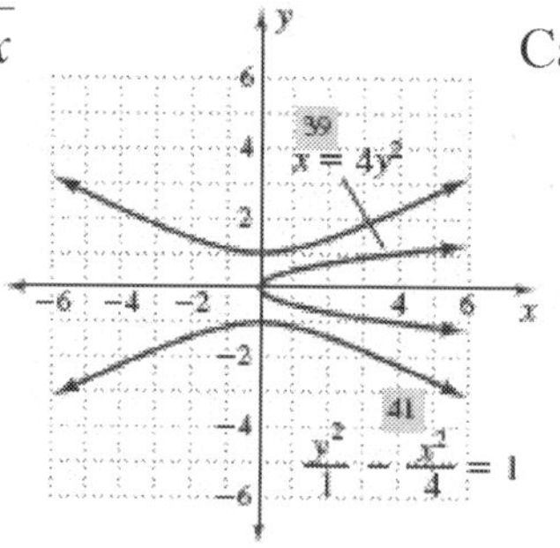

43. Parabola; vertex at (0, 0)

Calculator: y = $\pm\sqrt{-x}$

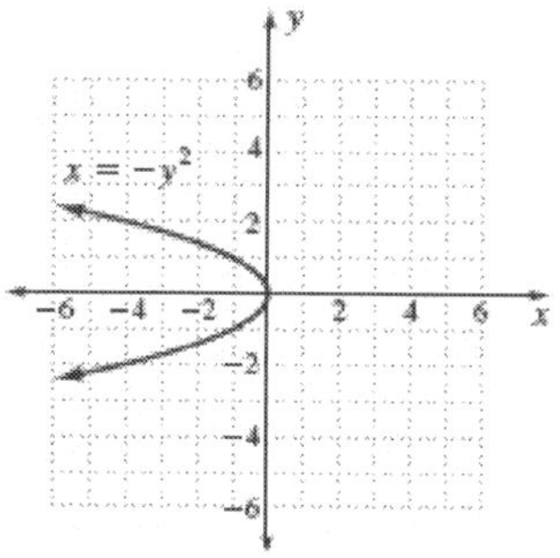

45. a. y/r; x/r; y/x

b. $x^2 + y^2 = r^2$

c. $\left(\dfrac{y}{r}\right)^2+\left(\dfrac{x}{r}\right)^2=1; \dfrac{y^2}{r^2}+\dfrac{x^2}{r^2}=1; y^2+x^2=r^2$

d. $x^2+y^2=r^2; \dfrac{y^2}{r^2}+\dfrac{x^2}{r^2}=1; \left(\dfrac{y}{r}\right)^2+\left(\dfrac{x}{r}\right)^2=1$

$(\sin A)^2+(\cos A)^2=1$

47. a. Assume smaller circles have r = 1; centers are at (−2, 2); (2, 2) and (0, −2). Equations are $(x + 2)^2 + (y - 2)^2 = 1$; $(x - 2)^2 + (y - 2)^2 = 1$ and $x^2 + (y + 2)^2 = 1$.

For calculator, solve each for y: $y=\pm\sqrt{1-(x+2)^2}+2$, $y=\pm\sqrt{1-(x-2)^2}+2$, and

$y=\pm\sqrt{1-x^2}-2$. Use a square window and experiment until you find which equation to use from each pair. In this case use equations with negative square root.

Large circle has r = 4, center at the origin; equation = $x^2 + y^2 = 16$; $y = \pm\sqrt{16-x^2}$.

b. $r=\sqrt{(x-h)^2+(y-k)^2}$; $r^2 = (x - h)^2 + (y - k)^2$

49. **b.** Conic section is a hyperbola. **c.** Conic section is an ellipse.

Exercises 7.2: Solving Systems of Nonlinear Equations

1. Parabola and a line; set equal to each other and solve.
$x^2 - 3x + 2 = -x + 5 \quad x^2 - 2x - 3 = 0 \quad (x - 3)(x + 1) = 0 \quad x - 3 = 0$ or $x + 1 = 0$
$x = 3$ or $x = -1 \quad y = -(3) + 5,\ y = 2, \quad y = -(-1) + 5,\ y = 6 \quad (-1, 6), (3, 2)$

3. Hyperbola and a parabola; solve the parabola equation for x^2 and substitute into the hyperbola equation.
$y = x^2 - 4,\ x^2 = y + 4;\ \dfrac{y+4}{4} - \dfrac{y^2}{9} = 1,\ 36\left(\dfrac{y+4}{4} - \dfrac{y^2}{9}\right) = 36,$

$9y + 36 - 4y^2 = 36,\ -4y^2 + 9y = 0 \quad y(-4y + 9) = 0,\ y = 0$, or
$-4y + 9 = 0,\ -4y = -9,\ y = \frac{9}{4};\ x^2 = 0 + 4,\ x = \pm 2 \quad x^2 = \frac{9}{4} + 4,$
$x^2 = \frac{25}{4},\ x = \pm\frac{5}{2} \quad (\pm\frac{5}{2}, \frac{9}{4}), (\pm 2, 0)$

5. Both equations are parabolas; set equal to each other and solve.
$5x^2 + 2x = -x^2 + x + 2,\ 6x^2 + x - 2 = 0,\ (3x + 2)(2x - 1) = 0$
$3x + 2 = 0,\ 3x = -2,\ x = -\frac{2}{3};\ 2x - 1 = 0,\ 2x = 1,\ x = \frac{1}{2};\ y = 5(-\frac{2}{3})^2 + 2(-\frac{2}{3}),$
$y = \frac{20}{9} - \frac{4}{3},\ y = \frac{8}{9} \quad y = 5(\frac{1}{2})^2 + 2(\frac{1}{2}),\ y = \frac{5}{4} + 1,\ y = \frac{9}{4};\ (-\frac{2}{3}, \frac{8}{9}), (\frac{1}{2}, \frac{9}{4})$

7. Equations are an ellipse and a parabola. Substitute one into the other.
$y = x^2 - 1,\ x^2 = y + 1 \quad y + 1 + \frac{y^2}{4} = 1 \quad 4y + 4 + y^2 = 4,\ y^2 + 4y = 0$
$y(y + 4) = 0 \quad y = 0$ or $y + 4 = 0,\ y = -4 \quad 0 = x^2 - 1,\ x^2 = 1,\ x = \pm 1$
$-4 = x^2 - 1,\ x^2 = -3$. Not possible; discard extraneous answer. $(\pm 1, 0)$

9. Circle and a parabola; substitute one into the other.
$y = x^2 - 4,\ x^2 = y + 4; \quad y + 4 + y^2 = 16, \quad y^2 + y - 12 = 0,\ (y + 4)(y - 3) = 0;$
$y + 4 = 0,\ y = -4; \quad y - 3 = 0,\ y = 3; \quad -4 = x^2 - 4,\ x^2 = 0,\ x = 0;$
$3 = x^2 - 4,\ x^2 = 7,\ x = \pm\sqrt{7}, \quad (0, -4), (\pm\sqrt{7}, 3)$

11. Equations are both parabolas; set equal to each other and solve.
$x^2 - 6 = -x^2 + 2 \quad 2x^2 = 8,\ x^2 = 4 \quad x = \pm 2 \quad y = 4 - 6,\ y = -2 \quad (\pm 2, -2)$

13. Ellipse and a parabola; solve parabola for x^2 and substitute into ellipse equation.
$y = x^2 - 1,\ x^2 = y + 1;\ \dfrac{y+1}{4} + y^2 = 1, \quad y + 1 + 4y^2 = 4, \quad 4y^2 + y - 3 = 0,$
$(4y - 3)(y + 1) = 0, \quad y + 1 = 0,\ y = -1,\ 4y - 3 = 0,\ 4y = 3,\ y = 0.75;\ -1 = x^2 - 1,\ x^2 = 0,$
$x = 0,\ 0.75 = x^2 - 1,\ x^2 = 1.75,\ x = \pm\sqrt{1.75};\ (0, -1), (\pm\sqrt{1.75} \approx \pm 1.323, 0.75)$

15. Equations are a circle and a hyperbola. Use elimination and add equations.
$(x^2 + x^2) + (y^2 - y^2) = 4 + 4 \quad 2x^2 = 8;\ x^2 = 4,\ x = \pm 2 \quad 4 + y^2 = 4,\ y^2 = 0,\ y = 0 \quad (\pm 2, 0)$

17. Hyperbola and a line.

$x + y = 4,\ y = 4 - x;\ 4 - x = \frac{1}{x},\ 4x - x^2 = 1,\ x^2 - 4x + 1 = 0,$

$x = \dfrac{-(-4) \pm \sqrt{(-4)^2 - 4(1)(1)}}{2(1)} = \dfrac{4 \pm \sqrt{12}}{2},\quad x = \dfrac{4 \pm 2\sqrt{3}}{2},\quad x = 2 \pm \sqrt{3};$

$y = 4 - (2 + \sqrt{3}),\ y = 2 - \sqrt{3},\qquad y = 4 - (2 - \sqrt{3}),\ y = 2 + \sqrt{3};$

$(2 \pm \sqrt{3},\ 2 \pm \sqrt{3}) \approx (3.732, 0.268), (0.268, 3.732)$

19. Equations are a circle and an ellipse.

$x^2 = 9 - y^2,\ \dfrac{9 - y^2}{16} + \dfrac{y^2}{4} = 1,\ \ 9 - y^2 + 4y^2 = 16,\ 3y^2 = 7,\ y^2 = \frac{7}{3}$

$y = \pm\sqrt{\frac{7}{3}},\ y = \pm\dfrac{\sqrt{7}}{\sqrt{3}},\ \ y = \pm\dfrac{\sqrt{21}}{3}\qquad x^2 = 9 - \frac{7}{3},\ x^2 = \frac{20}{3},\ x = \pm\sqrt{\frac{20}{3}},$

$x = \pm\dfrac{2\sqrt{5}}{\sqrt{3}},\ \ x = \pm\dfrac{2\sqrt{15}}{3};\qquad \left(\pm\dfrac{2\sqrt{15}}{3},\ \pm\dfrac{\sqrt{21}}{3}\right)$

21. $3^x = 2 - 3^x\ \ 2 \cdot 3^x = 2,\ 3^x = 1,\ x = 0;\qquad y = 3^0,\ y = 1;\ (0, 1)$

23. $2^{x+2} = 4 + 2^x;\ 2^2 \cdot 2^x = 4 + 2^x\ \ 4 \cdot 2^x - 2^x = 4;\ 3 \cdot 2^x = 4,\ 2^x = \frac{4}{3}\ \ \log 2^x = \log \frac{4}{3},$

$x\log 2 = \log \frac{4}{3}\ \ x = \dfrac{\log \frac{4}{3}}{\log 2},\ x \approx 0.415\ \ y \approx 4 + 2^{0.415},\ y \approx 5.333\qquad (0.415, 5.333)$

25. $\log x = 3 - \log x,\ 2\log x = 3,\ \log x = 1.5,\ 10^{1.5} = x,\ x \approx 31.6;$
$y = \log 10^{1.5},\ y = 1.5\ \ (31.6, 1.5)$

27. $\log_4 x = 2 - \log_4 x,\ 2\log_4 x = 2\ \ \log_4 x = 1,\ 4^1 = x,\ x = 4;\ \ y = \log_4 4,\ y = 1\qquad (4, 1)$

29. $-\log_3 x = -1 + 0.5\log_3 x; 1 = 1.5\log_3 x; \dfrac{2}{3} = \log_3 x; 3^{\left(\frac{2}{3}\right)} = x; x \approx 2.08; y = -\log_3 3^{2/3} = -\dfrac{2}{3}; (2.08, -2/3)$

31. $y = \pm 2\sqrt{1 - \dfrac{x^2}{9}},\quad y = \pm 2\sqrt{x^2 - 1}$

33. $2^{x+2} = 2^2 \cdot 2^x = 4 \cdot 2^x\ \ 4 \cdot \frac{1}{3} = \frac{4}{3}$

35. $3^{x+1} = 3^1 \cdot 3^x = 3 \cdot 3^x\ \ 3 \cdot 2 = 6$

37. $2^{x-1} = 2^{-1} \cdot 2^x = \frac{1}{2} \cdot 2^x\ \ \frac{1}{2} \cdot 5 = \frac{5}{2}$

39. $4^{x-2} = 4^{-2} \cdot 4^x = \frac{1}{16} \cdot 4^x\ \ \frac{1}{16} \cdot 3 = \frac{3}{16}$

41. For $d > 1$, no solutions. For $d = 1$, one solution. For $-1 < d < 1$, two solutions. For $d = -1$, three solutions, For $d = -1.25$, two solutions. For $d < -1.25$, no solutions.

Exercises 7.3: Solving Systems of Three or More Linear Equations

1. $2a + 3b + 2c = 1$ (1)

$3a + 2c = 10$ (2)

$a + b = 1$ (3)

Solve (3) for b: $b = 1 - a$

Solve (2) for c: $c = 5 - 1.5a$

Substitute into (1):

$2a + 3(1 - a) + 2(5 - 1.5a) = 1$

Solve for a:

$2a + 3 - 3a + 10 - 3a = 1$

$13 - 4a = 1$

$4a = 12$

$\mathbf{a = 3}$

$b = 1 - 3$

$\mathbf{b = -2}$

$c = 5 - 1.5(3)$

$\mathbf{c = 0.5}$

3. $2a + b - 4c = 17$

$5a - 2b = 27$

$b + 3c + 7 = 0$

$a = 0.4b + 5.4$

$c = -\frac{1}{3}b - \frac{7}{3}$

$2(0.4b + 5.4) + b - 4(-\frac{1}{3}b - \frac{7}{3}) = 17$

$0.8b + 10.8 + b + 1.3b + 9.3 = 17$

$3.1b + 20.1 = 17$

$3.1b = -3.1$

$\mathbf{b = -1}$

$a = 0.4(-1) + 5.4$

$\mathbf{a = 5}$

$c = -\frac{1}{3}(-1) - \frac{7}{3}$

$\mathbf{c = -2}$

5. $x + y + z = 2$

$x + 3z = 0 \quad x = -3z$

$2z = 7 + x + y$

$-3z + y + z = 2$

$y - 2z = 2$ (a)

$2z = 7 + (-3z) + y$

$-y + 5z = 7$ (b)

Add (a) and (b).

$3z = 9$

$\mathbf{z = 3}$

$x = -3(3)$

$\mathbf{x = -9}$

$-9 + y + 3 = 2$

$\mathbf{y = 8}$

7. $2x + 3y - z = 11$

$3x + y + 2z = 13$

$2y = z + 7$

$z = 2y - 7$

$2x + 3y - (2y - 7) = 11$

$2x + y = 4$ (a)

$3x + y + 2(2y - 7) = 13$

$3x + 5y - 14 = 13$

$3x + 5y = 27$ (b)

$-5(2x + y) = -5(4)$

$-10x - 5y = -20$ (c)

Add (b) and (c).

$-7x = 7$; $\mathbf{x = -1}$

$2(-1) + y = 4$; $\mathbf{y = 6}$

$z = 2(6) - 7$; $\mathbf{z = 5}$

9.

$x + y - z = 2$ (1)

$x - y + z = 3$ (2)

$-x + y + z = 4$ (3)

Add (1) and (2).

$2x = 5$

$\mathbf{x = 2.5}$

Add (2) and (3).

$2z = 7$

$\mathbf{z = 3.5}$

Substitute into (1).

$2.5 + y - 3.5 = 2$

$\mathbf{y = 3}$

11.

$2a + 3b + c = -4$ (1)

$3a + 2b + 4c = 19$ (2)

$a - b - 3c = -19$ (3)

$-3(a - b - 3c) = -3(-19)$

$-3a + 3b + 9c = 57$ (4)

Add (2) and (4): $5b + 13c = 76$ (a)

$-2(a - b - 3c) = -2(-19)$

$-2a + 2b + 6c = 38$ (5)

Add (1) and (5): $5b + 7c = 34$ (b)

Subtract (a) and (b): $6c = 42$; $\mathbf{c = 7}$

$5b + 7(7) = 34$; $5b = -15$; $\mathbf{b = -3}$

$a - (-3) - 3(7) = -19$; $a - 18 = -19$; $\mathbf{a = -1}$

13.

$a + b + c = 10$ (1)

$-a + b = 4$ (2)

$2a - b + c = 0$ (3)

Subtract (1) from (3).

$a - 2b = -10$ (4)

Add (2) and (4).

$-b = -6$; $\mathbf{b = 6}$

$-a + 6 = 4$; $\mathbf{a = 2}$

$2 + 6 + c = 10$; $\mathbf{c = 2}$

15.

$x + y - z = 2$ (1)

$x - z = 4$ (2)

$x + y = 3$ (3)

Subtract (1) and (3).

$-z = -1$; $\mathbf{z = 1}$

$x - 1 = 4$

$\mathbf{x = 5}$

$5 + y = 3$

$\mathbf{y = -2}$

17.

$a + 2b + 3c = 2$ (1)

$4a + 5b + 6c = 3$ (2)

$a + b + c = 1$ (3)

Subtract (3) from (1).

$b + 2c = 1$ (4)

$-4(a + b + c) = -4(1)$

$-4a - 4b - 4c = -4$ (5)

Add (2) and (5).

$b + 2c = -1$ (6)

From (4) and (6): $1 = -1$

contradiction; no solution

19.

$2a + b + c = 3$ (1)

$a + 2b - 2c = 2$ (2)

$a - b + 3c = 1$ (3)

Subtract (2) and (3).

$3b - 5c = 1$ (4)

$-2(a - b + 3c) = -2(1)$

$-2a + 2b - 6c = -2$ (5)

Add (1) and (5).

$3b - 5c = 1$ (6)

From (4) and (6): $1 = 1$

identity; infinite

21. $2x - y + z = 1$ (1)

$x + z = -2$ (2)

$x - y = 3$ (3)

Add (2) and (3).

$2x - y + z = 1$ (4)

From (1) and (4): $1 = 1$

identity; infinite

23. $2x + y - z = 3$ (1)

$5x - 4z = 0$ (2)

$5y + 3z = 0$ (3)

$5x = 4z;\ x = 0.8z$

$5y = -3z;\ y = -0.6z$

$2(0.8z) + (-0.6z) - z = 3$

$1.6z - 0.6z - z = 3;\ 0 = 3$

contradiction; no solution

25. a. Intersection is a plane.

b. Intersection is a plane.

c. No common intersection.

d. No common intersection.

27.

Class c	**Study** s	**Work** w	**Total**
2	2(3) = 6	2(2) = 4	12
5	5(3) = 15	5(2) = 10	30
c	3c	2c	c + s + w = 60

$s = 3c$; $w = 2c$; $c + s + w = 60$

$c + 3c + 2c = 60$

$6c = 60;\ c = 10$ hr $s = 3(10) = 30$ hr $w = 2(10) = 20$ hr

29.

Angle x	**Angle** y	**Angle** z	**Total** 180°
30	30	30 + 21 = 51	111
40	40	40 + 21 = 61	141
x	y = x	x + 21	x + y + z

$x = y$; $z = x + 21$ (or $x = z - 21$); $x + y + z = 180$

$x + x + x + 21 = 180$

$3x = 159;\ x = 53°;\ y = 53°;\ z = 53 + 21;\ z = 74°$

31. Number of grams is a quantity; number of calories per gram is a rate.

	Protein	Carbs	Fat	Total
	5	10	13	28
grams	p	c = 2p	f	28
	20	40	117	177
calories	4p	4c	9f	222

$p + c + f = 28$; $c = 2p$; $4p + 4c + 9f = 222$

$p + 2p + f = 28$; $3p + f = 28$; $f = 28 - 3p$

$4p + 4(2p) + 9f = 222$

$12p + 9f = 222$

$12p + 9(28 - 3p) = 222$

$12p + 252 - 27p = 222$

$252 - 15p = 222$

$30 = 15p$ **p = 2 g** $c = 2(2)$; **c = 4 g** $f = 28 - 3(2)$; **f = 22 g**

33.

$c + f + p = 29$

$p = f$

$4c + 9f + 4p = 121$

$c + p + p = 29$

$c + 2p = 29$ $c = 29 - 2p$

$4(29 - 2p) + 9p + 4p = 121$

$116 - 8p + 13p = 121$

$5p = 5$

p = 1 g

f = 1 g

$c + 1 + 1 = 29$

c = 27 g

35.

$n + d + q = 44$

$0.05n + 0.10d + 0.25q = 5.80$

$q = d + 3$

$n + d + (d + 3) = 44$

$n + 2d = 41$ $n = 41 - 2d$

$0.05(41 - 2d) + 0.10d + 0.25(d + 3) = 5.80$

$2.05 - 0.10d + 0.10d + 0.25d + 0.75 = 5.80$

$0.25d + 2.80 = 5.80$

$0.25d = 3.00$ **d = 12**

$q = 12 + 3$; **q = 15**

$n + 12 + 15 = 44$; **n = 17**

37. a. Using the first letter of each type of coffee for the variable:

$E + F + G + C = 1000$

$10.95(E) + 8.95(F) + 8.50(G) + 8.25(C) = 8.59(1000)$

$C = 2G;\ F = 4E$

$E + 4E + G + 2G = 1000 \qquad 4E + 3G = 1000$

$10.95E + 8.95(4E) + 8.50G + 8.25(2G) = 8590 \qquad 46.75E + 25G = 8590$

$25G = 8590 - 46.75E \qquad G = 343.6 - 1.87E$

$4E + 3(343.6 - 1.87E) = 1000 \qquad 4E + 1030.8 - 5.61E = 1000$

$30.8 = 1.61E$ **E = 19.1 Round to 50 lb. E = 50 lb**

$G = 343.6 - 1.87(50)$ **G = 250 lb**

$F = 4(50)$ **F = 200 lb;** $C = 2(250)$ **C = 500 lb**

b. Using the first letter of each ingredient as the variable:

$s + p + b + c + v = 20$

$p + b = v$

$c = 2b \qquad b = 0.5c$

$s + b + v = 12$

$s + c = 10 \qquad s = 10 - c$

$(10 - c) + (0.5c) + v = 12 \qquad 10 - 0.5c + v = 12 \qquad v = 2 + 0.5c$

$p + 0.5c = 2 + 0.5c$ **p = 2 oz**

$(10 - c) + 2 + (0.5c) + c + (2 + 0.5c) = 20$

$14 + c = 20$ **c = 6 oz**

$6 = 2b;$ **b = 3 oz**

$2 + 3 = v;$ **v = 5 oz**

$s + 3 + 5 = 12;$ **s = 4 oz**

Exercises 7.4: Solving Systems of Inequalities

1. $x < 0$ and $y > 0$ is in quadrant 2.

3. $x < 0$ and $y < 0$ is in quadrant 3.

5. $y < 0$ is in quadrants 3 and 4.

7. $x < 0$ is in quadrants 2 and 3.

9. In quadrant 2, $x < 0$ and $y > 0$.

11. In quadrants 2 and 4, $x < 0$ and $y > 0$ or $x > 0$ and $y < 0$.

13. $300 \le y \le 2800$ and $x \ge 0$.

15. $y \le x$ and $x \le 0$.

17. (2, -1)

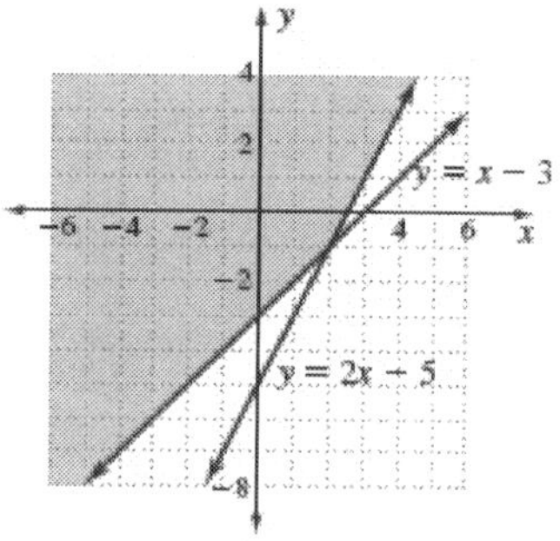

19. (11/5, 3/10)

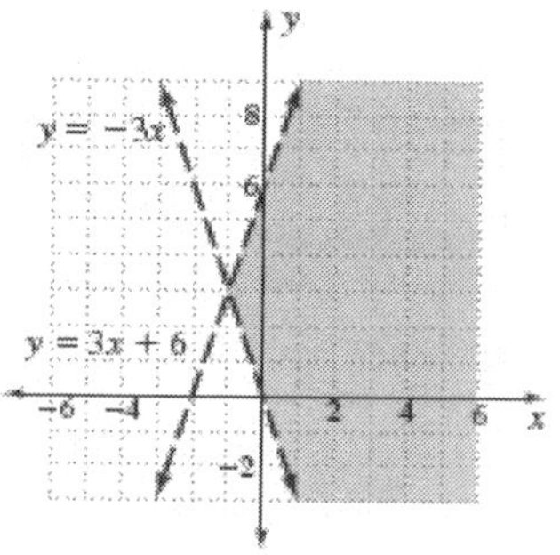

21.

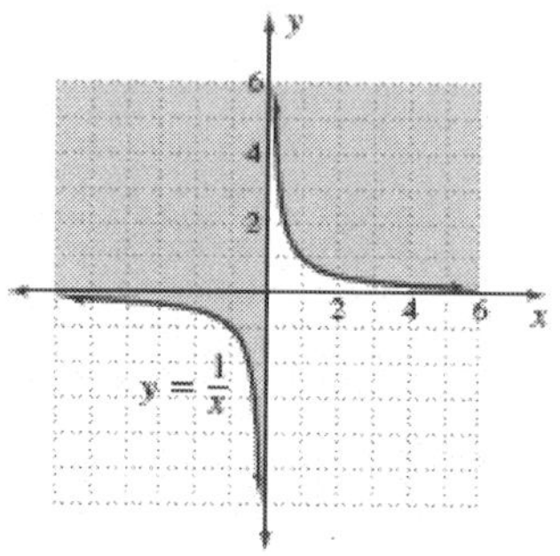

23.

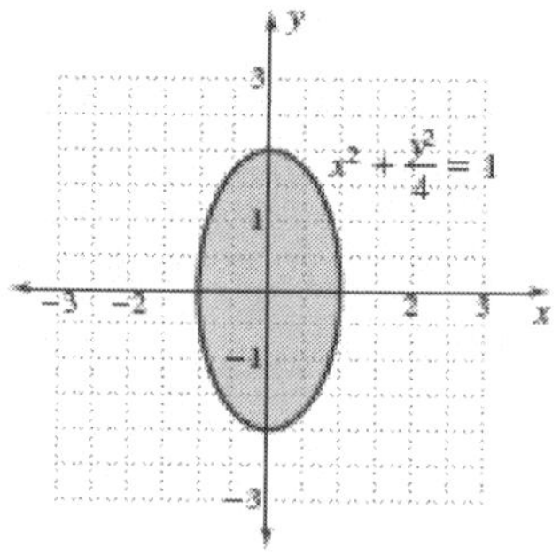

25.

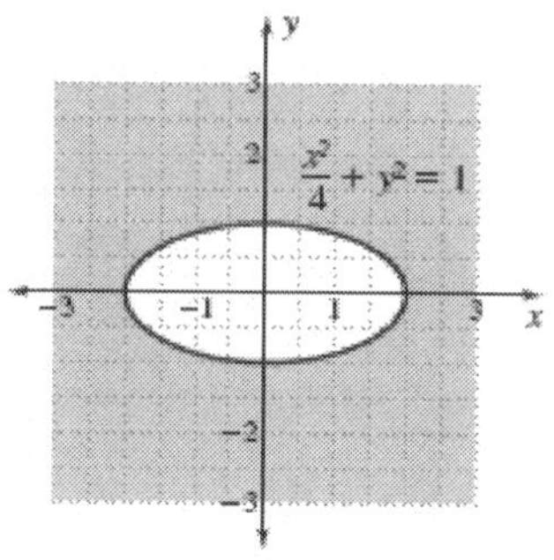

27.

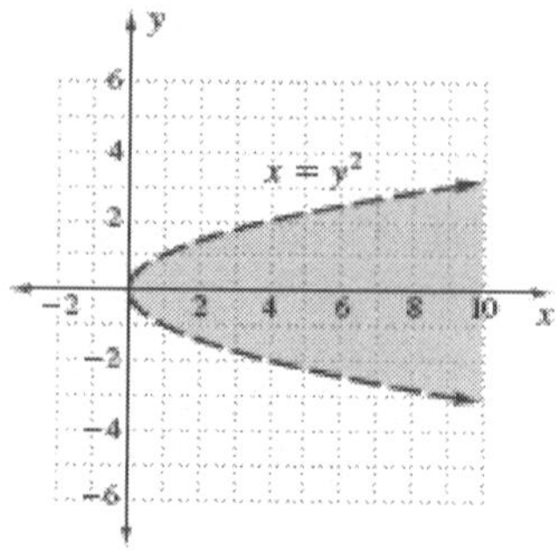

29. $y \le 0$ and $y \ge x^2 - 3$

31. $y < 0$, $y < -x^2 + 4$

33. x^2 is always ≥ 0, so if $y \ge x^2$, $y \ge 0$

35. (-1.2496, 0.7808); (1.2496, 0.7808)

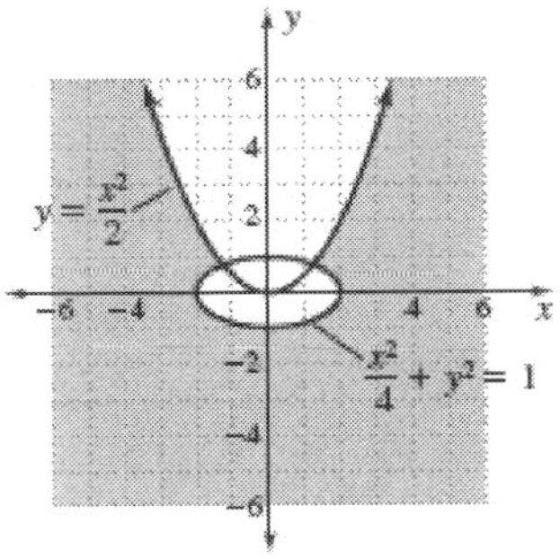

37. (-1, -1), (1, 1)

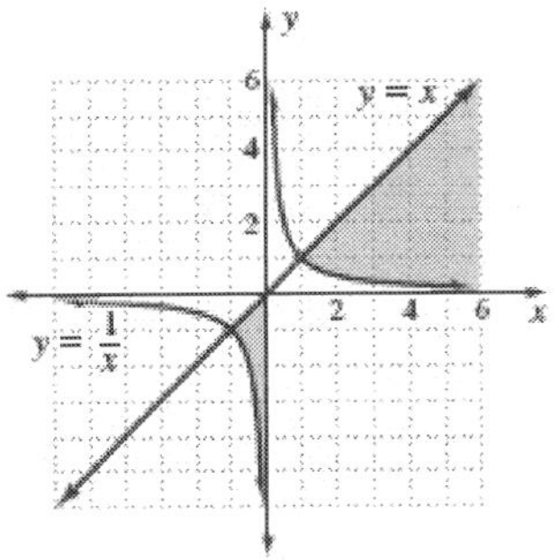

39. $\left(-\sqrt{2}, \frac{\sqrt{2}}{2}\right)$

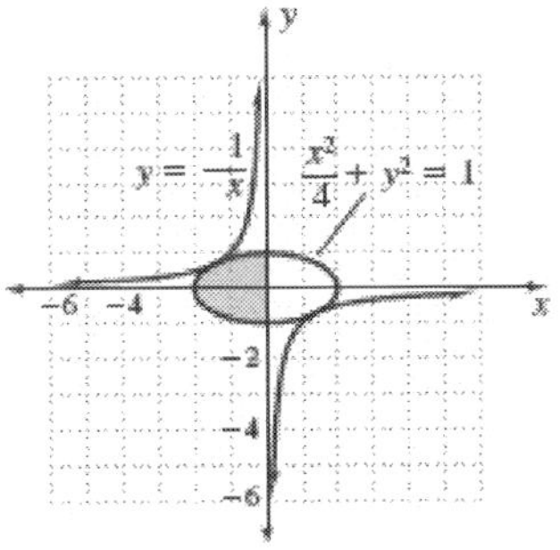

41. $150x + 500y \geq 5{,}000{,}000$

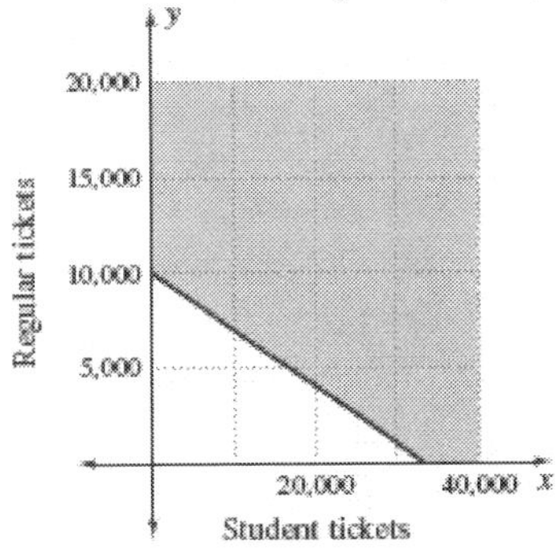

x = # of student tickets;
y = # of regular tickets

43. x = # of units IV; y = # of units fluid

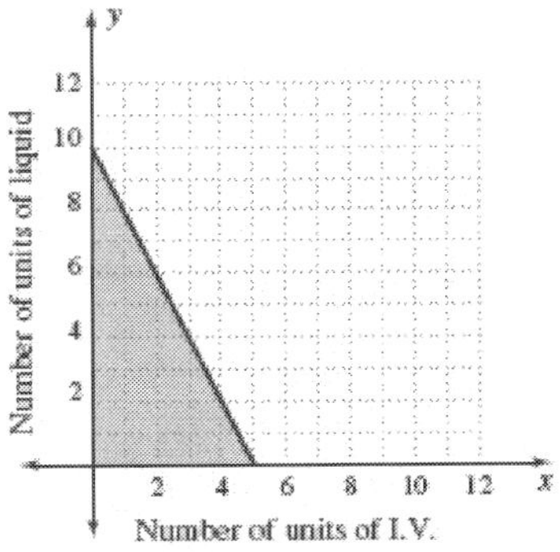

$500x + 250y \leq 2500$

45. Solid boundaries are included in the solution, while dashed boundaries are not.

47. a.

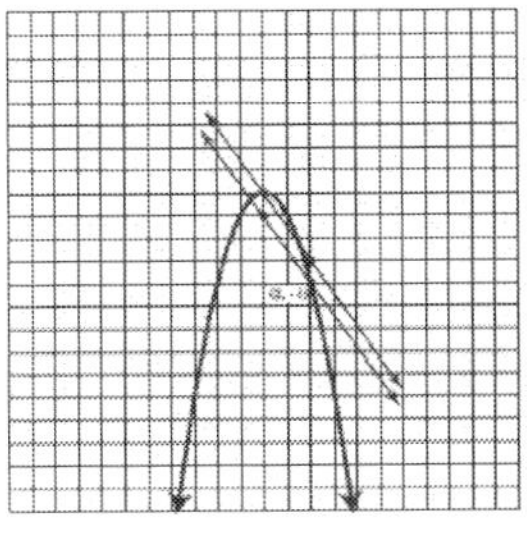

b. An area bounded by parallel lines top and bottom, and the parabola on the left and right sides.

c. Quadratic

d. $\frac{1.5 \pm \sqrt{1.5^2 - 4(1)(-1)}}{2(1)} = -0.5, 2; y = 2.75, 2$; (-0.5, 2.75), (2, -1);

$2x^2 - 3x = 0; x(2x-3) = 0; x = 0, 3/2; y = 3, 3/4$; (0,3), (1.5, .75)

Exercises 7.5: Sequences and Series

1. $a_{20} = 18 + (20-1)(-2) = -20; s_{20} = \frac{20}{2}(18 + -20) = -20$

3. $a_{20} = 10 + (20-1)(8) = 162; s_{20} = \frac{20}{2}(10 + 162) = 1720$

5. Not arithmetic

7. $a_1 = 18, a_n = -2 + a_{n-1}; a_1 = 10, a_n = 8 + a_{n-1}$; geometric: $a_1 = 3, a_n = 3a_{n-1}$

9. Not arithmetic; pairs have different sums. Quadratic: $a_n = \frac{n(n+1)}{2}$; $a_1 = 1, a_n = n + a_{n-1}$

11. $a_n = 47 + (n-1)$; 67 seats

13. $a_n = 1896 + (n-1)4$; 27 games; WWI; WWII

15. $a_n = 1 + (n-1)3$; 49 tests.

17. $s_{10} = \frac{9(1-3^{10})}{1-3} = 265716$

19. $s_{10} = \frac{64(1-0.5^{10})}{1-0.5} = 127.875$

21. $s_{10} = \frac{3(1-3^{10})}{1-3} = 88572$

23. $a_n = 3(3)^n = 9(3)^{n-1}$ and $a_1 = 9, a_n = 3a_{n-1}$;

$a_n = 64(0.5)^{n-1}$ and $a_1 = 64, a_n = 0.5a_{n-1}$;

$a_n = 4(3)^{n-1}$ and $a_1 = 4, a_n = 3a_{n-1}$

25. $s_{20} = \frac{135(1-(2/3)^{20})}{1-2/3} \approx 404.88''$

27. $s_\infty = \frac{1}{1-0.5} = 2$

29. $s_\infty = \frac{18}{1-5/6} = 108'$

31. a. $a_n = 2n-1$

b. $y = 2x-1$ or $a_n = 2n-1$

c. 1, 4, 9, 16, … Square numbers

d. The orner shapes fit together to form a square.

e. $s_n = \frac{n}{2}[1 + (2n-1)] = n^2$

33. a. $\frac{3}{10}, \frac{3}{100}, \frac{3}{1000}, \ldots; \frac{3}{10}; \frac{1}{10}; s_\infty = \frac{3/10}{1-1/10} = \frac{1}{3}$

b. $\frac{4}{10}, \frac{4}{100}, \frac{4}{1000}, \ldots; \frac{4}{10}; \frac{1}{10}; s_\infty = \frac{4/10}{1-1/10} = \frac{4}{9}$

c. $\frac{12}{100}, \frac{12}{10000}, \frac{12}{1000000} \ldots; \frac{12}{100}; \frac{1}{100}; s_\infty = \frac{12/100}{1-1/100} = \frac{4}{33}$

d. $\frac{42}{100}, \frac{42}{10000}, \frac{42}{1000000} \ldots; \frac{42}{100}; \frac{1}{100}; s_\infty = \frac{42/100}{1-1/100} = \frac{14}{33}$

e. $\frac{9}{10}, \frac{9}{100}, \frac{9}{1000}, \ldots; \frac{9}{10}; \frac{1}{10}; s_\infty = \frac{9/10}{1-1/10} = 1$

Chapter 7 Review

1. Opening to the left indicates $x = ay^2$ equation; $-3 = a(1)^2$; $a = -3$: $x = -3y^2$

2. $x^2 + y^2 = 16^2$ or $x^2 + y^2 = 256$

3. $\frac{x^2}{3^2} + \frac{y^2}{2^2} = 1$; $\frac{x^2}{9} + \frac{y^2}{4} = 1$

4. No x-intercepts indicates y^2 term comes first; $\frac{y^2}{16} - \frac{x^2}{9} = 1$.

In exercises 5 to 14, some graphs are combined.

5. Parabola, $x^2 = y - 5$

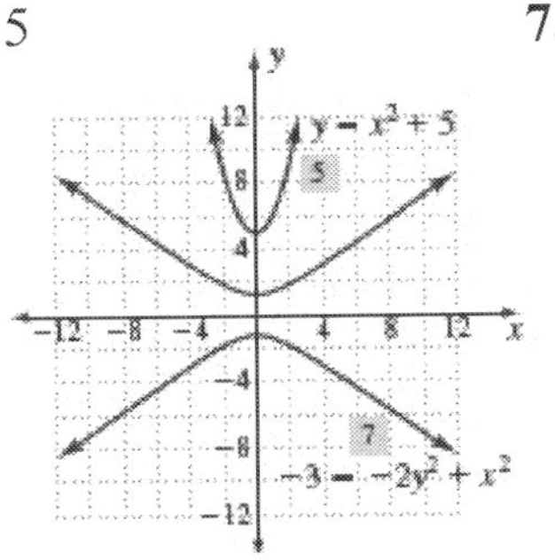

7. Divide both sides by −3:

$\frac{2y^2}{3} - \frac{x^2}{3} = 1$; hyperbola; $y^2 = \frac{1}{2}x^2 + \frac{3}{2}$

6. Parabola, $3y^2 = x + 4$;

$y^2 = \frac{1}{3}x + \frac{4}{3}$

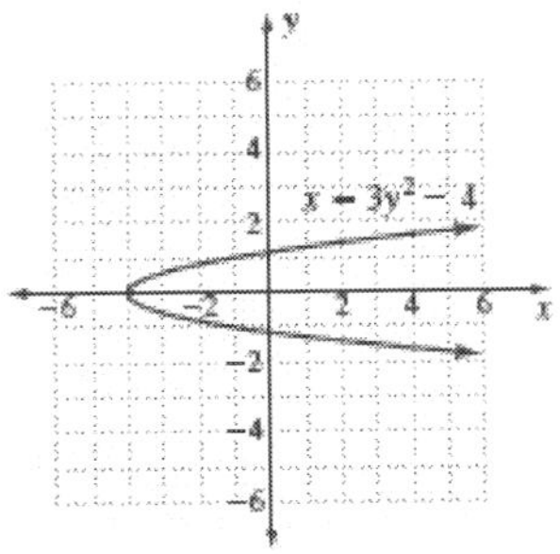

8. $x^2 = -y^2 - 5$; not a conic section, no graph

9. Circle, $y = \pm\sqrt{8 - x^2}$

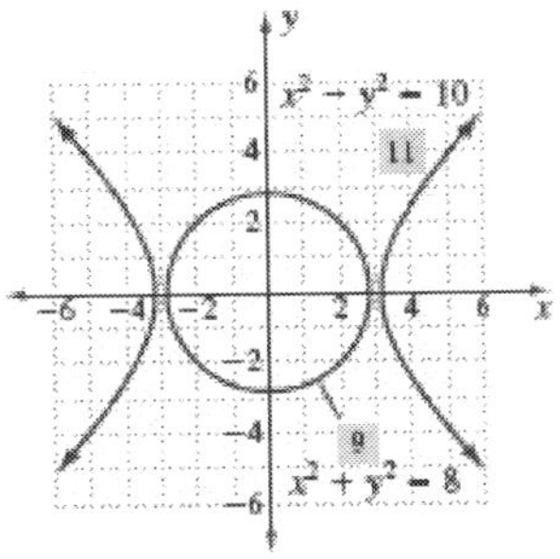

11. Divide both sides by 10: $\frac{x^2}{10} - \frac{y^2}{10} = 1$;

hyperbola; $x = \pm\sqrt{10 + y^2}$

10. Hyperbola; $y = \pm\sqrt{x^2 - 4}$

12. Circle; $x = \pm\sqrt{5 - y^2}$

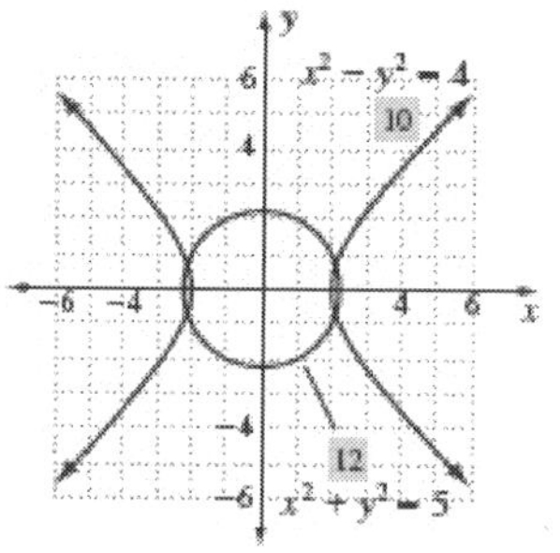

13. $y = \pm\sqrt{2 - \frac{x^2}{2}}$

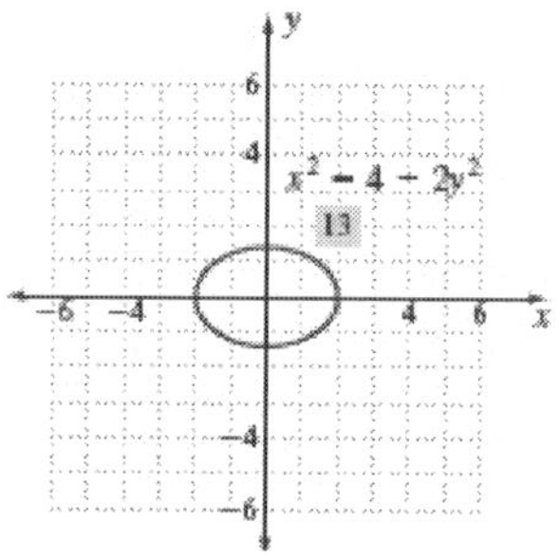

14. Ellipse; $y = \pm\sqrt{9 - 2x^2}$

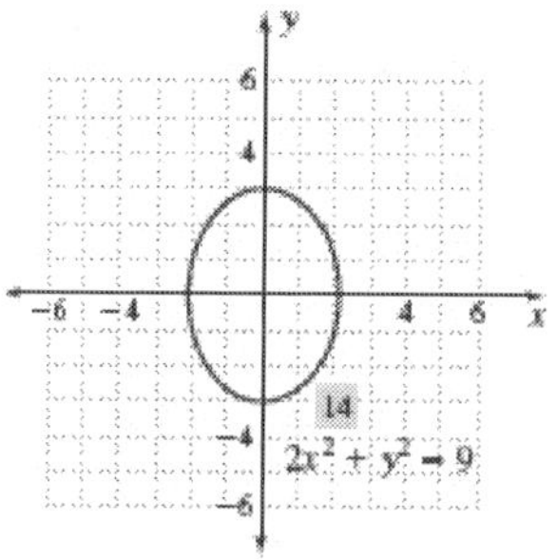

15. Parabola and line

$x^2 + x - 2 = -x - 3;$

$x^2 + 2x + 1 = 0$

$(x + 1)^2 = 0;\ x = -1$

$-(-1) - 3 = -2$

$(-1, -2)$

16. Parabola and line

$-x^2 - x - 2 = x + 2$

$-x^2 - 2x - 4 = 0$

$x^2 + 2x + 4 = 0$

$2^2 - 4(1)(4) = -12$, no real number solutions

$\{\ \}$ or $\varnothing$

17. Parabola and line

$x^2 - 4x - 1 = -x + 3$

$x^2 - 3x - 4 = 0$

$(x - 4)(x + 1) = 0;\ x = 4$ or $x = -1$

$-4 + 3 = -1;\ \ -(-1) + 3 = 4$

$(4, -1), (-1, 4)$

18. Parabola and line

$x^2 + 2x - 5 = -x - 1$

$x^2 + 3x - 4 = 0$

$(x + 4)(x - 1) = 0;\ x = -4$ or $x = 1$

$y = -(-4) - 1 = 3;\ \ y = -1 - 1 = -2$

$(-4, 3), (1, -2)$

19. Circle and parabola

$x^2 = y - 1;\; y - 1 + y^2 = 16$

$y^2 + y - 17 = 0$

$y = \dfrac{-1 \pm \sqrt{1^2 - 4(1)(-17)}}{2(1)}$

$y \approx -4.653$ or $y \approx 3.653$

$x^2 = -4.653 - 1$ (not possible)

$x^2 = 3.653 - 1 = 2.653$

$x = \pm\sqrt{2.653} \approx \pm 1.629$

$(\pm 1.629, 3.653)$

20. Hyperbola and circle

$x^2 + \frac{1}{x^2} = 2;\; x^4 + 1 = 2x^2$

$x^4 - 2x^2 + 1 = 0$

$(x^2 - 1)^2 = 0,\; x^2 = 1,\; x = \pm 1$

$y = \frac{1}{1} = 1;\; y = \frac{1}{-1} = -1$

$(1, 1), (-1, -1)$

21. Parabolas

$x^2 - x = -x^2 + 3$

$2x^2 - x - 3 = 0$

$(2x - 3)(x + 1) = 0$

$x = \frac{3}{2}$ or $x = -1$

$(\frac{3}{2})^2 - \frac{3}{2} = \frac{3}{4}$

$(-1)^2 - (-1) = 2$

$(-1, 2), (\frac{3}{2}, \frac{3}{4})$

22. Parabolas

$-x^2 + 1 = x^2 - 5$

$2x^2 = 6$

$x^2 = 3$

$x = \pm\sqrt{3} \approx \pm 1.732$

$y = 3 - 5 = -2$

$(\pm 1.732, -2)$ or $(\pm\sqrt{3}, -2)$

23. Parabolas

$x^2 + 1 = x^2 - 5$

$1 = -5$; inconsistent

$\{\ \}$ or $\varnothing$

24. Parabola and ellipse

$\dfrac{x^2}{4} + (x^2 - 1)^2 = 1;\; \dfrac{x^2}{4} + x^4 - 2x^2 + 1 = 1$

$x^2 + 4x^4 - 8x^2 + 4 = 4$

$4x^4 - 7x^2 = 0;\; x^2(4x^2 - 7) = 0$

$x^2 = 0, x = 0$ or $4x^2 = 7, x = \pm\sqrt{\frac{7}{4}} \approx \pm 1.323$

$y = 1.75 - 1 = 0.75$ or $y = 0 - 1 = -1$

$(\pm 1.323, 0.75), (0, -1)$

25. $2^x = 4 - 2^x$

$2 \cdot 2^x = 4$

$2^{x+1} = 2^2$

$x + 1 = 2;\; x = 1$

$y = 2^1;\; y = 2$

26. $3^x = 7 - 3^{x+1}$

$3^x = 7 - (3)3^x$

$(4)3^x = 7,\; 3^x = 1.75$

$x = \dfrac{\log 1.75}{\log 3} \approx 0.509$

$y = 3^{0.509} = 1.75$

27.

$\log x = 3 - 2\log x$

$3\log x = 3$

$\log x^3 = 3$

$10^3 = x^3;\ x = 10$

$y = \log 10;\ y = 1$

28.

$5 - \log x = \log x$

$5 = 2\log x$

$2.5 = \log x$

$x = 10^{2.5}$

$y = \log 10^{2.5} = 2.5$

29.

$a + b + c = 0$ (1)

$a - b + c = 6$ (2)

$a - 2b - c = 7$ (3)

Subtract (2) from (1).

$2b = -6;$ $\mathbf{b = -3}$

Substitute into (2) and (3).

$a + 3 + c = 6;\ a + c = 3$ (4)

$a - 2(-3) - c = 7;\ a - c = 1$ (5)

Add (4) and (5).

$2a = 4;$ $\mathbf{a = 2}$

$2 + (-3) + c = 0$

$\mathbf{c = 1}$

30.

$x + 2y - z = -1$ (1)

$-3x - y + z = -4$ (2)

$x + 3z = -3\quad x = -3 - 3z$ (3)

Substitute (3) into (1) and (2).

$(-3 - 3z) + 2y - z = -1$

$-3(-3 - 3z) - y + z = -4$

$2y - 4z = 2$ (4)

$-y + 10z = -13\quad y = 10z + 13$ (5)

Substitute (5) into (4).

$2(10z + 13) - 4z = 2$

$20z + 26 - 4z = 2$

$16z = -24;$ $\mathbf{z = -1.5}$

$x + 3(-1.5) = -3;$ $\mathbf{x = 1.5}$

$1.5 + 2y - (-1.5) = -1$

$2y = -4;$ $\mathbf{y = -2}$

31.

$3x - y + z = 4$

$-x + z = 2;\ x = z - 2$

$-y + 4z = 10;\ y = 4z - 10$

$3(z - 2) - (4z - 10) + z = 4$

$3z - 6 - 4z + 10 + z = 4$

$4 = 4$; identity

Infinite number of solutions

32.

$2x - 4z = 20;\ x = 2z + 10$

$5y + 3z = -19;\ y = -0.6z - 3.8$

$-2x + 3y = -14$

$-2(2z + 10) + 3(-0.6z - 3.8) = -14$

$-4z - 20 + -1.8z - 11.4 = -14$

$-5.8z = 17.4;$ $\mathbf{z = -3}$

$2x - 4(-3) = 20;$ $\mathbf{x = 4}$

$5y + 3(-3) = -19;$ $\mathbf{y = -2}$

33.

$2x + y - z = 3$ (1)

$x + y + 2z = 4$ (2)

$x - y - 3z = 6$ (3)

Subtract (1) and (2).

$x - 3z = -1$ $x = 3z - 1$ (4)

Subtract (4) from (3).

$-y = 7$; $\mathbf{y = -7}$

Substitute y and (4) into (2).

$(3z - 1) + (-7) + 2z = 4$

$5z - 8 = 4$; $\mathbf{z = 2.4}$

$x - (-7) - 3(2.4) = 6$; $\mathbf{x = 6.2}$

34.

$x + y + z = 180$

$x = 2y$; $y = 0.5x$

$x = z$

$x + 0.5x + x = 180$

$2.5x = 180$

$\mathbf{x = 72}$

$\mathbf{z = 72}$

$72 = 2y$

$\mathbf{y = 36}$

35.

$p + f + c = 49$

$4p + 9f + 4c = 201$

$c = p + 10$

$p + f + (p + 10) = 49$

$2p + f = 39$

$4p + 9f + 4(p + 10) = 201$

$8p + 9f = 161$ (1)

$-4(2p + f) = -4(39)$

$-8p - 4f = -156$ (2)

Add (1) and (2).

$5f = 5$; **f = 1 g**

$2p + 1 = 39$; **p = 19 g**

$c = 19 + 10$; **c = 29 g**

36.

$L + W + H = 19$

$W + H = L + 34$

$W = L - 2$

$L + (L - 2) + H = 19$

$2L + H = 21$; $H = 21 - 2L$

$(L - 2) + H = L + 3$

$L + H = L + 5$; **H = 5 in.**

$5 = 21 - 2L$

$2L = 16$; **L = 8 in.**

$W = 8 - 2$; **W = 6 in.**

37. $n + d + q = 74$

$0.05n + 0.10d + 0.25q = 12.40$

$q = 2n - 5$

$n + d + (2n - 5) = 74$

$3n + d = 79 \quad d = 79 - 3n$

$0.05n + 0.10d + 0.25(2n - 5) = 12.40$

$0.55n + 0.10d = 13.65$

$0.55n + 0.10(79 - 3n) = 13.65$

$0.55n + 7.9 - 0.30n = 13.65$

$0.25n = 5.75$

$\mathbf{n = 23}$

$q = 2(23) - 5$; $\mathbf{q = 41}$

$23 + d + 41 = 74$; $\mathbf{d = 10}$

38. $c + p + i = 215$

$p = 2c - 1$; $c = \frac{1}{2}(p + 1)$

$p = 47i$; $i = \frac{1}{47}p$

$\frac{1}{2}(p + 1) + p + \frac{1}{47}p = 215$

$94[\frac{1}{2}(p + 1) + p + \frac{1}{47}p] = 94(215)$

$47(p + 1) + 94p + 2p = 20{,}210$

$143p + 47 = 20{,}210$

$143p = 20{,}163$; $\mathbf{p = 141}$ **mg**

$i = \frac{1}{47}(141)$; $\mathbf{i = 3}$ **mg**

$c + 141 + 3 = 215$; $\mathbf{c = 71}$ **mg**

39. $2t + 1p + 0.5n = 17$

$26t + 3.50p + 1.50n = 152.50$

$n = t + p$

$2t + p + 0.5(t + p) = 17$

$2.5t + 1.5p = 17$

$26t + 3.50p + 1.50(t + p) = 152.50$

$27.5t + 5p = 152.50 \quad (1)$

$-11(2.5t + 1.5p) = -11(17)$

$-27.5t - 16.5p = -187$ *add to (1)*

$-11.5p = -34.5$; $\mathbf{p = 3}$

$2.5t + 1.5(3) = 17$; $\mathbf{t = 5}$

$n = 5 + 3$; $\mathbf{n = 8}$

40. a. The algebra yielded a contradiction – **true.**

b. The graphs intersect in a unique point – **false.**

c. The algebra yielded an identity – **false.**

41.

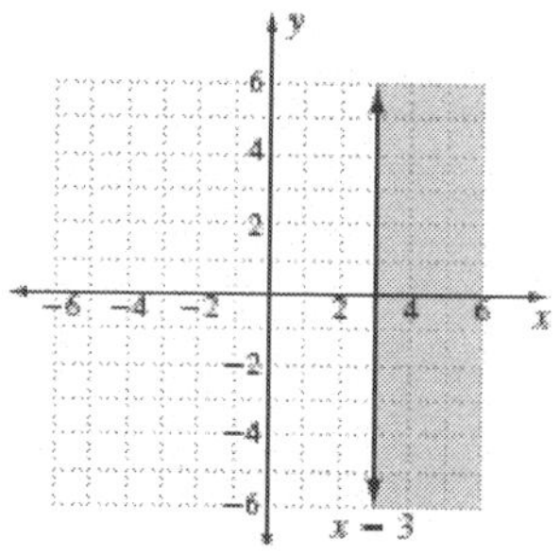

42.

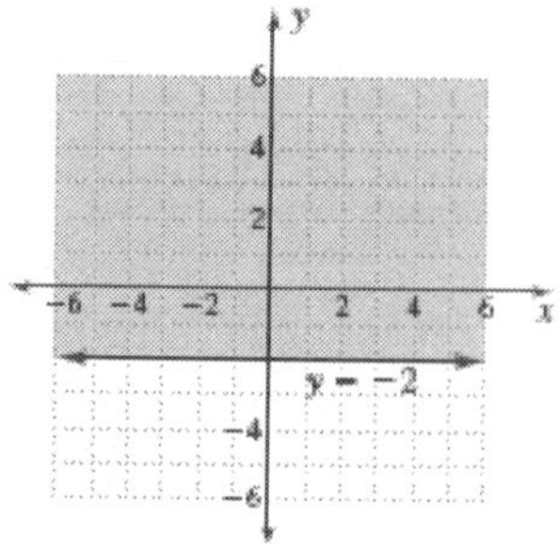

43.

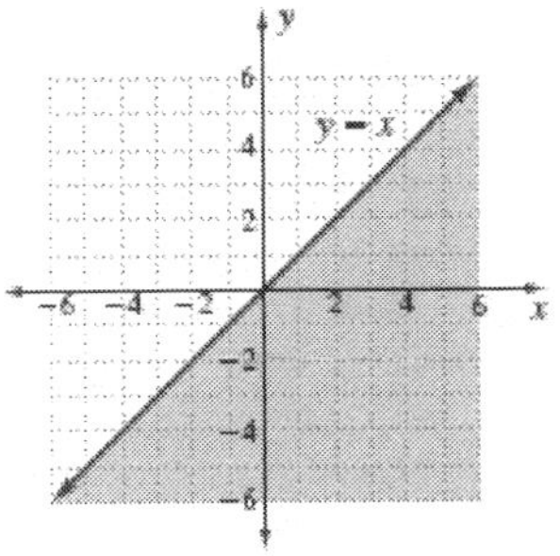

44.

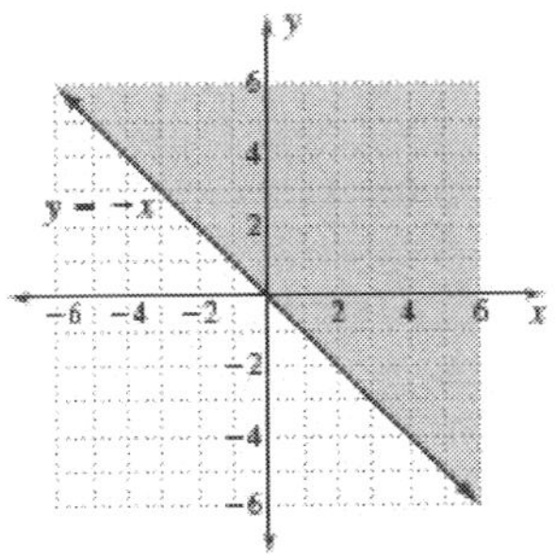

45.

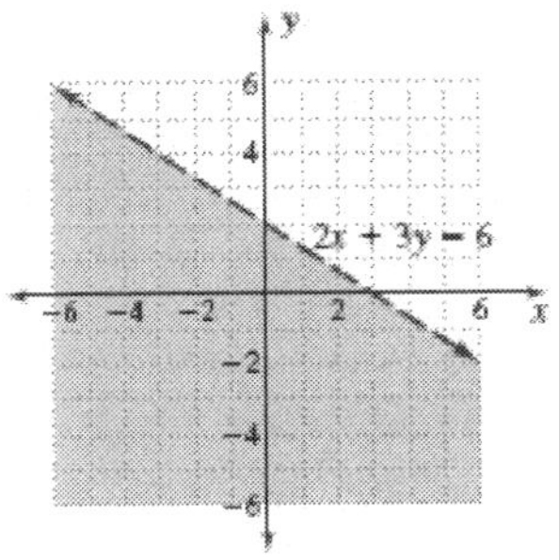

46.

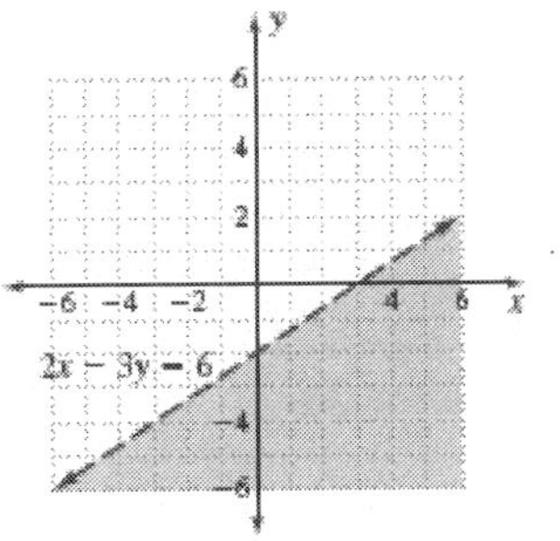

47.

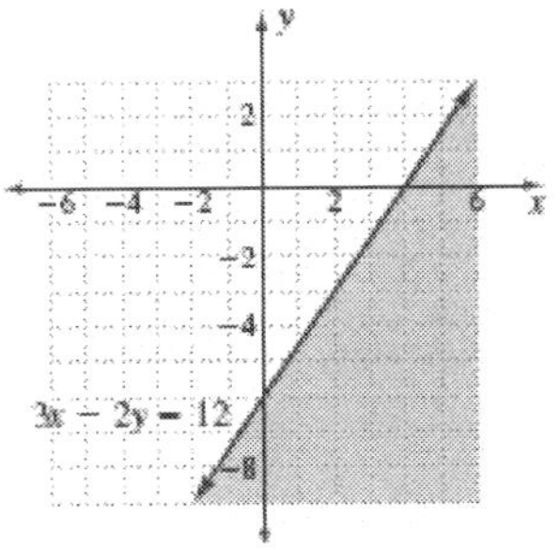

48.

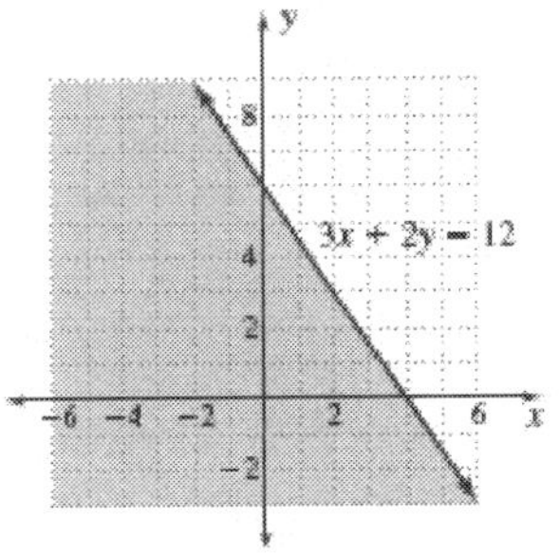

49.

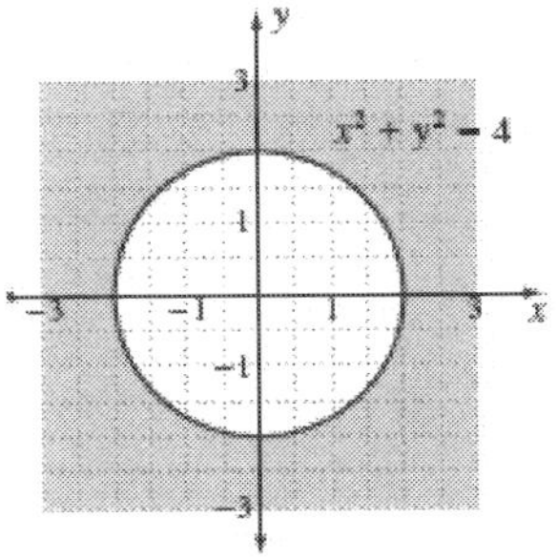

50.

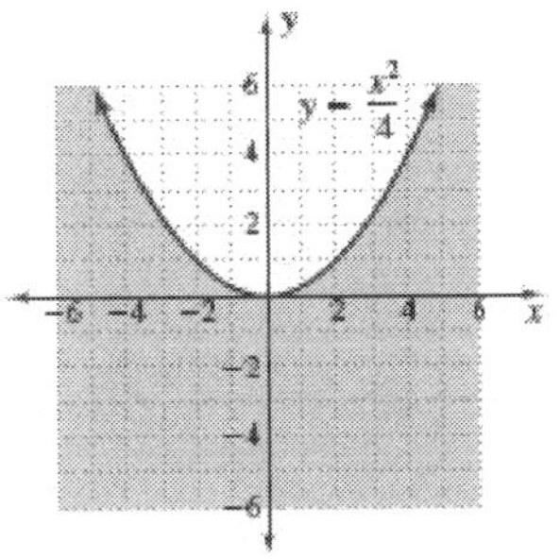

51. $\left(\dfrac{6\sqrt{13}}{13}, \dfrac{6\sqrt{13}}{13}\right), \left(\dfrac{-6\sqrt{13}}{13}, \dfrac{-6\sqrt{13}}{13}\right)$

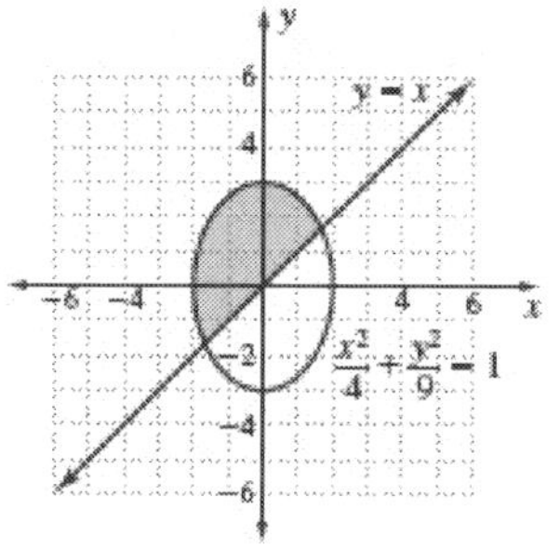

52. (0, 0), (-1, 1)

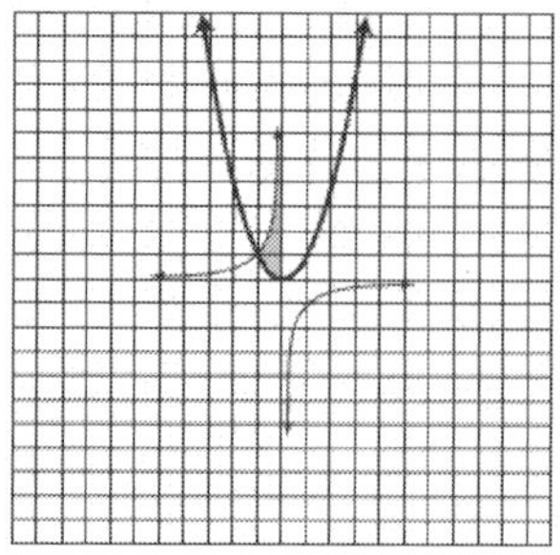

53. (-1.175, 2.760)

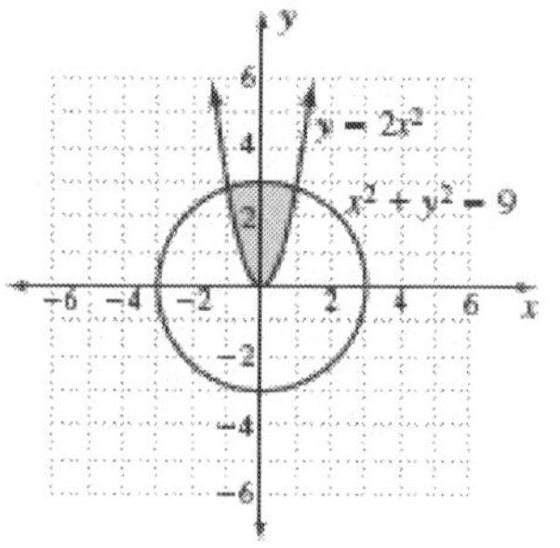

54. $\left(\dfrac{-1-\sqrt{5}}{2}, \dfrac{3-\sqrt{5}}{2}\right), \left(\dfrac{-1+\sqrt{5}}{2}, \dfrac{3+\sqrt{5}}{2}\right)$

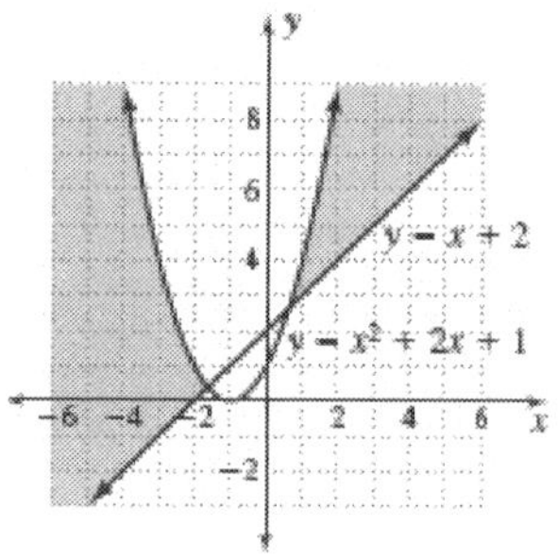

55. $y < 0$, $y > -x^2 + 4$

56. $y \geq 0$ and $y \leq -x^2 + 3$

57. $\begin{bmatrix} 2 & -3 & -4 \\ 3 & -2 & 1 \\ 4 & -1 & -3 \end{bmatrix} \cdot \begin{bmatrix} x \\ y \\ z \end{bmatrix} = \begin{bmatrix} -4 \\ -17 \\ -3 \end{bmatrix}$

(−2, 4, −3)

58. $\begin{bmatrix} 3 & -1 & -4 \\ 2 & -3 & 1 \\ -2 & 1 & 3 \end{bmatrix} \begin{bmatrix} x \\ y \\ z \end{bmatrix} = \begin{bmatrix} 9 \\ 19 \\ -7 \end{bmatrix}$

(3, −4, 1)

59. Determinant = 0
Add first 2 equations;
$x - y - z = 2$
No solution

60. Determinant = 0
Add 2nd and 3rd equations.
Equals 1st equation.
Infinite number of solutions

61. Geometric; 5, 2.5, 1.25; $y = 160(0.5)^x$ or $y = 80(0.5)^{x-1}; a_1 = 80, a_n = 0.5a_{n-1}$

62. Arithmetic; 22, 25, 28; $y = 3x + 4; a_1 = 7; a_n = 3 + a_{n-1}$

63. Neither, quadratic with constant second difference; 27, 35, 44;
$y = \frac{1}{2}x^2 + \frac{3}{2}x; a_1 = 2, a_n = n + 2 + a_{n-1}$

64. Geometric; 3125, 15625, 78125; $y = 5^x$; $a_1 = 5, a_n = 5a_{n-1}$

65. Arithmetic; 39, 46, 53; y = 7x+4; $a_1 = 4, a_n = 7 + a_{n-1}$

66. Geometric; $6\frac{2}{3}, 2\frac{2}{9}, \frac{20}{27}; y = 540\left(\frac{1}{3}\right)^x, y = 180\left(\frac{1}{3}\right)^{x-1}; a_1 = 540, a_n = \frac{1}{3}a_{n-1}$

67. $a_{10} = 160(0.5)^{10} = 0.15625; S_{10} = \frac{80(1-0.5^{10})}{1-0.5} = 159.84375, S_\infty = \frac{80}{1-0.5} = 160;$

By addition: 275; $a_{10} = 4 + (10-1)7 = 67; S_{10} = \frac{10}{2}(4+67) = 355$

68. $a_{10} = 7 + (10-1)3 = 34; S_{10} = \frac{10}{2}(7+34) = 205;$

$a_{10} = (5)^{10} = 9765625; S_{10} = \frac{5(1-5^{10})}{1-5} = 12207030;$

$S_{10} = \frac{540(1-(1/3)^{10})}{1-(1/3)} = 809.986, S_\infty = \frac{540}{1-(1/3)} = 810$

69. a. $a(-1)^2 + b(-1) + c = -10$
$a(0)^2 + b(0) + c = -5$
$a(1)^2 + b(1) + c = -4$

$$\begin{bmatrix} (-1)^2 & -1 & 1 \\ 0 & 0 & 1 \\ 1^2 & 1 & 1 \end{bmatrix} \cdot \begin{bmatrix} a \\ b \\ c \end{bmatrix} = \begin{bmatrix} -10 \\ -5 \\ -4 \end{bmatrix}$$

$y = -2x^2 + 3x - 5$

b. $a(0)^2 + b(0) + c = -2$
$a(1)^2 + b(1) + c = -1$
$a(2)^2 + b(2) + c = -8$

$$\begin{bmatrix} 0 & 0 & 1 \\ 1 & 1 & 1 \\ 2^2 & 2 & 1 \end{bmatrix} \cdot \begin{bmatrix} a \\ b \\ c \end{bmatrix} = \begin{bmatrix} -2 \\ -1 \\ -8 \end{bmatrix}$$

$y = -4x^2 + 5x - 2$

c.
$$a(-2)^3 + b(-2)^2 + c(-2) + d = 0$$
$$a(-1)^3 + b(-1)^2 + c(-1) + d = -5$$
$$a(1)^3 + b(1)^2 + c(1) + d = -3$$
$$a(2)^3 + b(2)^2 + c(2) + d = -8$$
$$\begin{bmatrix} (-2)^3 & (-2)^2 & -2 & 1 \\ (-1)^3 & (-1)^2 & -1 & 1 \\ 1^3 & 1^2 & 1 & 1 \\ 2^3 & 2^2 & 2 & 1 \end{bmatrix} \cdot \begin{bmatrix} a \\ b \\ c \\ d \end{bmatrix} = \begin{bmatrix} 0 \\ -5 \\ -3 \\ -4 \end{bmatrix}$$
$$y = -x^3 + 2x - 4$$

d.
$$a(-1)^3 + b(-1)^2 + c(-1) + d = 0$$
$$a(1)^3 + b(1)^2 + c(1) + d = 2$$
$$a(2)^3 + b(2)^2 + c(2) + d = 0$$
$$a(3)^3 + b(3)^2 + c(3) + d = 4$$
$$\begin{bmatrix} (-1)^3 & (-1)^2 & -1 & 1 \\ 1^3 & 1^2 & 1 & 1 \\ 2^3 & 2^2 & 2 & 1 \\ 3^3 & 3^2 & 3 & 1 \end{bmatrix} \cdot \begin{bmatrix} a \\ b \\ c \\ d \end{bmatrix} = \begin{bmatrix} 0 \\ 2 \\ 0 \\ 4 \end{bmatrix}$$
$$y = x^3 - 3x^2 + 4$$

e.
$$a(-3)^4 + b(-3)^3 + c(-2)^2 + d(-3) + e = 50$$
$$a(-1)^4 + b(-1)^3 + c(-1)^2 + d(-1) + e = -6$$
$$a(0)^4 + b(0)^3 + c(0)^2 + d(0) + e = -1$$
$$a(1)^4 + b(1)^3 + c(1)^2 + d(1) + e = 2$$
$$a(2)^4 + b(2)^3 + c(2)^2 + d(2) + e = 15$$
$$\begin{bmatrix} (-3)^4 & (-3)^3 & (-3)^2 & -3 & 1 \\ (-1)^4 & (-1)^3 & (-1)^2 & -1 & 1 \\ 0 & 0 & 0 & 0 & 1 \\ 1^4 & 1^3 & 1^2 & 1 & 1 \\ 2^4 & 2^3 & 2^2 & 2 & 1 \end{bmatrix} \cdot \begin{bmatrix} a \\ b \\ c \\ d \\ e \end{bmatrix} = \begin{bmatrix} 50 \\ -6 \\ -1 \\ 2 \\ 15 \end{bmatrix}$$
$$y = x^4 - 2x^2 + 4x - 1$$

f.
$$a(-2)^4 + b(-2)^3 + c(-2)^2 + d(-2) + e = 8$$
$$a(-1)^4 + b(-1)^3 + c(-1)^2 + d(-1) + e = 1$$
$$a(0)^4 + b(0)^3 + c(0)^2 + d(0) + e = 2$$
$$a(2)^4 + b(2)^3 + c(2)^2 + d(2) + e = 4$$
$$a(3)^4 + b(3)^3 + c(3)^2 + d(3) + e = 53$$
$$\begin{bmatrix} (-2)^4 & (-2)^3 & (-2)^2 & -2 & 1 \\ (-1)^4 & (-1)^3 & (-1)^2 & -1 & 1 \\ 0 & 0 & 0 & 0' & 1 \\ 2^4 & 2^3 & 2^2 & 2 & 1 \\ 3^4 & 3^3 & 3^2 & 3 & 1 \end{bmatrix} \cdot \begin{bmatrix} a \\ b \\ c \\ d \\ e \end{bmatrix} = \begin{bmatrix} 8 \\ 1 \\ 2 \\ 4 \\ 53 \end{bmatrix}$$
$$y = x^4 - 3x^2 - x + 2$$

g. 2, 3, $n + 1$

70. a. $(-12)^2 + 5^2 + (-12)D + 5(E) + F = 0$; $-12D + 5E + F = -169$

$12^2 + 5^2 + 12D + 5E + F = 0$; $12D + 5E + F = -169$

$5^2 + (-12)^2 + 5D + (-12)E + F = 0$ $5D - 12E + F = -169$

$\begin{bmatrix} -12 & 5 & 1 \\ 12 & 5 & 1 \\ 5 & -12 & 1 \end{bmatrix}\begin{bmatrix} D \\ E \\ F \end{bmatrix} = \begin{bmatrix} -169 \\ -169 \\ -169 \end{bmatrix}$, $D = 0, E = 0, F = -169$; $x^2 + y^2 - 169 = 0$, $x^2 + y^2 = 169$

b. $6^2 + (-8)^2 + 6D + (-8)E + F = 0$ $6D - 8E + F = -100$

$6^2 + 8^2 + 6D + 8E + F = 0$ $6D + 8E + F = -100$

$(-8)^2 + 6^2 + (-8)D + 6E + F = 0$ $-8D + 6E + F = -100$

$\begin{bmatrix} 6 & -8 & 1 \\ 6 & 8 & 1 \\ -8 & 6 & 1 \end{bmatrix}\begin{bmatrix} D \\ E \\ F \end{bmatrix} = \begin{bmatrix} -100 \\ -100 \\ -100 \end{bmatrix}$, $D = 0, E = 0, F = -100$; $x^2 + y^2 - 100 = 0$, $x^2 + y^2 = 100$

c. $(-4)^2 + (-1)^2 + (-4)D + (-1)E + F = 0$ $-4D - 1E + F = -15$

$(-5)^2 + 6^2 + (-5)D + 6E + F = 0$ $-5D + 6E + F = -61$

$3^2 + 6^2 + 3D + 6E + F = 0$ $3D + 6E + F = -45$

$\begin{bmatrix} -4 & -1 & 1 \\ -5 & 6 & 1 \\ 3 & 6 & 1 \end{bmatrix}\begin{bmatrix} D \\ E \\ F \end{bmatrix} = \begin{bmatrix} -15 \\ -61 \\ -45 \end{bmatrix}$, $D = 2, E = -6, F = -15$; $x^2 + y^2 + 2x - 6y - 15 = 0$

d. $(-1)^2 + (-5)^2 + (-1)D + (-5)E + F = 0$ $-D - 5E + F = -26$

$6^2 + 2^2 + 6D + 2E + F = 0$ $6D + 2E + F = -40$

$(-1)^2 + 3^2 + (-1)D + 3E + F = 0$ $-D + 3E + F = -10$

$\begin{bmatrix} -1 & -5 & 1 \\ 6 & 2 & 1 \\ -1 & 3 & 1 \end{bmatrix}\begin{bmatrix} D \\ E \\ F \end{bmatrix} = \begin{bmatrix} -26 \\ -40 \\ -10 \end{bmatrix}$, $D = -4, E = 2, F = -20$; $x^2 + y^2 - 4x + 2y - 20 = 0$

Chapter 7 Test

1. $x^2 + y^2 = 16^2,\ x^2 + y^2 = 256$

2. $\dfrac{x^2}{1^2} + \dfrac{y^2}{4^2} = x^2 + \dfrac{y^2}{16} = 1$

3. Opening to the left indicates $x = ay^2$ equation; $-4 = a(1)^2$, $a = -4$; $x = -4y^2$

4. No y-intercepts indicates $\dfrac{x^2}{a^2} - \dfrac{y^2}{b^2} = 1$ equation; $\dfrac{x^2}{3^2} - \dfrac{y^2}{2^2} = \dfrac{x^2}{9} - \dfrac{y^2}{4} = 1$

5. Parabola and line

$3x^2 - 4x - 5 = x - 3$

$3x^2 - 5x - 2 = 0$

$(3x + 1)(x - 2) = 0$

$x = -\frac{1}{3}$ or $x = 2$

$y = -\frac{1}{3} - 3 = -3\frac{1}{3} = -\frac{10}{3}$

$y = 2 - 3 = -1 \Rightarrow (-\frac{1}{3}, -\frac{10}{3}), (2, -1)$

6. Parabola and line

$x^2 - 3x + 2 = 0.5x - 2$

$x^2 - 3.5x + 4 = 0$

$$\frac{-(-3.5) \pm \sqrt{(-3.5)^2 - 4(1)(4)}}{2(1)} = \frac{3.5 \pm \sqrt{-3.75}}{2}$$

Not a real number; no solution.

7. Circle and hyperbola

$x^2 + y^2 = -x^2 + y^2$

$2x^2 = 0;\ x = 0$

$0 + y^2 = 4;\ y = \pm 2$

$(0, \pm 2)$

8. Parabola and circle

$x^2 = 4 - y^2;\ y = 4 - y^2 + 1$

$y^2 + y - 5 = 0$

$$\frac{-1 \pm \sqrt{1^2 - 4(1)(-5)}}{2(1)} = \frac{-1 \pm \sqrt{21}}{2}$$

$y = 1.791$ or $y = -2.791$

$-2.791 = x^2 + 1$ (not possible, discard soln.)

$1.791 = x^2 + 1;\ x^2 = 0.791;$

$x = \pm\sqrt{0.791} \approx \pm 0.890 \Rightarrow (\pm 0.890, 1.791)$

9. $3^x = 5 - 3^x$

$2(3^x) = 5$

$3^x = 2.5$

$\log 3^x = \log 2.5$

$x \log 3 = \log 2.5$

$x = \dfrac{\log 2.5}{\log 3} \approx 0.834$

$y = 3^{0.834} \approx 2.5$

$(0.834, 2.5)$

10. $\log_3 x = 4 - \log_3 x$

$2\log_3 x = 4$

$\log_3 x^2 = 4$

$3^4 = x^2$

$81 = x^2;\ x = \pm 9$; (discard negative)

$y = \log_3 9;\ 3^y = 9;\ 3^y = 3^2;\ y = 2$

$(9, 2)$

11.

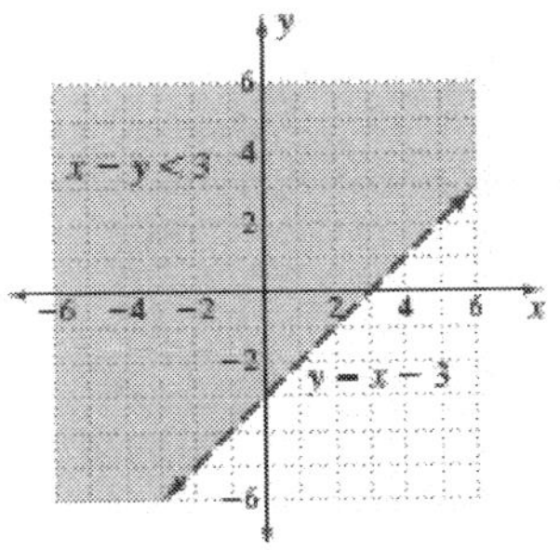

12.

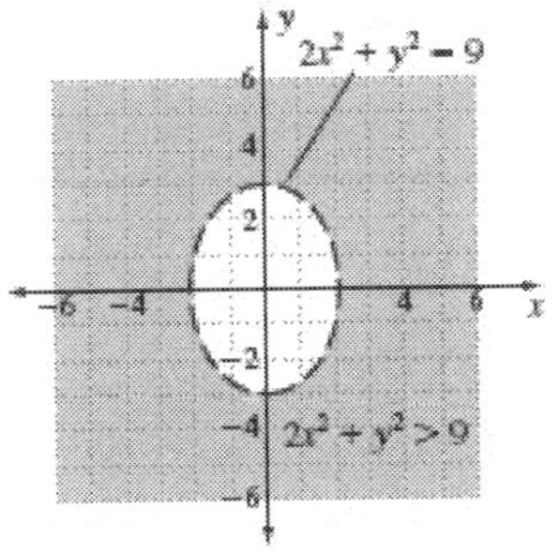

13. $\left(-2, 2\sqrt{3}\right), \left(-2, -2\sqrt{3}\right)$

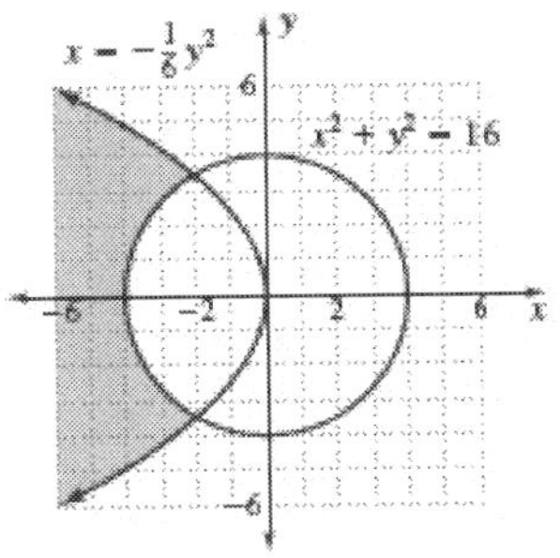

14. (-1, -1)

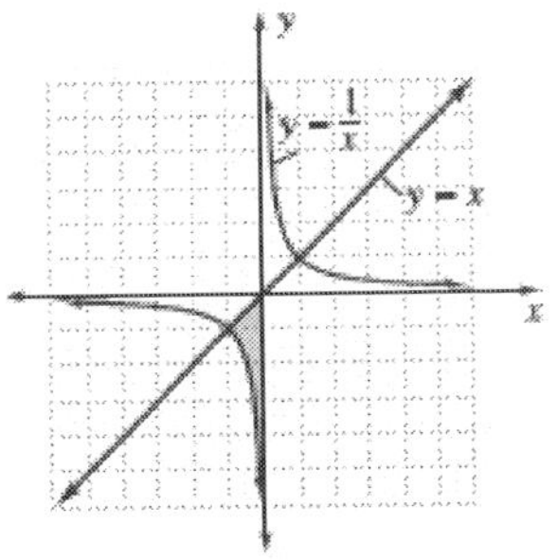

15. line, ellipse, circle, parabola, hyperbola, line, respectively.

16.

$2x + y + 2z = 5$ (1)

$x + 3z = 4$; $x = 4 - 3z$ (2)

$4x + 3y = 0$ (3)

$-3(2x + y + 2z) = -3(5)$

$-6x - 3y - 6z = -15$ (4)

Add (3) and (4).

$-2x - 6z = -15$ (5)

Substitute (2) into (5).

$-2(4 - 3z) - 6z = -15$

$-8 + 6z - 6z = -15$

$-8 = -15$; contradiction

No solution; inconsistent

17.

$3x + y + z = 8$ (1)

$2x - y - 2z = 17$ (2)

$-x - y + 3z = -9$ (3)

Add (1) and (2), then add (1) and (3).

$5x - z = 25$ $z = 5x - 25$ (4)

$2x + 4z = -1$ (5)

Substitute (4) into (5).

$2x + 4(5x - 25) = -1$

$2x + 20x - 100 = -1$

$22x = 99$; $\mathbf{x = 4.5}$

$z = 5(4.5) - 25$; $\mathbf{z = -2.5}$

$3(4.5) + y + (-2.5) = 8$ $\mathbf{y = -3}$

18. $L = W + H;\ L - W - H = 0$

$W = H + 8$

$L + W + H = 24$

Add 1st and 3rd equations.

$2L = 24;$ $\mathbf{L = 12}$ **in.**

$12 + (H + 8) + H = 24$

$2H + 20 = 24;$ $\mathbf{H = 2}$ **in.**

$W = 2 + 8;$ $\mathbf{W = 10}$ **in.**

19. $p + n + q = 59$

$q = n + 10$

$0.01p + 0.05n + 0.25q = 6.35$

$p + n + (n + 10) = 59$

$p + 2n = 49;\ p = 49 - 2n$

$0.01p + 0.05n + 0.25(n + 10) = 6.35$

$0.01p + 0.30n = 3.85$

$0.01(49 - 2n) + 0.30n = 3.85$

$0.49 - 0.02n + 0.30n = 3.85$

$0.28n = 3.36;$ $\mathbf{n = 12}$

$q = 12 + 10;$ $\mathbf{q = 22}$

$p + 12 + 22 = 59;$ $\mathbf{p = 25}$

20. $j + a + c = 11$ (1)

$0.89j + 0.59a + 3.98c = 15.07$ (2)

$17j + 8a + 16c = 158$ (3)

$-0.89(j + a + c) = -0.89(11)$

$-0.89j - 0.89a - 0.89c = -9.79$ (4) Add (2) and (4).

$-0.30a + 3.09c = 5.28$ (5)

$-17(j + a + c) = -17(11)$

$-17j - 17a - 17c = -187$ (6) Add (3) and (6).

$-9a - c = -29;\ c = -9a + 29$ Substitute into (5).

$-0.30a + 3.09(-9a + 29) = 5.28$ $\quad -0.30a - 27.81a + 89.61 = 5.28$

$-28.11a = -84.33;$ $\mathbf{a = 3}$ $\quad c = -9(3) + 29;$ $\mathbf{c = 2}$ $\quad j + 3 + 2 = 11;$ $\mathbf{j = 6}$

21.
- **a.** The graphs intersect in a plane or in a line – **true.**
- **b.** The algebra yielded an identity – **true.**
- **c.** The system contains parallel planes – **false.**
- **d.** The algebra yielded a contradiction – **false.**

22. 52; arithmetic

23. $a_n = 243\left(\frac{2}{3}\right)^{n-1};a_1 = 243, a_n = \left(\frac{2}{3}\right)a_{n-1}$

24. $a_{15} = 16 + (15 - 1)9 = 142; S_{15} = \frac{15}{2}(16 + 142) = 1185$

25. $S_\infty = \frac{243}{1-(2/3)} = 729$

CHAPTER 8

PROBABILITY AND COUNTING

Exercises 8.1: Probability: Definitions and Single-Event Probability

1. **a.** GG, GB, BG, BB **b.** GGG, GGB, GBG, GBB, BGG, BGB, BBG, BBB
c. GGGG, GGGB, GGBG, GGBB, GBGG, GBGB, GBBG, GBBB, BGGG, BGGB, BGBG, BGBB, BBGG, BBGB, BBBG, BBBB

3. BCG, BGC, CBG, CGB, GCB, GBC

5. **a.** {4, 5, 6} **b.** {1, 2, 3, 4, 5, 6} **c.** {1, 4} **d.** {2, 3, 4, 5, 6}

7. **a.** {HH, HT, TH} **b.** {TT, TH, HT} **c.** {HH} **d.** {TH, HT, TT}

9. 2/5

11. **a.** 8/17 **b.** 7/17 (one is not prime) **c.** 2/17
d. 4/17 **e.** 7/17

13. **a.** 1/11 **b.** 2/11 **c.** 5/11
d. 4/11 **e.** 5/11

15. **Row 1:** Boy, ½, ½; **Row 2:** GG, ¼, ¾; **Row 3:** BB, ¾, ¼; **Row 4:** 2, 4; 3/5; **Row 5:** z, 2/3; **Row 6:** Odd sum, ½, ½.

17. **a.** 30/36 = 5/6 **b.** 36/36 = 1 **c.** 30/36 = 5/6
d. 0 **e.** 5/36

19. **a.** ¼ **b.** ¼ **c.** 1/7
d. 4/7 (zero is even) **e.** ¾ **f.** 3/14
g. 0

21. a. 4/52 **b.** 13/52 = ¼ **c.** 12/52 = 3/13 **d.** 40/52 = 10/13

23. a. {1, 2, 3, 4, 5, 6} **b.** 7/36 **c.** 27/36 = ¾ **d.** 21/36 = 7/12

25. 49/139; 8/139

27. a. 16.8/29.3 **b.** 4.2/29.3 **c.** 5.3/29.3 **d.** 1.3/29.3

29. 58/127; 27/127

31. a. 1/5
b. P(A) = 21/100; P(B) = 22/100; P(C) = 16/100; P(D) = 21/100; P(E) = 20/100
c. P(A) = 105/500; P(B) = 104/500; P(C) = 86/500; P(D) = 96/500; P(E) = 109/500
d. Drawing 500 times has slightly less variation from the theortical probability of 1/5 = 0.2.

33. Any event; any rational number p, $0 \leq p \leq 1$.

35. 1/10000

Exercises 8.2: Multiple-Event Probability

1. a. 11/36 **b.** 11/36 **c.** 2/36 = 1/18
d. 20/36 = 5/9 **e.** 16/36 = 4/9

3. a. 3/36 = 1/12 **b.** 12/36 = 1/3 **c.** 10/36 = 5/18 **d.** 14/36 = 7/18

5. a. 22/52 = 11/26 **b.** 16/52 = 4/13 **c.** ½ **d.** 25/52

7. 0

9. a.

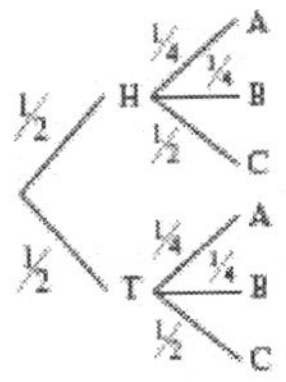

b. HA, HB, TA, TB each have probability: (1/2)(1/4) = 1/8; HC, TC each have probability: (1/2)(1/2) = ¼
c. ½
d. ½ + 1/8 = 5/8; Second addition principle
e. ¼ + ½ = ¾
f. P(TB) = 1/8 (see part b)

11. a.

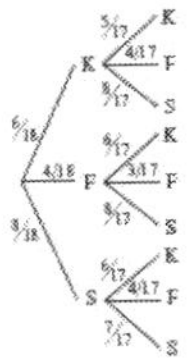

b. P(Spoon and Fork) = 32/306 + 32/306 = 32/153
c. P(Knife and Fork) = 24/306 + 24/306 = 8/51
d. P(2 Spoons) = (8/18)(7/17) = 28/153
e. P(2 forks) = (4/18)(3/17) = 2/51

13. a. Independent **b.** Dependent **c.** Independent
d. Dependent **e.** Independent

15. a. independent **b.** (1/2)(2/3) = 1/3 **c.** (1/2)(1/3) = 1/6

17. a. Independent; Outcome of spinner 1 does not affect outcome of spinner 2.
b. (5/8)(1/4) = 5/32 **c.** (3/8)(3/4) = 9/32

19. a. Dependent; Contents of the box changed. **b.** (4/7)(3/6)(3/5) = 4/35
c. (3/7)(2/6)(1/5) = 1/35

21. Answers will vary

Exercises 8.3: Using Counting Principles

1. 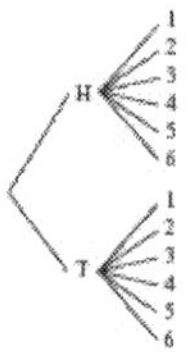

3. 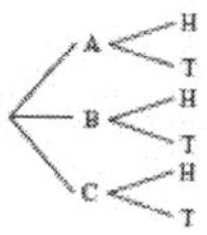

5. $10^9 = 1{,}000{,}000{,}000$

7. 4^{20}

9. a. $40^3 = 64{,}000$

b. $(40)(39)(38) = 59{,}280$

11. a. $\frac{1}{10^4} = \frac{1}{10000}$

b. No; Neither combination nor permutation.

Although order counts, numbers are repeated so only the multiplication principle applies.

13. a. $(2)(26)(26) = 1352$

b. $(2)(26)(26)(26) = 35{,}152$

c. $1352 + 35{,}152 = 36{,}504$

15. a. $5 + 12 = 17$

b. $15(3) = 45$

c. $12(11)(10)(15) = 19{,}800$

d. $12 + 15 = 27$

e. $15(14)(13) = 2730$

f. $5(12 + 3) = 75$

17. a. $8(2)(10) = 160$

b. $8(10)(10) = 800$

c. $(8^2)(10^8)$

d. $800(799)(10^4)$

19. a. 40,320

b. 3,628,800

c. 1

d. 3,628,800

e. 40,322

f. 720

21. $6! = 720$ (also could be MP)

23. $7^7 = 823{,}543$; MP

25. ${}_4P_2$; ab, ac, ad, ba, bc, bd, ca, cb, cd, da, db, dc

27. ${}_3C_2$; bc, bd, cd

29. a. 120 **b.** 120 **c.** 12

31. a. 15 **b.** 6 **c.** 495

33. a. ${}_5P_3 = 60$ **b.** $5^3 = 125$

35. $10^3 = 1000$

37. Different Sets; ${}_{10}C_3 = 120$

39. ${}_{25}C_4 = 12{,}650$

41. a. MP; $36^4 = 1{,}679{,}616$ **b.** Permutations; ${}_{36}P_4 = 1{,}413{,}720$
c. Combinations; ${}_{36}C_4 = 58{,}905$ **d.** Permutations; $4! = 24$

43. a. Combintations; ${}_{18}C_2 = 153$ **b.** Permutations; $9! = 362{,}880$
c. Permutations; 18!

45. a. Combinations; ${}_{35}C_6 = 1{,}623{,}160$ **b.** Combinations and MP; $3({}_{12}C_3)({}_{20}C_2) = 125{,}400$

47. $48(13)/{}_{52}C_5 = 624/{}_{52}C_5$; Example 10

49. a. ${}_{13}C_6 = 1716$ **b.** ${}_{25}C_6 = 177{,}100$
c. $({}_{13}C_2)({}_{12}C_4) = 38{,}610$ **d.** $1716/{}_{30}C_6 = 0.00289$

51. a. $2(2)(2)(2) = 16$ **b.** HHH, HHT, HTH, THH, HTT, THT, TTH, TTT
c. 1, 3, 3, 1 **d.** 0, 1, 2, or 3 tails.
e. The events cannot all happen at the same time.
f. $1 + 3 + 3 + 1 = 8$
g. 1, 5, 10, 10, 5, 1; $32 = 2^5$
h. Flipping 0, 1, 2, 3, 4, 5, or 6 tails (or heads).
i. yes

Exercises 8.4: Pascal's Triangle and Combinations

1.

```
                    1
                  1   1
                1   2   1
              1   3   3   1
            1   4   6   4   1
          1   5  10  10   5   1
        1   6  15  20  15   6   1
      1   7  21  35  35  21   7   1
    1   8  28  56  70  56  28   8   1
  1   9  36  84 126 126  84  36   9   1
1  10  45 120 210 252 210 120  45  10   1
```

3. $128 = 2^7$; $256 = 2^8$

5. 1, 5, 10, 10, 5, 1 respectively.

7. 1, 12, 66, 220, 495, 792, 924, 792, …

9. a. 35 **b.** 15 **c.** 6

11. a. 20 **b.** 1 **c.** 1 **d.** 16

13. The first two are 1 and n and the last two are n and 1.

15. r + 1; because the first number represents r = 0 not r = 1.

17. HH, HT, TH, TT; 1, 2, 1

19. Yes; 0 head in 1 way, 1 head in 4 ways, 2 heads in 6 ways, 3 heads in 4 ways, 4 heads in 1 way.

21. 3/8; 4/8 = ½ because TTT is also included.

23. 5/32 **25.** 1/32 **27.** 6/64 = 3/32

29. $2^{12} = 4096$ **31.** 66/4096 = 33/2048 **33.** 792/4096 = 99/512

35. 66/4096 = 33/2048 **37.** Because ${}_{12}C_{10}$ and ${}_{12}C_2$ are the same.

39. a. 10 ways **b.** 20 ways
c. 21 ways **d.** 66 ways

41. $\frac{n(n+1)}{2}, n = 1,2,3...$; Answers will vary

Review Exercises Chapter 8

1. **a.** {O, E} **b.** {P, R, B, L, M} **c.** 2/7
d. 5/7 **e.** 5/7

2. **a.** {A, E, I, O} **b.** {B, D, L, R, T, V} **c.** 7/17
d. 10/17 **e.** 5/17 **f.** 7/17

3. **a.** 2 **b.** 2(2) = 4 **c.** 2(2)(2) = 8
d. 2(2)(2)(2) = 16

4. The outcomes in Exercise 2 are not equally likely. There are multiple ways to obtain outcomes such as A, E, R.

5. **Row 1:** HT, TH; ½; ½; **Row 2:** HH, HT, TH; ¼; ¾; **Row 3:** Odd numbers, 2/5, 3/5; **Row 4:** White, blue; 2/3; 1/3; **Row 5:** b, d; ½; ½; **Row 6:** a, b, c; ¾; ¼;

6. **Row 1:** GB, BG, BB; ¾; ¼; **Row 2:** GG, BB; ½, ½; **Row 3:** 2, 3, 5; 2/5;3/5; **Row 4:** green, yellow; 2/3; 1/3; **Row 5:** j; ¾; ¼; **Row 6:** j; ¼; ¾;

7. $\frac{32}{120} \approx 0.267; \frac{48}{120} = 0.4; \frac{40}{120} \approx 0.333$; Empiracal; No, should maybe try 1000 trials.

8. **a.** 9/30 = 3/10 **b.** 13/30 **c.** 8/30 = 4/15
This is a small sample and shows a wide distribution of loads. It is an empirical study that probably requires many more classes to study student load.

9. **a.** AP; 1 (will always roll a 1-6 on a standard die) **b.** MP; (½)(½) = ¼

10. **a.** AP; ½ **b.** MP; (1/2)(2/6) = 1/6

11. (1/2)(1/6) + ½ = 7/12;

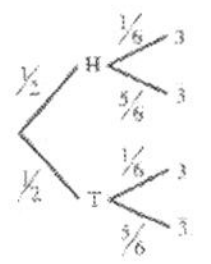

12. (1/2)(1/6) + ½ = 7/12;

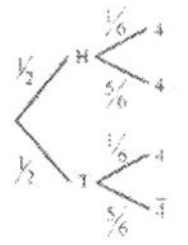

13. **a.** Overlap; 15 **b.** Overlap; 14
c. Overlap; 20 **d.** No Overlap; 16/36 = 4/9
e. Overlap; 8/36 = 2/9 **f.** No Overlap; 24/36 = 2/3

14. **a.** Overlap; 13 **b.** Overlap; 15
c. Overlap; 20 **d.** Overlap; 13/36
e. No Overlap; 12/36 = 1/3 **f.** Overlap; 18/36 = ½

15. **a.**

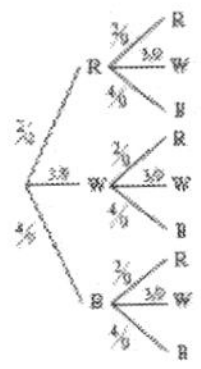

b. (2/9)(1/3) (2) = 4/27
c. 2/9 + 1/3 + (4/9)(2/9) + (4/9)(3/9) = 65/81
d. 1 – (4/9)(4/9) = 65/81
e. (5/9)(5/9) = 25/81

16. a.

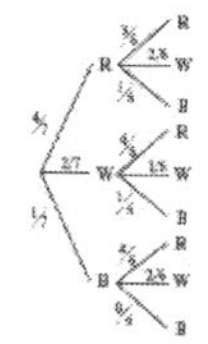

b. (1/7)(2/6) + (2/7)(1/6) = 2/21
c. 1 (would have to get BB for that to no happen and BB is impossible – only 1 blue).
d. 1 (see c)
e. (4/7) + (3/7)(4/6) = 6/7

17. a.

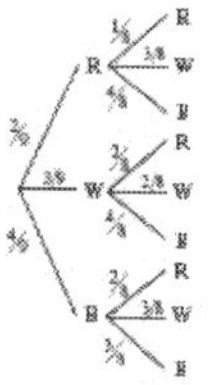

b. (2/9)(3/8) + (1/3)(1/4) = 1/6
c. 2/9 + 1/3 + (4/9)(1/4) + (4/9)(3/8) = 5/6
d. 1 – (4/9)(3/8) = 5/6
e. (5/9)(1/2) = 5/18

18. a.

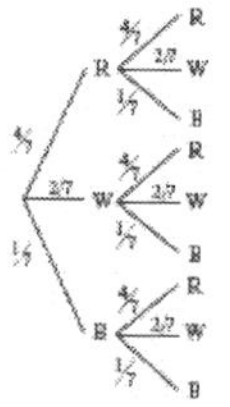

b. (1/7)(2/7) + (2/7)(1/7) = 4/49
c. 4/7 + 2/7 + (1/7)(4/7) + (1/7)(2/7) = 48/49
d. 1 – (1/7)(1/7) = 48/49
e. 4/7 + (2/7)(4/7) + (1/7)(4/7) = 40/49

19. a. 8 students with a 61 or higher and only 1 passed (D+): 1/8
b. 13 students between 46 and 61 and 10 had a C or better: 10/13
c. No students with those hobbies passed: 0
d. All students with an A had a load of 61 or less: 1
e. 13 of the 30 students got a C or better: 13/30

20. a.

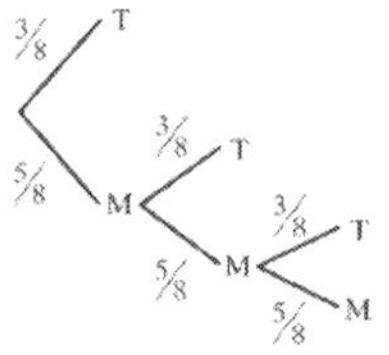

b. (5/8)(5/8)(5/8) = 125/512
c. 3/8 + (5/8)(3/8) + (5/8)(5/8)(3/8) = 387/512

21. 10^{13}

22. a. 362,880 **b.** 35 **c.** 6
d. 840 **e.** 125,970 **f.** 20,160

23. a. Answers will vary – there are 5,040 different possibilities. For example: "Hard to see the dark side is"; "To hard see the dark side is"; "see hard to the dark side is"; "the hard to see dark side is"; "Dark hard to see the side is"; "side hard to see the dark is"; "is hard to see the dark side"
b. ${}_7P_7 = 5040$

24. a. Answers will vary. There are 24 different possibilities. For example: "Revealed your opinion is"; "Your revealed opinion is"; "Opinion revealed your is"; "Is revealed your opinion"
b. ${}_4P_4 = 24$

25. a. P; ${}_{15}P_4 = 32{,}760$ **b.** C; ${}_{15}C_3 = 455$

26. a. P; 15!

b. C; ${}_{15}C_3 = 455$

27. a. C; ${}_{10}C_3 = 120$

b. P; ${}_{10}P_{10} = 3{,}628{,}800$

c. P; ${}_{10}P_3 = 720$

d. C; ${}_{10}C_4 = 210$

e. MP; $10^5 = 100{,}000$

28. a. P; ${}_8P_8 = 40{,}320$ **b.** C; ${}_8C_3 = 56$ **c.** MP; $8^3 = 512$ **d.** P; ${}_8P_3 = 336$

29. a. $\dfrac{1}{10^3} = \dfrac{1}{1000}$ **b.** $\dfrac{1}{{}_{10}C_3} = \dfrac{1}{120}$ **c.** no

30. a. $\dfrac{1}{8^3} = \dfrac{1}{512}$ **b.** $\dfrac{1}{8 \cdot 7 \cdot 6} = \dfrac{1}{336}$ **c.** no

31. ${}_{218}C_8 = 1.11 \times 10^{14}$

32. a. ${}_{12}C_4 = 495$

b. $\left({}_{12}C_4\right) \cdot 4! = 495 \cdot 4! = 11{,}880$; Combination has a factor of 4! in the denominator compared with a permutation.

33. a. 1, 7, 20, 36, 36, 20, 7, 1

b. yes; $128 = 2^7$

c. 14

d. $\frac{22}{2^6} = \frac{11}{32}$

e. Number of ways of obtaining 0 items from a set of *n* items.

34. a. 1, 9, 38, 83, 125, 125, 83, 38, 9, 1

b. yes; $512 = 2^9$

c. 53

d. $\frac{72}{2^8} = \frac{9}{32}$

e. value of $(n - r)!r!$ in formula for ${}_nC_r$ is the same for 3 and 5

35. ${}_aC_b = \dfrac{a!}{(a-b)!b!}$;1; a, $\dfrac{a!}{(a-2)!2!}$, ..., $\dfrac{a!}{(a-2)!2!}$, a,1

36. ${}_xC_y = \dfrac{x!}{(x-y)!y!}$;1; x, $\dfrac{x!}{(x-2)!2!}$, ..., $\dfrac{x!}{(x-2)!2!}$, x,1

37. a.

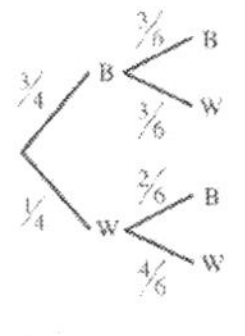

b. (3/4)(1/2) + (1/4)(2/3) = 13/24

c. (3/4)(1/2) + (1/4)(1/3) = 11/24

38. Answers will vary depending on experiment.

39. a. Regular once a week menu.

b. Know I get pizza every week.

c. Once out of 5.

d. 1/5

e. Answers will vary.

f. Answers will vary

g. Answers will vary.

h. Many holidays occur on Monday and Friday so Tuesday and Wednesday are the most frequent and because of Thanksgiving, Thursday one day less common than Tuesday and Wednesday.

Chapter 8 Test

1. **a.** 3/8 **b.** 5/8 **c.** 3/8

2. No; Not all months have 31 days.

3. **a.** 1/6 **b.** 2/6 = 1/3 **c.** 3/6 = ½
d. 3/6 = ½ **e.** 0

4. **a.**

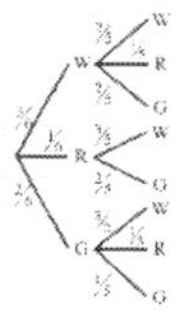

b. 1/6 + (1/2)(1/5) + (1/3)(1/5) = 1/3
c. 0
d. (1/3) +(1/6)(2/5) + (1/2)(2/5) = 3/5

5. **a.**

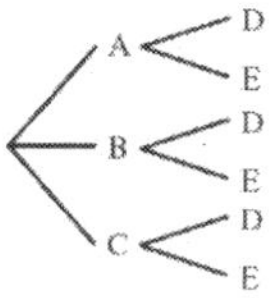

b. {AD, AE, BD, BE, CD, CE}
c. AD or BD
d. BD or BE
e. 1/3 + (1/3)(1/2) + (1/3)(1/2) = 2/3
f. (1/3)(1/2) = 1/6

6. **a.** 10 + 15 = 25 **b.** 10(15) = 150 **c.** 15(14) = 210
d. 15/25 = 3/5 **e.** 1 **f.** (15/25)(14/25) = 7/20

7. **a.** 10(25)(25) = 6250 **b.** 10^5 = 100,000

8. **a.** 52 **b.** $\frac{52!}{0!}$ **c.** 1
d. 1 **e.** 1 **f.** 52

9. **a.** ${}_{10}C_7$ **b.** ${}_{10}P_5$ **c.** 10^5
d. ${}_{10}P_7$ **e.** ${}_{10}P_3$

10. $\dfrac{1}{{}_{30}C_5} = \dfrac{1}{142506}$

11. **a.** 1, 8, 27, 56, 72, 56, 27, 8, 1 **b.** yes, $256 = 2^8$ **c.** 20
d. $36/2^7 = 9/32$ **e.** Represents 0, 1, 2, 3, … boys; the 4th number: 36.
f. The total of the row is $2^7 = 128$

12. **a.** careful was I, careful I was, I was careful, I careful was, was I careful, was careful I
b. Answers vary with explanation

Made in United States
Orlando, FL
04 November 2026

90205880R00155